The Cambridge Encyclopedia of Amateur Astronomy

Being an amateur astronomer is great fun, with many different and interesting areas to get involved in. This complete reference provides a wealth of practical information covering all aspects of amateur astronomy. Organized thematically for ease of use, it covers observing techniques, telescopes and observatories, internet resources, and the objects that can be studied. Those new to the field will find tips, techniques and plans on how to begin their quest, and more advanced observers will find lots of useful advice on how to get more out of their hobby. Containing the most recent data, the book is highly accurate, and is illustrated throughout with stunning color images and graphics. It is an essential guide for both beginning stargazers and more advanced observers.

MICHAEL E. BAKICH obtained a Bachelor's degree in Astronomy from Ohio State University in 1975, and a Master's degree in Planetarium Education from Michigan State University in 1977. He has written numerous original planetarium programs, and gives lectures on astronomy to groups of all ages. He is also a tour guide to eclipses and astro-archaeological sites. Bakich has written two previous books – The Cambridge Guide to the Constellations and The Cambridge Planetary Handbook – both published by Cambridge University Press.

THE CAMBRIDGE

ENCYCLOPEDIA OF

AMATEUR ASTRONOMY

Michael E. Bakich

CAMBRIDGE
UNIVERSITY PRESS

PUBLISHED BY THE PRESS SYNDICATE OF THE UNIVERSITY OF CAMBRIDGE
The Pitt Building, Trumpington Street, Cambridge, United Kingdom

CAMBRIDGE UNIVERSITY PRESS
The Edinburgh Building, Cambridge CB2 2RU, UK
40 West 20th Street, New York, NY 10011-4211, USA
477 Williamstown Road, Port Melbourne, VIC 3207, Australia
Ruiz de Alarcón 13, 28014 Madrid, Spain
Dock House, The Waterfront, Cape Town 8001, South Africa

http://www.cambridge.org

First published 2003

Printed in the United Kingdom at the University Press, Cambridge

Typeface Joanna 10.25/12.5pt.   *System* QuarkXpress   [HMCL]

A catalogue record for this book is available from the British Library

Library of Congress cataloging in Publication data

Michael E. Bakich
The Cambridge Encyclopedia of Amateur Astronomy/Michael E. Bakich

.

Includes bibliographical references and index.

ISBN 0 521 81298 4 hardback

Contents

Foreword

Jeff Medkeff
Contributing Editor, *Sky & Telescope*

When I got my start as an amateur astronomer at about age nine, I proceeded with a planisphere and my grandmother's opera glasses. From the front yard of my childhood home in Cuyahoga Falls, Ohio, I was able to identify Vega and the summer triangle, and find many of the brighter summer Milky Way clusters. Over the years, I gained proficiency in recognizing the constellations and observing the sky, and I found myself wanting to spend time under the stars more and more. I acquired a telescope and a few eyepieces, and some spiffy star charts. I began to enjoy hunting down obscure deep-sky objects and striving to see fine detail in tiny planetary images on nights of steady seeing. But more than this, I made friends of my fellow astronomers – people who proved to be important mentors to me.

And yet nothing has proved more enduring and closer to my soul than the night sky itself. As the years have passed, the night sky has more and more proven to be my natural element. Nowhere do I feel more comfortable or more natural than out in the countryside, alone or with a few good friends, in the dark, looking at the night sky. Urban paranoia may not sympathize with my enjoyment of nighttime walks in the country, spent mostly looking upward; nor do corporate managers relate well to the special circumstances required for my nocturnal pursuits. It doesn't bother me much at all. Every clear, dark night, I can plunge deep into the wilderness – a wilderness untouched by humans, and utterly isolated from every nuisance of society. I lay my eyes on things and places that few, if any, humans have ever seen, and where none have ever been. And only there – even with all the special places in this world of ours, and even with all its people to spend time with – only there, in the night sky, do I find my true element. The night sky is my true home.

Today, over twenty years since I first directed opera glasses toward the summer Milky Way, I sleep soundly while a robotic telescope and electronic camera automatically take images of variable stars for later analysis. With this equipment, my backyard has turned into a fully fledged research observatory, complete with published scientific papers and recognition from the IAU.

I have not lost my passion for looking at the night sky with my eyes, binoculars, and telescopes. But the twenty years I have spent in this hobby have been years of change. When I got my start, an eight inch scope on standard German equatorial mount was a good-sized scope – with the now-venerable C-8 a new-fangled alternative for those who drove subcompact or sports cars. I was observing with a 4.5" reflector that my uncle had picked up for me at an astronomy convention – a scope whose optics are still in use today. Within a few years, an inexpensive alt-az design swept through the ranks of amateur astronomy like a wildfire – the Dobsonian revolution had hit us hard. By the mid-1980s I upgraded to a ten inch Coulter dob – which, judging by the weight, was built of neutron star material. In short order, wide-field eyepieces were replacing the narrow-field Plossls as the observer's ocular of choice. Soon Dobsonians didn't weigh as much as a semi truck. By the early 1990s, the computer revolution took off in earnest, and we began to see computerized telescopes, CCD cameras, and online amateur astronomy discussion forums.

The hobby, in short, has not been static. The changes have shocked some of the old timers, who increasingly retreat to their observatories and star party sites for secluded observing in the old style. (I am shocked, in my turn, to be identified by some as one of the "old timers" myself.) Others have embraced the changes and have pursued new facets of the hobby with gusto. The changes have attracted a new crop of amateur astronomer as well – some of whom adopt the latest technology for its own sake, or as a tool; while others prefer to "go retro," keeping the heritage of amateur astronomy alive.

This book serves as a snapshot of where we are as amateur astronomers today. For many of us, it will also show where we've been. There is no facet of the hobby, as it has existed in the last quarter century or more, which isn't at least hinted at here. Those looking for current information will find much more than hints. I'm convinced that both beginner and veteran will find plenty of material here to instruct and entertain. I expect I'll get a lot of use out of my copy. I hope you do too.

The past twenty years have been quite a ride. It will be interesting to see where we are in another twenty years. When we get there, be sure to look me up. You'll find me somewhere dark.

Preface

You want to observe the sky, but you don't know where to start. You are ready to buy equipment, but you don't know which would be the best for you. You want to produce results, but you don't know how. To top it all off, you are in a hurry. Purchasing this book may be the first step you have taken in your quest to become an amateur astronomer.

Practical. If I could only use one word to sum up what I am trying to accomplish with this book, it would be the word "practical." There are so many books that present great quantities of facts about astronomy. I'm not saying there's anything wrong with that. After all, I wrote two of them! And while this book will indeed present many facts related to astronomy, its main purpose is to help the reader learn more about how to *do* astronomy.

I started learning about the sky 40 years ago. I suppose when one begins that young and stays that long with any subject it no longer seems as difficult. As an educator in planetariums for half of those years, however, I have been forced to take a step back and see the reactions of those who are just being introduced to what I like to call "the wonders of the heavens." In most cases, it isn't easy for them. Things that I now take for granted are met with wide-eyed puzzlement. "How do you know that's Jupiter?", "Is this a good telescope for me?" and "When can I start taking pictures of what I see?" are only three of a myriad of questions asked by budding amateur astronomers.

I intend that this book provide some answers. Those new to the field of amateur astronomy will find tips, techniques, and plans on how to begin their quest. The first few chapters will provide a working knowledge of some of the terminology and concepts involved. More enlightened amateurs may find a review of this information helpful. Advanced amateurs will, I am certain, pick and choose from the many sections within this book.

This book is intended as a beginning, not an end. There is so much information on each of the subjects covered. If you view this work as a personal tour through what I consider "the highlights," you will be on the right track. The "highlights" of others may differ somewhat. Because of this, I have provided an annotated bibliography, as well as information on catalogs, software and websites, for those wishing to go into more depth on a particular subject. Each of us has a higher interest in one or two of the topics covered here. Whatever the subject, I hope that this book will help you not only to understand astronomy better, but, in the most practical sense, to *do* astronomy better as well.

The telescope of the author's dreams. A 500 mm StarMaster with GoTo drive. This is the personal "travel scope" of Rick Singmaster, owner of StarMaster Telescopes. (Photo by the author)

Acknowledgements

First, I again want to thank my wife, Holley. Her level of understanding, compassion and forbearance was part of the fuel that helped to move this project along. Her willingness to drop everything because "I have another idea for a graphic!" is, here, acknowledged with humble thanks. Holley created all the non-photographic illustrations in the book. There may be no graphic-based software program which she has not mastered. But aside from the technical help, her support of me has been more valuable than the Sun and Moon (or any other celestial object covered in this text). Holley, I love you.

When this project was conceived, my thought was to include contributions from as many amateur astronomers as possible. I therefore put out a "call for images" and I also contacted a number of individuals, asking if they would like to contribute an image or more. The response of the amateur astronomy community was heartening. The next paragraph lists, alphabetically, the people to whom I am indebted and to whom I extend my sincere gratitude. Each is also credited in the caption beneath all images they contributed or with their direct quotation.

Leonard B. Abbey, FRAS, Mark Abraham (Olathe, Kansas), Paul Alsing (Poway, California), Chris Anderson (western Kentucky), Thomas M. Back (Cleveland, Ohio), Ulrich Beinert (Kronberg, Germany), Steve and Susan Carroll (Fort Scott, Kansas), Roland Christen (Rockford, Illinois), Steven Coe (Phoenix, Arizona), A. J. Crayon (Phoenix, Arizona), Mark Cunningham (Craig, Colorado), Richard Dibon-Smith (Toronto, Canada), Eugene Dolphin (San Diego, California), Jim Gamble (El Paso, Texas), Robert Gendler (Avon, Connecticut), Ed Grafton (Houston, Texas), Robert Haler (Kansas City, Missouri), Jeffrey R. Hapeman (Madison, Wisconsin), David Healy (Sierra Vista, Arizona), Carlos E. Hernandez (Houston, Texas), Jane Houston Jones (San Francisco, California), Mick Hradek (El Paso, Texas), Tim Hunter (Tucson, Arizona), Steven Juchnowski (Balliang East, Victoria, Australia), Jere Kahanpaa (Jyväskylä, Finland), Al Kelly (Danciger, Texas), David W. Knisely, (Lincoln, Nebraska), Arpad Kovacsy (Mt. Vernon, Virginia), Ron Lambert (El Paso, Texas), Shane Larson (Bozeman, Montana), Eugene Lawson (El Paso, Texas), Tan Wei Leong (Singapore), Charles Manske (Watsonville, California), Mark Marcotte (El Paso, Texas), James McGaha (Tucson, Arizona), Arild Moland (Oslo, Norway), Craig Molstad (Onamia, Minnesota), Mike Murray (Bozeman, Montana), Larry Robinson (Olathe, Kansas), Ray Rochelle (Chico, California), Jim Sheets (McPherson, Kansas), Raymond Shubinski (Prestonsburg, Kentucky), Rick Singmaster (Arcadia, Kansas), Brian Skiff (Flagstaff, Arizona), Shay Stephens (Seattle, Washington), Rick Thurmond (Mayhill, New Mexico), Alin Tolea (Baltimore, Maryland), John Wagoner (Cleveland, Texas), Kent Wallace (Palominas, Arizona), Chris Woodruff (Valencia, California)

I have saved the mention of three individuals for special recognition. These are the people who went above and beyond the call of duty in the provision of images or the offer of comments and critiques related to specific sections of this book. Alphabetically, they are: Adam Block, of Tucson, Arizona, who runs the Advanced Observers Program at Kitt Peak. With few exceptions, if you see a galaxy in this book, Adam had a hand in providing it. Early on, the many high-quality images and information that Adam provided helped to set my mind at ease and assure me that the book would look great. Thanks, Adam.

The second person I have singled out for special recognition is Robert Kuberek, of Valencia, California. No request I made was too much for Bob. Whether it was for an image of a celestial object or an image of a piece of equipment (Bob is known for the high quality of "stuff" that he has accumulated), I never heard anything other than, "Sure, I'll get that off to you right away." Thanks, Bob.

The last (but not least) of the notables is Jeff Medkeff, of Sierra Vista, Arizona. I have learned a great deal about a number of areas of amateur astronomy from Jeff, who was always willing to share information and answer questions. More than simple knowledge, however, Jeff has inspired me in areas (CCD, the Moon, etc.) where I really had no interest before. Thank you, Jeff.

On the corporate end of things, there are also many individuals and companies to thank. First, I extend my gratitude to Meade Instruments Corporation for the generous loan of a 300 mm LX200 GPS model Schmidt–Cassegrain telescope. I used this telescope to test several pieces of equipment, filters, etc., reports of which are included in this book.

Thanks to Thomas M. Bisque of Software Bisque for providing a brand new, full-blown copy of their excellent software TheSky. Whether he gets credit or not, Tom has been a great supporter from my very first book.

Al Misiuk, of Sirius Optics in Kirkland, Washington, provided a number of filters for evaluation and to aid in observation from my light-polluted backyard. My only regret was that my deadline was such that I was unable to test several exciting upcoming products. And thanks for some enlightening phone calls, Al.

Thanks to Glenn Eaton, the Advertising & Marketing

Coordinator for the Astronomical Society of the Pacific, for providing a copy of RealSky. In conjunction with several of the star charting software packages, RealSky provides a level of realism that is incredible.

Finally, my thanks go out to the staff of Cambridge University Press as well. Once again their professionalism and expertise have guided me, nearly effortlessly, through the complex world of book publishing. Specifically, I wish to thank Dr Simon Mitton, who started the process and saw this book through the acceptance process, and Miss Jacqueline Garget, who has taken over the lead editor's spot for Dr Mitton. Both have been kind and helpful, answering all of my questions and providing much valuable advice.

But my final, and highest, praise goes to my copy editor, Fiona Chapman. If you like this book – I mean, really *like* it – Fiona is the one to thank. Yes, mine was the idea and mine were the words, but hers was the craft which molded these words into this book. Thank you, Fiona.

CHAPTER ONE

Background

Being an amateur astronomer is a lot of fun. There are so many different and interesting areas. But before we *do* astronomy, let's learn a little about what we're going to be observing by taking a little tour. We will start right here on Earth. (Note: Experienced observers can proceed beyond this section.)

Earth

The Earth is a planet. Planets are different from stars because they do not give off their own light (although large planets like Jupiter may radiate some energy). Planets are also generally much smaller than stars, so they seem to orbit stars (although technically both the star and the planet are orbiting a common point known as the center of mass). For our purposes, we will be referring to planets as objects which are orbiting the Sun.

The Earth is the third planet out from the Sun. Closest to the Sun is Mercury, then Venus. These are known as the inner or inferior planets. Further than the Earth are the planets (in order) Mars, Jupiter, Saturn, Uranus, Neptune and Pluto. These are the outer or superior planets. Four planets (Jupiter, Saturn, Uranus, Neptune) are larger than the Earth and the other four are smaller. The four larger planets are also different

Gil Machin, of Kansas City, Missouri, with his superb homemade 400 mm Newtonian. Gil ground the mirror as well and the views must be experienced to be believed. (Photo by the author)

The Moon. (Photo by Robert Kuberek of Valencia, California)

in that they are mainly composed of gas, with relatively small rocky or metallic cores deep within. And although Pluto may be mainly composed of frozen gases, the other three small planets are much more Earth-like in composition.

Moon

The first stop out from the Earth is the Moon. The Moon is about one-quarter of the diameter of our Earth, but much less massive. It orbits the Earth every 27 1/3 days, approximately. However, since the Earth is also moving through space (as it orbits the Sun) the Moon takes about 29 1/2 days to go through a complete cycle of phases. The lunar cycle begins at New Moon (which cannot be seen from Earth). New Moon occurs when the Moon is between the Sun and the Earth.

Eclipses

Once in a while, we see the Moon pass in front of the Sun. This is called a solar eclipse. Eclipses can be total, if the Moon covers the Sun entirely, or partial, if the Moon covers only part of the Sun, or annular, if the Moon is far from Earth and (although it passes directly in front of the Sun) does not cover the entire disk of the Sun but leaves a ring of the Sun's surface visible. Eclipses do not occur every 29 1/2 days because the Moon's orbit is tilted to the orbit of the Earth around the Sun. So at times it is above the Sun at New Moon and at other times it is below.

The two intersections of the orbit of the Moon and the apparent path of the Sun (the ecliptic) are known as nodes. Only when the Full Moon is at one of the nodes is an eclipse possible. When the Sun is at the same node, the eclipse will be solar. When the Sun and Moon lie at opposite nodes, the eclipse will be lunar. (Illustration by Holley Y. Bakich)

There are also eclipses of the Moon (lunar eclipses). These events happen at Full Moon, when the Earth is between the Moon and the Sun. Lunar eclipses can be total or partial. Also, since the Earth's shadow has two parts, a dark inner part (called the umbra) and a lighter outer part (the penumbra), there are times when the Moon does not enter the umbra. Such an eclipse is called a penumbral eclipse. All total or partial eclipses of the Moon occur when the Moon passes through the umbra of the Earth's shadow.

Observers who are not interested in the Moon generally schedule their observing sessions around the time of New Moon. Actually, during the roughly two weeks between Last Quarter and First Quarter, the light of the Moon is diminished and many good observations may be obtained. Also, a Quarter Moon is in the sky only half the night (First Quarter from sunset to midnight and Last Quarter from midnight to sunrise). One further point. Due to the angle of the Sun's light striking the Moon's surface, Quarter Moons are only 10% as bright as the Full Moon.

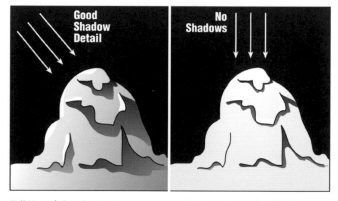

Full Moon (when the Sun's rays are perpendicular to the surface) is the worst time to observe the Moon, as the detail is reduced. (Illustration by Holley Y. Bakich)

Angle of sunlight is also the reason why those interested in observing the Moon do not generally observe at Full Moon. During that phase, the Sun is directly behind the observer. Thus, as seen from the Moon, it would be overhead and any shadows would be at their shortest. The best location on the lunar surface for observing detail is where the sunlit part of the Moon meets the dark portion. This line is the point of sunrise (when the phase is between New Moon and Full Moon) or sunset (when the phase is between Full Moon and New Moon) and is called the terminator.

Mercury and Venus

As we move outward from the Moon (in the direction of the Sun), we encounter the planets. Mercury and Venus, because their orbits are smaller than the orbit of the Earth, are never seen very far from the Sun. At most, Mercury can be 28° from the Sun and Venus 47°. A degree, by the way, is a unit of measuring distance across the sky. This type of distance is called angular distance, as you are really measuring the angle observed, and not the true distance in miles or kilometers. To get an idea of degree measurement, make a fist and hold it at arm's length. From the top to the bottom of your fist is roughly 10°. If you extend one finger, that's approximately 2°. (Rough measurements in the sky include the distance between the two stars in the bowl of the Big Dipper (called the Pointer Stars). They are a little over 5° apart. From the tip of the bowl to the end of the handle the distance is a little over 25°. Other distance "helpers" will be discussed later.)

Mercury is small and essentially no detail has ever been seen on it by amateur observers. Venus is larger and closer, but its surface is totally obscured by a thick layer of cloud. Occasionally, and by employing the right equipment, cloud structure can be seen when observing Venus. This will be discussed in more detail in the section on observing Venus.

Mercury and Venus can line up with the Earth and Sun in two ways. Each can be between the Sun and Earth (this is called inferior conjunction) or on the opposite side of the Sun from the Earth (superior conjunction). Just like eclipses, usually the three bodies are not directly in line. However, if either Mercury or Venus is directly in line with the Earth, the disk of the planet will appear as a black dot on the face of the Sun. This is called a transit. Transits of Mercury are infrequent, but much more common than transits of Venus (which occur in pairs separated by more than 100 years).

Sun

Since we've been moving in the direction of the inferior planets, let me say a few introductory words about the Sun. The Sun is a star, similar to those that we see at night. It is an

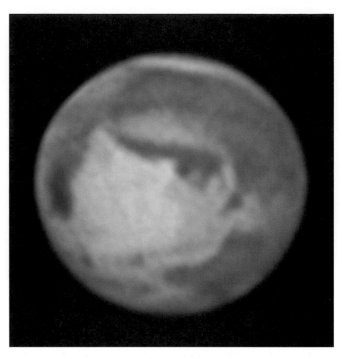

Mars. Tan Wei Leong, of Singapore, took this image at 16:18 UT on 20 Jun 2001. (Celestron C11 and an SBIG ST7E CCD camera)

average star. About half the stars in space are larger, half are smaller. Half the stars are hotter, half are cooler. Half the stars are more massive, half are less massive. But more importantly than being average, the Sun is also a *stable* star.

Two major forces are at work within the Sun. The first is gravity. Gravity causes the material of the Sun (and that of all other celestial objects) to remain a single body. But gravity wants to pull all the mass of the Sun to its center. Opposing the force of gravity is the force created by the energy being produced in the Sun's core. There, at temperatures of 15 million degrees Celsius, the centers of hydrogen atoms are being smashed into one another with such force that a helium atom (at least its center, called the nucleus) can form. In this process, some mass is lost as energy is released. This is the reason the Sun shines. All of us have probably at least heard of Albert Einstein's famous equation $E = mc^2$. This is a simple way of stating how mass (m in the equation) can be transformed into energy (E).

Mars

Continuing our journey outward, we come to Mars. Of all the planets, Mars most resembles Earth. Its day is just over 24 hours long. The tilt of its axis (that is, how much the poles point from straight up – the Earth's tilt is 23½ degrees) is similar to that of the Earth. This means that, like the Earth, Mars has four distinct seasons. Mars' temperature range is also closer to that of Earth than any of the other planets, although it still is very cold there on average. Mars is also the only planet where the telescopes of amateur astronomers can resolve details on its surface.

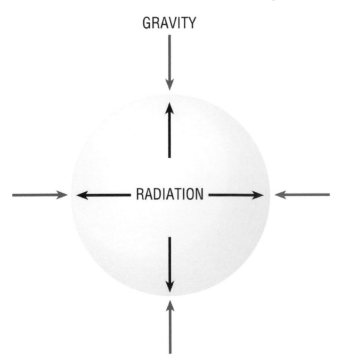

GRAVITY

RADIATION

In a stable star such as the Sun, the forces of gravity and radiation are in balance. (Illustration by Holley Y. Bakich)

Asteroid belt

Between Mars and the next planet, Jupiter, lies the asteroid belt. Asteroids (also called minor planets) are rocky bodies which also orbit the Sun, taking from three to six years to do so. Hundreds of thousands of asteroids inhabit this region. The largest of these, Ceres, has a diameter of only 930 km. Some asteroids do not lie within this area of the solar system, but range far and wide. Those few that approach our region of space are called NEOs, or Near-Earth Objects.

Jupiter, Saturn, Uranus, and Neptune

The planets Jupiter, Saturn, Uranus, and Neptune are called either the Jovian planets (for their resemblance to Jove, that is, Jupiter) or the gas giants. In composition, these planets are much more like the Sun than they are like the Earth. They are made mostly of hydrogen and helium, with other gases in smaller amounts. When we observe these planets, we are viewing only the tops of their atmospheres. Uranus and Neptune reveal little detail (except for color) in amateur telescopes. Jupiter and Saturn are showpieces.

Even through small telescopes, Jupiter reveals several belts and four bright moons. Jupiter also rotates (spins) quickly, so features on it change a lot. Saturn is best known for its magnificent system of rings. However, Saturn is not the only planet with rings. The other three gas giants also have ring systems. In contrast, their rings are thin and composed of dark material which makes them impossible to observe directly from Earth. Saturn's rings, on the other hand, are big and bright, being composed mainly of ice crystals and ice-covered rock.

Pluto

The last planet in our journey out from the Sun is Pluto. Pluto has successfully been observed through medium-sized amateur telescopes. It looks just like a star, however, and a faint one at that. You also need a good star chart to find it.

Kuiper Belt

After Pluto, there is still plenty of solar system left. A disk of comets called the Kuiper Belt extends from the area of Pluto out to a distance of several thousand times as far as the Earth is from the Sun. By the way, the average Earth–Sun distance has a name. Astronomers call it the astronomical unit; it is often abbreviated AU.

Oort Cloud

The extent of our solar system is defined by the Oort Cloud. This is a much larger group of comets lying much further from the Sun. The Oort Cloud lies between 30 000 and 100 000 astronomical units from the Sun. This places the furthest comets in the Oort Cloud slightly over one-third the distance to the next nearest star.

Comets

The previous two paragraphs have discussed comets, but only as icy objects lying far from the Sun. Often, however, comets approach the Sun and we find them in our region of space. When this happens, heat from the Sun begins to evaporate the ice of which the comet is mostly made. (The other main ingredient of comets is dust.) At this point the comet may develop a coma (a cloud of gas surrounding the central body of the comet; once the coma forms, that central body is called the nucleus) and it may even develop a tail. Here is where observers start getting excited. The large area of glowing gas makes the comet much easier to observe. How bright any comet gets depends on three factors: (1) the composition of the comet itself; (2) distance from the Sun; and (3) distance from the Earth. The closer a comet gets to the Sun, the brighter it becomes. However, if the comet is on

Halley's Comet, 4 Jan 1986, Naco, Arizona. (8" f/1.5 Celestron Schmidt camera, 5 minutes on hypered TP 2415 film. Image by David Healy, Sierra Vista, Arizona)

the other side of the Sun from the Earth at that time, it will not look as bright as if it were also near the Earth. This is the reason that Halley's Comet (among others) sometimes appears brilliant while at other times it is barely visible.

Stars

Moving out from our solar system, we encounter the stars. The stars are so far away that a new scale of distance is now required. An example will help. The brightest star in the night sky is Sirius. Sirius is easy to see from late fall to mid-

The constellation Canis Major, featuring the brilliant star Sirius. On the original slide, no less than five Messier objects are visible: M41, M46, M47, M50, and M93. Messier objects are discussed later in the text. (Photo by the author)

A light year may sound like a unit of time but it is actually a unit of distance. It is the distance that light (moving at the approximate speed of 300 000 km/s) travels in one year. After performing a little math, we discover that a light year corresponds to nearly 9.5 trillion kilometers! Sirius, lies 8.65 light years distant.

Parsecs are slightly more difficult to understand as they are more of a mathematical construction. One parsec is defined as the distance of an object if its parallax equals one second of arc. Parallax is defined as one-half the angle that an object makes when measured from both sides of the Earth's orbit (see diagram). A second of arc is 1/3600 of a degree (a very tiny angle). When you are looking at the sky, it is impossible to measure kilometers from one object to another, so astronomers use angular measurements. There are 360 degrees in a circle, 60 minutes of arc in each degree and 60 seconds of arc in each minute of arc. Thus, the distance of any object lying at one parsec is a little over three and one-quarter light years (3.2616 to be exact).

Stars come in different varieties. All stars, however, go through the same stages of "birth," "life" and "death." (Describing it like this is an easy way to picture the various parts of a star's existence.) Stars are born in vast clouds of gas, almost entirely composed of hydrogen and helium. Such clouds are found within galaxies. Gravity causes areas of the gas to collapse inward, raising the temperature and pressure. This contraction continues until the temperature of the star reaches about 10 000 000 °C. At this point, a process called nuclear fusion can occur. Lighter atoms, sped up by the tremendous energy available, collide with one another, "fusing" into heavier elements. The most common transformation happens when hydrogen atoms collide and combine, finally fusing to become helium. As this happens, a

spring and can be located by using the belt of the constellation Orion by simply drawing a line down from the belt to brilliant Sirius. When astronomers describe the distance to this star they do not use kilometers but rather light years or parsecs.

The solar antapex lies to the south and west of Sirius, within Columba. This is the point away from which our solar system is moving, due to the rotation of the Milky Way. (Illustration by Holley Y. Bakich)

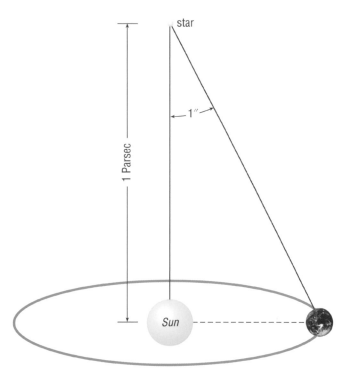

No star is as near as on parsec, where its parallax would be one second of arc. Not to scale. (Illustration by Holley Y. Bakich)

The Pleiades, a very young open star cluster in Taurus. SBIG ST7e NABG camera with the SBIG CLA Nikon lens adapter on a Vixen GP-DX EQ mount. 80 mm Nikon lens, 10-minute exposure. A set of 2" RGB filters were manually switched between shots. (Image by Chris Woodruff, of Valencia, California)

small part of the mass of the hydrogen is converted into energy. We have to thank Albert Einstein for helping us understand this process. $E = mc^2$ is his famous formula. Simply, it says that energy equals mass multiplied by a large number, the speed of light squared. This explains why a little mass can produce a large amount of energy. Nuclear fusion, then, is what makes the Sun and most of the other stars shine.

When a star begins to fuse hydrogen (and this only happens in its center, or core) it quickly achieves a balance between two forces. One is the force of the energy (flowing outward) and the other is the force of gravity (pulling inward). When these are in balance, a star is said to be stable. This is by far the most important attribute of our Sun – it is stable. If there were fluctuations in its temperature, size, etc., well, you can easily imagine that life on Earth would be impossible.

The middle part of a star's life (sometimes called the hydrogen-burning stage because that is what the star is using for fuel) is where most visible stars are. How long a star stays in this stage depends only on its mass. Massive, hot stars, with very high temperatures in their cores, use up their hydrogen fuel quickly (at least quickly in astronomical terms). Less massive stars, like our Sun, may be in this stage for billions of years. (The Sun's estimated middle lifespan is 10 billion years. It has been in this stage for approximately half that time so far.)

Massive (notice that I'm not saying "big" because it is mass, not size, that is the important factor) stars are very

M27. (Tri-color image with SBIG ST-8 CCD camera, R: 20 minutes, G: 30 minutes, B: 60 minutes, full resolution, Meade 16" LX200 @ f/6, 18 Jul 1998, Sierra Vista, Arizona. Image by David Healy)

The apex of the Sun's way, or solar apex, lies approximately 3° from the star θ Her. The nearest bright star is Vega. The solar apex is the direction in space toward which the Sun and our solar system are heading, due to our rotation in the Milky Way galaxy. (Illustration by Holley Y. Bakich)

bright, but due to the extremely high temperatures in their cores they only live for a short time. Once the energy output starts to diminish, the force of gravity begins to win the "tug-of-war" with energy and the star begins to contract, but this only causes a new change to come about. The temperature and pressure in the core increase and soon helium becomes the fuel used to produce energy. Because helium fuses at a much higher temperature than hydrogen, more energy is released and the star's outer layers expand until the balance is once again restored between energy and gravity. The star has now become a red giant. Other, heavier elements may be fused at higher temperatures to turn the star into a red supergiant, but this process soon ends. Different things can happen to the star now.

Planetary nebulae

One scenario is that the star will continue to make energy by using hydrogen and helium outside of the core; its surface will rise and fall and the star will become a variable star. When out of control, the layers of gas will pull away, forming a shell of gas known as a planetary nebula. As an amateur astronomer, you will have lots of opportunities to observe planetary nebulae. Many of these objects are reasonably bright and easy to observe.

White dwarf

As the outer layers are puffed away, the core contracts to roughly Earth-like size. The star is now known as a white dwarf. Further contraction is prevented because the star

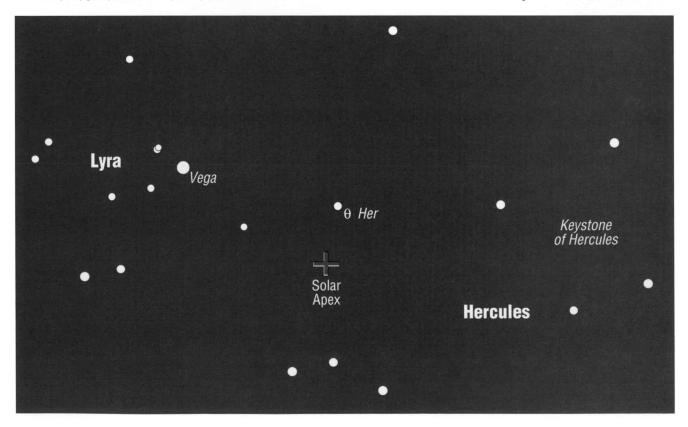

doesn't have enough mass to overcome the force caused by the repulsion of electrons in its matter. Probably the most famous white dwarf star is the first to be discovered, Sirius B. As the name implies, it is a companion to the brightest nighttime star, Sirius. It was discovered in 1862 by a famous American telescope maker named Alvan Clark. The overwhelming glare of Sirius makes this companion difficult to observe unless it lies far enough away in its orbit. The section on double stars later in this book has a diagram showing when Sirius B is best for observation.

Supernova

Very massive stars will continue to fuse heavy elements in order to produce more energy. However, once iron is formed, more fusion cannot occur since iron is very stable and its fusion would require more energy than even the largest stars have available. Rather quickly after the formation of iron the core will collapse under its own gravity and huge amounts of gas on the surface of the star will explode out. The star has become a supernova. Supernovae are the largest explosions in the universe and really bright ones often outshine the entire galaxy in which they are located!

Neutron star

During a supernova explosion, tremendous amounts of energy are available. The core of the star is crushed to such a degree that protons and electrons are made to combine, leaving only neutrons, and those with very little of the original space that separated them. Neutron stars consist of matter that is 100 million times denser than white dwarf matter. I once heard it put this way: a teaspoonful of neutron star material would weigh more than every automobile ever produced on Earth!

Pulsars

If the original star which became a supernova and produced a neutron star was rotating, the law of "conservation of angular momentum" insures that the neutron star is

The constellation Lyra, the Harp. Just to the upper left of brilliant Vega you can see the Double-Double resolved into "two" stars. (Kodak Elite Chrome 200, pushed 1 stop. Nikon F2 with a 50 mm Nikkor lens. Photo by Ulrich Beinert of Kronberg, Germany)

spinning very fast. The usual analogy given here is that of a skater whose arms are pulled close to their body while spinning, increasing the speed of the spin. Hot spots, caused by magnetic fields, may form on the neutron star. If these brighter spots pass by our field of view, the star will appear to blink on and off, or pulse. Such a star is known as a pulsar. Pulsars cannot be observed with equipment available to amateur astronomers.

Black holes

If the original star was massive – say, six to eight times the mass of our Sun or more – an object even stranger than a neutron star will be produced. Astronomers have coined the term "black hole" for such an object, but "invisible star" would be equally descriptive. Black holes are produced by the most powerful supernova explosions which force the matter within the star's core to an incredible, no, unbelievable density. In the tug-of-war between gravity and energy, a black hole represents the ultimate victory for gravity. Such an object has a gravitational field so strong that nothing – not even light – can escape. Black holes are not observed directly even by professional astronomers using the largest telescopes. They simply cannot be seen. Their presence is revealed by the effect they have upon nearby stars, or upon gas which they are consuming.

Constellations

All visible stars are grouped into constellations. There are 88 constellations that cover the sky. (For a complete list, see Appendix A.) There is no overlap among constellations and no gaps between them. The boundaries of the constellations were officialized in 1928 (published in 1930). Today, when we talk about celestial objects being "in" a particular constellation, this means that the object is to be found within the official boundary established long ago.

I think that all amateur astronomers should be familiar with at least the most prominent constellations. At gatherings, questions of location are common as are answers such as, "It is in Bootes." You still are clueless if you don't know where Bootes is. Also, it is good to know which are the "constell-ations of the seasons," usually defined as the constellations visible after sunset during the middle of each season.

While you take a little time to learn the locations and general figures of the constellations in the sky, also take a look at their correct pronunciations (see Appendix A). Oh, and while I'm on this subject, there are two constellations of the zodiac whose names are often butchered. They are Scorpius and Capricornus. If you are heard to say "Scorpio" or "Capricorn," prepare yourself for some abuse, because that's how *astrologers* refer to those star groups.

In addition to the 88 "official" constellations, there are a number of unofficial groups of stars, or *asterisms* in the sky. Two examples are the Big Dipper in Ursa Major and the Teapot in Sagittarius. Often, asterisms are made of stars which come from more than a single constellation. One example is the Summer Triangle, composed of three stars, one each from the constellations Lyra, Aquila and Cygnus.

The Eta Carinae Nebula. Truly one of the wonders of the sky. (This 20-minute exposure was on hypered Fuji HG 400 through a Celestron 5 at f/6.3. Image by Steven Juchnowski, Balliang East in the State of Victoria, Australia)

Nebulae

The word "nebula" comes from the Latin for "cloud." So when I speak of a nebula I mean a cloud of gas and dust in space. Two main types of nebulae exist: emission and reflection. In an emission nebula, the atoms are giving off light because they are being excited by massive, hot stars within the nebula itself. These nebulae are generally red in color because they are made almost entirely of hydrogen. When hydrogen is excited it gives off light, red light being the strongest color. Emission nebulae often have dark areas caused by clouds of dust which block the light. The combination of "red" hydrogen gas and dust gives us some very interesting objects. A great example of this is the North American Nebula, in the constellation Cygnus.

The other main type of nebula is known as a reflection nebula. Such an object is produced when dust in a cloud of gas reflects light from stars not within the cloud itself. Reflection nebulae are often blue since the blue light is scattered throughout the cloud by dust particles. Scattering of blue light is the same phenomenon that gives us a blue daytime sky. Some nebulae are made up of both reflection and emission components. A good example of this is the Trifid Nebula in the constellation Sagittarius.

Star clusters

Many stars occur in groups called star clusters. Astronomers divide star clusters into three main types: associations, open clusters and globular clusters. Associations are the loosest groupings of stars, usually containing a few stars to a few

Emission and reflection nebula in Orion (NGC 1973, 1975, 1977). Sometimes called the "Running Man" nebula. (IMG1024 camera (FLI), 12.5" Ritchey-Chretien at f/7.5, total exposure = 100 minutes. Image by Robert Gendler, Avon, Connecticut)

hundred. Open clusters (also called galactic clusters), have more stars than associations, on the order of hundreds or thousands. The stars in both of these are mainly young stars. Globular clusters (so named because they are round like globes) are much older and contain many more stars, usually tens of thousands up to a million. So, in essence, there are really two types of star clusters . . . those that are round and those that aren't. In addition, associations and open star clusters are found within the disk of our galaxy, the Milky Way. Globular clusters, on the other hand, are located just outside the galaxy in a spherical distribution around the galaxy's core.

The open cluster NGC 188. (Celestron Fastar 8 at f/1.95, PixCel 237 CCD. Unguided 60-second exposure atop the AP900 mount at −26.81°C. Image by Chris Anderson of Kentucky)

Galaxies

Every amateur astronomer wants to observe galaxies. Unfortunately, galaxies do not lend themselves to being viewed well with small telescopes. The wonderful pictures shown in many places (this work included) may provide an impression that galaxies are easy to observe. They are not. The details and colors which can be brought out in digital or photographic images simply are impossible to observe (except in the rarest cases) with the telescope/eye combination.

Experience helps to bring out the finer points, so all is not lost. We will go into more detail about this in the section on galaxies. As a rule of thumb, however, the most important thing to remember when observing galaxies is that the size of the telescope is what is important, and the bigger, the better.

Our own galaxy, called the Milky Way, consists of more than 250 billion stars, with our Sun being a typical specimen. It is a fairly large spiral galaxy and it has three main components: a disk, of which the solar system is a very tiny part, a central bulge at the core, and an all encompassing halo.

M82. (Image by Adam Block/NOAO/AURA/NSF, using a 0.4 m Meade LX200 telescope)

M61. (Image by Adam Block/NOAO/AURA/NSF, using a 0.4 m Meade LX200 telescope)

A magnificent spiral, the colorful Whirlpool Galaxy, M51. (Image by Adam Block/NOAO/AURA/NSF, using a 0.4 m Meade LX200 telescope)

The disk of the Milky Way has four spiral arms and it is approximately 300 parsecs (pc) thick and 30 kiloparsecs (kpc) in diameter. It is made up predominantly of Population I stars which tend to be blue and are reasonably young, spanning an age range between a million and ten billion years.

The bulge, at the center of the galaxy is a flattened spheroid, measuring approximately 1 kpc by 6 kpc. This is a high-density region where Population II stars predominate – stars which tend toward red and are very old, about 10 billion years. There is growing evidence for a very massive black hole at the center of the Milky Way.

The halo, which is a diffuse spherical region, surrounds the disk. It has a low density of old stars mainly in globular clusters, discussed above. The halo is believed to be composed mainly of dark matter which may extend well beyond the edge of the disk.

Galaxies are preferentially found in groups or larger agglomerations called clusters. The cluster in which our Milky Way galaxy is located is called the Local Group. It consists of the following galaxies. The approximate distance from the Milky Way to each galaxy, in kiloparsecs, is given in parentheses.

Wolf–Lundmark–Melotte Galaxy (1300)

IC 10 (1300)
Cetus Dwarf (925)
NGC 147 (750)
Andromeda III (900)
NGC 185 (775)
M110 (900)
Andromeda IV (900)
M32 (900)
M31 (900)
And I (900)
Small Magellanic Cloud (65)
Sculptor Dwarf (90)
LGS 3 3000
IC 1613 (900)
Andromeda V (900)
Andromeda II (900)

M33 (925)
Phoenix Dwarf (500)
Fornax Dwarf (160)
UGCA 86 (1900)
UGCA 92 (925)
Large Magellanic Cloud (55)
Carina Dwarf (90)
Leo A (2150)
Sextans B (1225)
NGC 3109 (1250)
Antlia Dwarf (1250)
Leo I (270)
Sextans A (1225)
Sextans Dwarf (90)
Leo II (250)
GR 8 (1550)
Ursa Minor Dwarf (75)
Draco Dwarf (85)
Sagittarius Dwarf Elliptical Galaxy (25)
Sagittarius Dwarf Irregular Galaxy (600)
NGC 6822 (525)
Aquarius Dwarf (600)
IC 5152 (925)
Tucana Dwarf (925)
Andromeda VII (900)
Pegasus Dwarf (1850)
Andromeda VI (900)

Regular clusters of galaxies have a concentrated central core and a well-defined spherical structure. These are subdivided according to their richness, which is defined by the number of galaxies within 1.5 megaparsecs (Mpc) of the center. This distance is known as the Abell radius. Typically, they have a size in the range 1–10 Mpc. A good example of a regular galaxy cluster is the Coma cluster, a very rich cluster with thousands of elliptical galaxies inside the Abell radius.

Irregular clusters of galaxies have no well-defined center, but are roughly the same size. They generally contain far fewer galaxies and have a mass one-tenth to one-thousandth

Abell 426, a cluster of galaxies in Perseus. (Image by Ed Grafton, Houston, Texas, using a Celestron 14 and an ST5c CCD camera)

of that of a regular galaxy cluster. An example is the nearby Virgo cluster.

The largest structures in the universe are superclusters of galaxies. Superclusters usually consist of chains of about a dozen galaxy clusters which have a mass approximately ten times that of a regular galaxy cluster. Our own Local Supercluster is centered in the direction of the constellation Virgo and is relatively poor, having a size of 15 Mpc. The largest superclusters, like that associated with Coma, are up to 100 Mpc in extent.

The most distant objects visible may be quasars (quasi-stellar radio sources). Some unimaginable source is powering their incredible energy output. They seem to be fairly small, on the order of 0.1% the size of an average

Not a very impressive picture? Think again. The arrow points to the brightest quasar discovered, 3C273, in Virgo. (300 mm Takumar lens, f/4, 20 minutes on Kodak 103a-E, North Sandwich, NH, April 1977. Image by David Healy, Sierra Vista, Arizona)

galaxy, but are apparently emitting a thousand times more energy than a galaxy.

Measurements of the velocities of galaxies and their deviations away from the general expansion of the universe can be made. Such studies have revealed enormous, similar motions of large numbers of galaxies on scales in excess of 60 Mpc. Consistent with these motions, our Milky Way is moving at about 600 km/s toward a distant object astronomers have named the "Great Attractor." The Great Attractor lies at a distance of 65 Mpc in the direction of the constellation Centaurus and has a mass approaching 5×10^{16} solar masses. Detailed investigation of that region of the sky finds ten times too little visible matter to account for this movement, implying a dominant role for what astronomers call "dark matter," discussed below.

Other astronomical surveys reveal a very bubbly structure to the universe with galaxies primarily confined to regions which resemble sheets and filaments. The areas between these structures, called voids, are the dominant feature and have a typical diameter of about 25 Mpc. They fill about 90% of space and the largest observed, called the Bootes void, has a diameter of about 124 Mpc. Other features that have been observed are the "Great Wall," a "sheet" of galaxies 100 Mpc long at a distance of about 100 Mpc.

Dark matter

Many astronomers believe that the universe contains a very large amount of what is called dark matter. This matter either gives off no light of its own or is of such a scale that our present observational techniques cannot detect it. Many forms of dark matter may exist. There could, for example, be a huge number of Jupiter-type planets and/or many low-luminosity stars (red and brown dwarves). There are indications that there is up to ten times more dark matter associated with each galaxy than previously estimated by astronomers.

There you have it. From the (relatively) small scale of objects which inhabit our solar system all the way out to superclusters of galaxies. As an amateur astronomer on the road to observation, imaging or more, your choices are certainly not limited.

Now that we have briefly described the objects which populate our incredible universe, you can be on your way to studying them for yourself. As a step along the journey, let's briefly touch upon three areas an amateur astronomer must be at least passively familiar with: positional astronomy, time and the calendar, and the magnitude system.

Dark matter? Well, at least a dark nebula. This image shows Crux, the Southern Cross, and the smallest of the 88 constellations. The dark region known as the Coal Sack is very well-defined. (Photo by Steve Coe, Phoenix, Arizona)

1.2 Positional astronomy

What good are Mercator's
North Poles and Equators
Tropics, Zones, and Meridian Lines?
So the Bellman would cry,
and the crew would reply
"They are merely conventional signs"

Lewis Carroll – *The Hunting of the Snark*

Positions and coordinate systems

One of the first steps that we, as observers, need to do is to be able to give reasonable descriptions for the positions of objects. This is done by assigning numbers to each position in space. These numbers are called coordinates and the system defined by this procedure a coordinate system.

Coordinate systems are based at a reference point in space from which the positions are measured. We call this the origin of the reference frame. Origins can be such things as the location of the observer, the center of Earth, the Sun, or the Milky Way. Any location in space is then described by its distance and its direction from the origin. The direction is given by following a straight line from the origin through the location (and on to infinity). In the coordinate systems used in astronomy, the direction is fixed by two angles, based on a defining plane of reference and a defining axis of reference. Let's look at an example.

For us living on the surface of the Earth, the coordinate system used is defined by longitude and latitude. The natural reference plane used is the one defined by the Earth's equator, and the natural reference axis used is the rotational axis which is determined by an imaginary line joining the planet's north and south poles. (To be fair, the poles are also defined as the two points on the Earth's surface equidistant from every point on the equator.) We can then define circles along the Earth's surface which are parallel to the equator – latitude circles. From any of these circles the angle at the planet's center is constant for all points on the circle. Half circles from pole to pole, which are perpendicular to the equator, are called meridians. One of the meridians, long ago defined as the one passing through the Greenwich Observatory in London, England, is taken as the reference meridian. Longitude is the angle between this and whichever meridian is under consideration, also measured from the center of the Earth.

A quick aside to talk about angles. There are 360 degrees in a circle, with the degree sign noted as °. Each degree is divided into 60 minutes, designated ′. Each minute is also divided into 60 seconds, shown as ″.

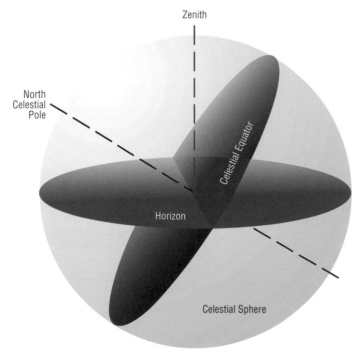

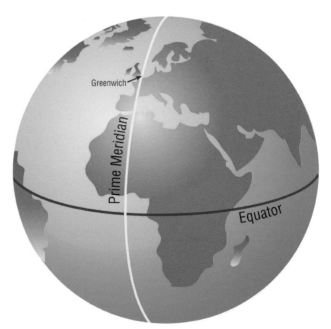

The Prime Meridian runs through Greenwich, England. Along this line the longitude is 0°. Longitude increases both east and west to 180°. (Illustration by Holley Y. Bakich)

The celestial sphere as it appears from the latitude of El Paso, Texas. The angle between the celestial pole (north or south) and the horizon is equal to the latitude. (Illustration by Holley Y. Bakich)

This is a real angle of 1°. Imagine splitting it into 3600 equal parts and then selecting only one of those. That is what an arcsecond is. (Illustration by Holley Y. Bakich)

There are four basic systems of astronomical coordinates: the altazimuth coordinate system, the equatorial coordinate system, the ecliptic coordinate system, and the galactic coordinate system. These systems have a common thread: all celestial objects are considered to be located on the inner surface of something called the celestial sphere.

Up to about 400 years ago, the common belief was that the heavens were made of the Sun, Moon, planets and a solid sphere to which the stars were attached. Although this idea was dead wrong, it is helpful for us to consider what astronomers call the celestial sphere. It is an imaginary sphere of infinite size centered on the Earth and representing the entire sky. This concept works because the distances to planets, stars, etc., are not discernible to the eye, so they appear to be positioned on a great sphere very far away.

The celestial sphere is used for describing the positions and motions of astronomical objects. So, any one of these can be thought of as being located at the point where the line of sight intersects the surface of the celestial sphere. The beauty of this is that it can be done without actually knowing the distance to any of the objects. In astronomical coordinate systems, the coordinate axes are great circles on the celestial sphere.

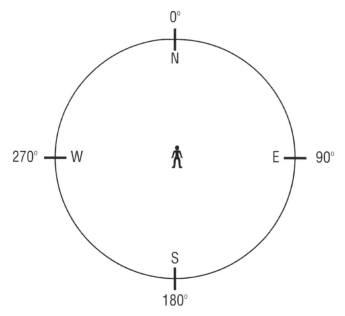

The azimuth scale. North is 0°. Azimuth increases toward the east. (Illustration by Holley Y. Bakich)

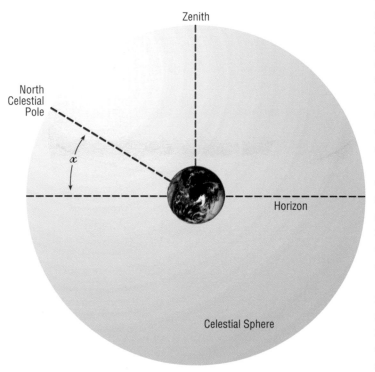

The height of either celestial pole, in degrees, is equal to the latitude of the observer. (Illustration by Holley Y. Bakich)

System 1. The altazimuth coordinate system

This coordinate system is often called the altazimuth co-ordinate system, or sometimes even the horizon coordinate system. Using this system, the position of an object on the celestial sphere is described relative to an observer's zenith and horizon. The coordinates of an object in this system are its altitude and azimuth. Altitude is the number of degrees from the horizon to the object. It ranges from 0° for an object on the horizon, to 90°, for an object at the zenith. If the true horizon cannot be seen (a common occurrence due to trees, building, mountains, etc.), then the altitude, in degrees, is 90 minus the distance from the zenith. So, if an object is 40° from the zenith, its altitude would be 90 − 40, or 50°.

To understand azimuth, let me define the term vertical circle. A vertical circle actually can be thought of as one-quarter of a circle that starts at the horizon and ends at the zenith. Azimuth, then, is measured along the horizon from north to the point where the body's vertical circle intersects the horizon. Since the azimuth coordinate can cover a full circle, it varies between 0° and 360°. The north point is 0° (or 360°), east is 90°, south is 180° and west is 270°. The main disadvantage of the horizon system is that the altitude and azimuth of a celestial body are constantly changing. This is due to the continuous rotation of the Earth. This problem can be removed by using a coordinate system which is fixed on the celestial sphere. So as the Earth rotates and the sky appears to move overhead, the coordinate system moves with it.

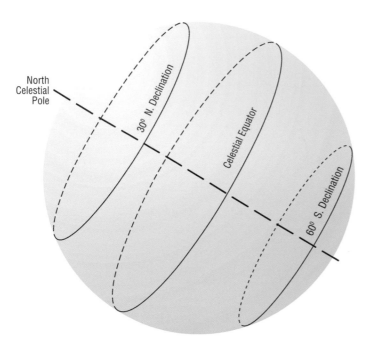

Declination is measured north or south of the celestial equator, which lies at declination 0°. All circles of declination are parallel to the celestial equator. (Illustration by Holley Y. Bakich)

System 2. The equatorial coordinate system

Another frequently used coordinate system is the equatorial coordinate system. In looking at this system, let us pose the question: what if we were able to project the Earth coordinates of longitude and latitude into the sky? The "latitude" would not be a problem, but the "longitude" in our new system would continue to change moment by moment due to the Earth's rotation. Somehow, the "longitude" needs to be attached to the sky.

Imagine projecting Earth's equator and poles to the celestial sphere. This produces the celestial equator as well as the north and the south celestial pole. Great circles through the celestial poles are always perpendicular to the celestial equator and are called hour circles. To designate the position of a star, consider one of these imaginary great circles passing through the celestial poles and through the star in question. This is the star's hour circle, and it corresponds to a meridian of longitude on Earth.

The first coordinate in the equatorial system, corresponding to the latitude, is called declination, and is the angle between the position of an object and the celestial equator (measured, as always, along the object's hour circle). It varies from 0°, for an object on the celestial equator to 90° north or south. Sometimes, + is used for objects having north declination and − is used for objects south of the celestial equator.

All that's left is to set the zero point of the "longitude" coordinate, which has come to be called right ascension. The origin of this term is somewhat obscure. The term "ascension" is just what it says: a noun that comes from the same root as the verb "to ascend." This is what stars appear to do if you observe them in the east. Then, I am told, if you measure that ascension with respect to a right celestial sphere (i.e. one on which the fixed plane of reference is the plane through the celestial equator), it is right ascension.

Back now to setting the zero point of the right ascension coordinate. For this, the intersection point of the Earth's equator and its orbital plane, the ecliptic, is used. This is called the March (or vernal) equinox, but is sometimes referred to as the first point of Aries. As the Earth orbits the Sun, the Sun appears to move through this point each year around 21 March, crossing the celestial equator moving from south to north.

We can now measure the angle between the vernal equinox and the point where the hour circle of the object intersects the celestial equator. This angle is called the object's right ascension and is measured in hours, minutes, and seconds rather than in the more familiar degrees, minutes, and seconds. (Sometimes, astronomers do things just to be different.) A circle contains 360°. Since the day is divided into 24 hours, the circle of right ascension also has 24 hours, each hour corresponding to 15°. It follows, then, that one minute of right ascension equals 15 arcminutes (15′) of angular measure and one second of right ascension equals 15 arcseconds (15″). Right ascension is always measured from west to east starting at the vernal equinox, which becomes the starting point of 0 hours. Also, since the vernal equinox lies on the celestial equator, its declination is also 0°.

The two intersections of the ecliptic and the celestial equator occur at the March equinox (labeled 0ʰ), when the Sun is headed north, and the September equinox, when the Sun is headed south. Right ascension increases to the east from the March equinox. Any celestial object lying on the solid curve will have a right ascension of 3 hours. (Illustration by Holley Y. Bakich)

Regarding the right ascensions and declinations of celestial objects for the purposes of amateur astronomy, some coordinates change and some do not (except over very long periods of time). Objects whose coordinates change include the Sun, Moon, planets, asteroids, and comets. Objects whose coordinates may be said not to change include stars, nebulae, clusters, and galaxies.

Because of small but constant changes of the rotation axis of Earth, mainly caused by the gravitational pulls of the Sun and Moon, the vernal equinox is not always in the same place. Its position changes slowly, so that the whole equatorial coordinate system is also slowly changing with time. Therefore, it is necessary to specify an epoch (a moment of time) to which the coordinate system is referenced. Currently, most sources use epoch 2000.0, the beginning of the year 2000 AD. In the past, star charts have reflected a change of epoch every 50 years. This may be reduced to every 25 years, however, due to increased calls for more accuracy. Also, it sells more star charts!

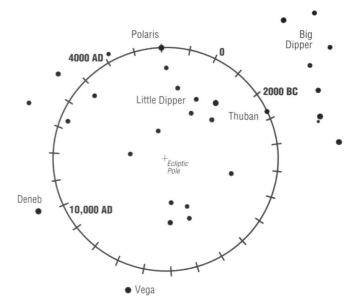

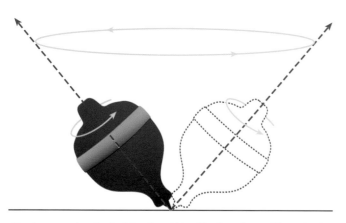

Due to precession, the Earth's axis wobbles, like that of a top. (Illustration by Holley Y. Bakich)

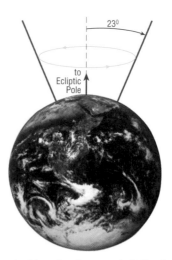

The precessional cycle. A long time from now, both Deneb and Vega will be near the north celestial pole, but not as close as either Thuban or Polaris. (Illustration by Holley Y. Bakich)

For completeness, this change of position of the vernal equinox is due to something called precession. Because of the gravitational attraction of the Sun and Moon, the Earth's axis wobbles, like that of a dying top. This is a very slow motion. One complete precessional cycle takes approximately 26 000 years. The effect of precession is twofold. (1) Precession changes the position of the north (and south) celestial pole. Throughout history, different "pole stars" have been recognized. In the days of the early pharaohs of Egypt, Thuban, the alpha star of the constellation Draco, was the north star. (2) Precession also moves the position of the vernal equinox slowly westward through the stars.

Above, I mentioned that the vernal equinox is sometimes referred to as the first point of Aries. This is because, 2000 years ago, the position of the vernal equinox lay within the confines of the constellation Aries. Today, it is in Pisces, the Fishes. In 1969, the American singing group The Fifth Dimension released a song entitled "The Age of Aquarius." It

spoke of love, peace, and harmony at the dawning of the age of Aquarius. (Here, we are defining "age of . . . " as the constellation containing the vernal equinox.) We are, indeed, moving toward the age of Aquarius, but not because the Moon is in the seventh house or because Jupiter is aligning with Mars, as the song says. We are at the dawning of the age of Aquarius because of our old friend, precession. But don't look for worldwide peace and understanding just yet. This momentous event is still roughly 800 years away.

Effects of Earth's rotation

As mentioned above, Earth's (or another celestial body's) rotation has remarkable effects on the appearance of the sky: stars and other celestial bodies appear to rotate around the celestial poles (as actually Earth rotates and carries the observer away below them), i.e. move along circles of constant declination in the co-rotating equatorial system.

An interesting relationship exists between your earthly

latitude and a star's declination. First, the northern hemisphere rule: if a star's declination is greater than 90° minus your latitude, that star will never set. It is known as a circumpolar star. In the southern hemisphere, if a star's declination is less than the latitude (taken as a positive number) minus 90° the star will be circumpolar. Any number in the southern hemisphere will be, by necessity, negative.

System 3. The ecliptical coordinate system

In the ecliptical coordinate system, the fundamental reference plane is chosen to be the ecliptic, which as we learned above, is the orbital plane of the Earth around the Sun. If we define two points which are 90° away from every point on the ecliptic, we will have the north and the south ecliptic pole. The north ecliptic pole lies in the constellation Draco and the south ecliptic pole is found within the boundaries of the constellation Dorado.

Ecliptic latitude is defined as the angle between any object and the ecliptic and, like declination, has values between −90° and +90°. Ecliptic longitude, like right ascension, begins at the vernal equinox and runs from 0° to 360°, also measured from west to east.

Ecliptical coordinates are most frequently used for solar system calculations such as planetary and cometary orbits and appearances. For this purpose, two ecliptical systems are used: the heliocentric coordinate system with the Sun at its center, and the geocentric one with the Earth at its center. The ecliptical coordinate system is almost never used by amateur astronomers.

System 4. The galactic coordinate system

This coordinate system is most useful for considerations of objects beyond the solar system, especially for considerations of objects of our Milky Way galaxy, and sometimes beyond. Again, it is rarely used by amateur astronomers and, like the ecliptical system, will only be considered briefly here.

In the galactic coordinate system, the galactic equator is used as reference plane. This is the great circle of the celestial sphere which is pretty well defined by the visible Milky Way. The zero point for galactic longitude is defined as being in the direction of the galactic center, which lies within the constellation Sagittarius. Galactic latitude is the angle between any object and the galactic equator and (again like declination) runs from −90° to +90°. Galactic longitude runs from 0° to 360°. The galactic north pole lies within the boundaries of the constellation Coma Berenices. The galactic south pole is found in Sculptor.

Like the ecliptical system, the galactic coordinate system alternately uses either the Sun or the Earth as its center. As if those weren't enough, for some calculations the center of the Milky Way galaxy itself is the origin point for this coordinate system. In such cases, the interesting term galactocentric galactic coordinates is used!

1.3 Time and the calendar

Time

Time may be defined as the continuous and irreversible progression of existence. In amateur astronomy, it is the basis for correlating all our observations. It is the one element of all observations that immediately translates as understandable to all astronomers, amateur and professional alike. As such, it is very important. It is not enough to enter in your log, "Saw the occultation of Saturn by the Moon tonight." Other questions must be answered. What was the date? What time did the event begin? How long did it last? An observation without these facts is essentially useless. But we need to be even more specific. For amateur astronomers, it is important to understand and use what is called Universal Time (abbreviated UT).

Sidereal time

The basic unit of astronomical time measurement is the day, one rotation of the Earth on its axis. But there is more than one way to define "day." We can define it based on the Sun. One solar day is the time for the Sun to leave and return to your local meridian.

Sidereal time is referenced to the stars. It is defined as the right ascension of a star (real or imagined) on the observer's meridian. (Another way to think of sidereal time is the

to distant star

Sun

extra

Earth's motion in 24 hours

A solar day is slightly longer than a sidereal day because the Earth is not stationary. As we orbit the Sun, the Earth must rotate the "extra" amount to place the Sun back on the meridian. (Illustration by Holley Y. Bakich)

length of time since the vernal equinox has crossed the local celestial meridian.) Since the meridian differs for each observer, sidereal time is a concept good only for one location. Another way to define sidereal time is as the hour angle (the number of hours from the meridian, measured westward) of the vernal equinox. Our everyday (civil) time is referenced to the (average) motion of the Sun, and not to the stars. Thus, sidereal time generally does not coincide with the everyday (clock) time. To be precise, the sidereal time agrees with the solar time only at the autumnal equinox; at any other time, they differ (they are exactly 12 hours apart at the time of the vernal equinox).

The sidereal day is defined to be the length of time for the vernal equinox to return to your celestial meridian. Compare this to the solar day, which is defined to be the length of time for the Sun to return to your celestial meridian. The two are not the same, and here's why.

Because the Earth is revolving around the Sun as well as rotating on its axis, in the course of a day the Earth must turn 3 minutes and 56 seconds longer to bring the Sun back to the celestial meridian than to bring the vernal equinox back to the celestial meridian. So the solar day is 3 minutes and 56 seconds longer than the sidereal day. It is this discrepancy that causes the difference between sidereal and solar time. This time difference between the sidereal and solar days is also responsible for the fact that different constellations are overhead at a given time of day throughout the year.

When the vernal equinox is on your local meridian, the sidereal time is 0^h (0^h is the right ascension of the vernal equinox). As time passes, the stars overhead seem to move. Later, you may note that a star on your local meridian has a right ascension of 1^h. The sidereal time at your location is 1^h. So sidereal time is defined as the right ascension of a "star" (real or imagined) which lies on your local meridian. Sidereal time is not like time zones, where an entire area has the same time. The sidereal time at your observing location is different from your friend's across town.

Universal Time

Using sidereal time worldwide is impractical. That is why the times of various events, particularly astronomical and weather phenomena, are often given in UT. This practice started in 1928, when the International Astronomical Union recommended that the time used in the compilation of astronomical almanacs be referred to as Universal Time. Sometimes you might still hear the old term "Greenwich Mean Time" (abbreviated GMT) used, but that is passing out of style. Either of these terms is used to refer to time kept on the zero degree longitude meridian, the one running through the old Greenwich Observatory in England. This meridian is five hours ahead of US Eastern Standard Time. Times given in UT are given in terms of a 24-hour clock. Thus, 14:02 (often written simply 1402) is 2:02 p.m., and 21:47 (2147) is 9:17 p.m. Sometimes a Z is appended to a time to indicate UT, as in 0539Z.

When a precision of one second or better is needed, however, it is necessary to be more specific about the exact meaning of UT. For that purpose different designations of UT have been adopted. In astronomical and navigational usage, UT often refers to a specific time called UT1, which is a measure of the rotation angle of the Earth as observed astronomically. It is affected by small variations in the way the Earth spins, and can differ slightly from the time on the Greenwich meridian. Times labeled UT in data found in astronomical almanacs produced by the US Naval Observatory are UT1.

However, in the most common civil usage, UT refers to a time scale called "Coordinated Universal Time" (abbreviated UTC), which is the basis for the system of time used worldwide. This time scale is set by time laboratories around the world and is determined using highly precise atomic clocks. These clocks provide the international standard UTC which is accurate to approximately a nanosecond (billionth of a second) per day. The length of a UTC second is defined in terms of an atomic transition of the element cesium under specific conditions, and is not directly related to any astronomical phenomena.

UTC is the time distributed by standard radio stations that broadcast time, such as WWV and WWVH. It can also be obtained readily from the Global Positioning System (GPS) satellites. UTC is the basis for civil standard time in the US and its territories. Standard time within US time zones is an integral number of hours offset from UTC.

Historical note

Prior to 1948, the observatory at Greenwich was known as the Royal Observatory. It was set up on a hill near the River Thames and commanded a view of the London Docks. In 1948, the observatory moved to Herstmonceux Castle in Sussex, becoming the Royal Greenwich Observatory. That was the name given to it despite the somewhat obvious fact that it was no longer at Greenwich. The (former) site at Greenwich became known as the Old Greenwich Observatory. Over time, the historic buildings and instruments were incorporated into the National Maritime Museum, the main buildings of which are located at the foot of Observatory Hill, close to the river. Following the closing of the RGO in the fall of 1998, the Old Greenwich Observatory was renamed the Royal Observatory Greenwich.

Greenwich Mean Time

Greenwich Mean Time is a time scale based on the apparent motion of the "mean" Sun with respect to the meridian through the Old Greenwich Observatory (zero degrees longitude). The "mean" Sun is used because time based on the actual or true apparent motion of the Sun doesn't "tick" at a constant rate. The Earth's orbit is slightly eccentric and the plane of the Earth's orbit is inclined with respect to the equator (about $23\frac{1}{2}$ degrees) hence at different times of the year the Sun appears to move faster or slower in the sky. That's why an uncorrected sundial can be "wrong" (if it is supposed to be telling mean time) by up to 16 minutes. So if the mean (i.e. corrected) Sun is directly over the meridian through Greenwich, it is exactly 12 noon GMT or 12:00 GMT. (Prior to 1925, astronomers reckoned mean solar time from noon so that when the mean Sun was on the meridian, it was actually 00:00 GMT. This practice arose so that astronomers wouldn't have a change in date during a night's observing. Some in the astronomical community still use the pre-1925 definition of GMT in the analysis of old data although it is recommended that the term Greenwich Mean Astronomical Time now be used to refer to time reckoned from noon.) Mean time on selected meridians 15° apart is generally known as standard time. For example, Eastern Standard Time (EST) is the mean solar time of the meridian at 75° W.

GMT and the BBC

The BBC began transmitting time signals in 1924. Since that time, the major global news headlines of the day have been preceded by the six Greenwich Time "pips." The pips were suggested after the chimes of Big Ben were first broadcast at midnight, 1 January 1924. Not long after that, on 5 February 1924, at the recommendation of the then Astronomer Royal, Frank Dyson, the six-pip time signal (officially known as the Greenwich time signal) was inaugurated. The six-pip time signal (pips for seconds 55, 56, 57, 58, 59, 00) was Dyson's brainchild, devised in discussion with Frank Hope-Jones, inventor of the free pendulum clock. The sixth pip signals the start of the next minute.

Control of the BBC's six pips was taken over by the Royal Observatory in 1949 from Abinger to where the time service had moved during the Second World War. The time service moved to Herstmonceux in 1957. The time service at Herstmonceux closed down during February 1990 when the BBC took over generation of the six pips. Since 5 February 1990, the 66th anniversary of the start of the Greenwich Time Service, the six pips have been synchronized to UTC by using the GPS satellite signals which are picked up by a pair of GPS receivers atop Broadcasting House in London.

For more on time, see the "Keeping time" section in the chapter on "Telescope accessories".

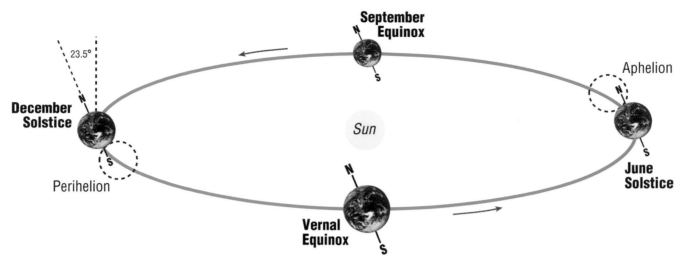

The Earth's orbit. Positions of equinoxes, solstices, perihelion, and aphelion are marked. (Illustration by Holley Y. Bakich)

The calendar

In addition to the day, there are two other naturally occurring time cycles of longer period. The first, based on the phases of the Moon, is the month. The second, based on the Sun, is the year.

Today's "month" is a convenience. It was found too troublesome to keep track of the exact time of New Moons, so months became roughly equal groups of 30 or 31 days (except for poor February) that filled up a year.

The year, like the day, has a basis in astronomy. It is the time it takes the Earth to orbit the Sun once. And although the year may be defined many different ways, for amateur astronomy the basic definition of the year as 365.2422 days is enough.

A great reference regarding all things calendric is *Calendar: Humanity's Epic Struggle to Determine a True and Accurate Year* by David Ewing Duncan (Avon Books, New York, 1999). For a definitive reference, very detailed, with equations, I recommend *Mapping Time: The Calendar and Its History* by E. G. Richards (Oxford University Press).

Julian date

The Julian day number system of time reckoning is a convenient way to determine the number of days between one event and another. You simply subtract days.

In 1583, the Italian philologist Joseph Justus Scaliger (1540–1609), combined three periods of time into what he called the Julian period. These were: (1) the solar cycle of 28 years, when the days of the week and the days of the month in the Julian calendar coincide; (2) the Metonic cycle of 19 years, as 19 solar years are roughly equal to 235 lunar months, and (3) the Roman indiction of 15 years, which had been decreed long before by the Emperor Constantine. The last time these periods coincided was on 1 January 4713 BC.

Incidentally, Scaliger's father's name was Julius Caesar Scaliger, so it has been conjectured that he called it the Julian period in honor of his father. However in his *De Emandatione Temporum* (Geneva, 1583) Scaliger says "We have termed it Julian because it fits the Julian year."

Although Scaliger did invent the Julian period, the British astronomer John F. W. Herschel (1792–1871) was the one who turned the idea into a complete time system (see *Outlines of Astronomy* by J. F. W. Herschel, Longmans, Brown, Green & Longmans, London, 1849, sections 928–931). Thus, 1 January 4713 BC became Julian Day (JD) number 1.

Julian days begin at noon, Universal Time. 1 January 2000 UT was Julian day number 2 451 544.5 (at noon on that date, the JD was 2 451 545). Because Julian dates are so large, astronomers often make use of a modified Julian date (MJD). The conversion is a simple subtraction

$$MJD = JD - 2\,400\,000.5$$

1.4 The magnitude system

The first known observer to describe and catalog differences in brightnesses of stars was the Greek astronomer Hipparchus, who lived in the second century BC. He divided his listing of approximately 850 visible stars into six brightness ranges, or magnitudes. The brightest he classed as stars of the first magnitude, and the faintest as stars of the sixth magnitude. His system was used, almost unchanged, for more than 1800 years.

But then came Galileo. In addition to discovering the phases of Venus, the large moons of Jupiter and many other things, he noted that his telescope did not just simply magnify, but also revealed that which was hitherto invisible. Writing in the *Sidereus Nuncius* in 1610, Galileo stated, "Indeed, with the glass you will detect below stars of the sixth magnitude such a crowd of others that escape natural sight that it is hardly believable." Then he coined a term which had not been used before. He called the brightest of the stars below naked eye visibility "seventh magnitude."

Many stars too faint to be detected by the unaided eye are visible in this photograph of h and χ Persei. (8" Celestron Schmidt camera, f/1.5, 15 minutes on TP 2415 film, hypersensitized 9.5 days in forming gas at 30° C, 8–9 Oct 1980, Naco, Arizona. Image by David Healy, Sierra Vista, Arizona)

So, after the invention of the telescope, it was deemed necessary to expand the magnitude system. Many stars fainter than those listed as sixth magnitude by Hipparchus were now visible. In addition, it was noted that stars of first magnitude varied greatly in brightness. Around the time of the great observational astronomer Sir William Herschel, a loose system was adopted that defined two stars differing by one magnitude as having a brightness difference of approximately two and a half. An important, albeit somewhat strange, point to always remember is that the smaller the number, the brighter the object. Or, fainter things have larger, more positive, magnitudes. The 30 brightest stars are listed in Appendix B.

William R. Dawes (1799–1868) proposed, in 1851, a simple and effective method of photometric comparison, depending upon the principle of equalization by limiting apertures, to the problem of a fixed standard of stellar magnitude. (*Monthly Notices of the Royal Astronomical Society*, vol. xi, p.187.)

In 1856, Norman R. Pogson suggested that all observations be calibrated by using the constant $10^{2/5}$. The ratio between magnitudes thus becomes approximately 2.511 886 5. That is, a star of a given magnitude is 2.511 886 5 times brighter than a star one magnitude fainter. At that time, the concept of using magnitudes equal to and less than zero also came into being. The rationale involved keeping some semblance of the original system, where the general limiting magnitude of the human eye was approximately sixth magnitude. With this limitation and Pogson's mathematical formula, it became evident that the brightest stars were much brighter than first magnitude, to say nothing of the bright planets, the Moon, and, of course, the Sun.

At the time, these intervals of magnitude were based on the nineteenth century belief of how the human eye perceives differences in brightnesses. It was thought that the eye sensed differences in brightness on a logarithmic scale. Thus, a star's magnitude is not directly proportional to the actual amount of energy we receive. Today, we know that the eye is not quite a logarithmic detector. Our eyes perceive equal ratios of intensity as equal intervals of brightness. What this means is that a star of fifth magnitude does not appear to the eye to be exactly halfway in brightness between stars of fourth and sixth magnitudes. Close, but not exactly.

The number given above − 2.511 886 5 − is the fifth root of 100. Therefore, a difference of five magnitudes is equal to a 100-fold difference in actual brightness. Thus Sirius (α CMa) at magnitude −1.46 is 100 times as bright as Wasat (δ Gem) at magnitude +3.53.

The magnitude system

Magnitude difference	Brightness ratio
0.1	1.0964782
0.2	1.2022644
0.25	1.2589254
0.3	1.3182567
0.333	1.3593563
0.4	1.4454397
0.5	1.5848932
0.6	1.7378008
0.666	1.8478497
0.7	1.9054607
0.75	1.9952623
0.8	2.0892961
0.9	2.2908677
1.0	2.5118865
1.5	3.9810719
2.0	6.3095738
2.5	10.000000
3.0	15.848932
3.5	25.118865
4.0	39.810719
4.5	63.095738
5.0	100.00000
5.5	158.48932
6.0	251.18865
6.5	398.10719
7.0	630.95738
7.5	1000.0000
8.0	1584.8932
8.5	2511.8865
9.0	3981.0719
9.5	6309.5738
10.0	10000.000
11.0	25118.865
12.0	63095.738
12.5	100000
13.0	158489.32
14.0	398107.19
15.0	1000000
16.0	2511886.5
17.0	6309573.8
17.5	10000000
18.0	15848932
19.0	39810719
20.0	100000000

In the table given opposite detailing the magnitude system, if values are sought which are not listed, simply multiply the ratios of the magnitude differences which, when added, give the desired difference. For example, to find the brightness ratio between Antares (α Sco) at magnitude 1.2 and Ras Algethi (α Her) at magnitude 3.5 (difference = 2.3), simply multiply the ratios of 6.309 573 8 (for a magnitude difference of 2) and 1.318 256 7 (for a magnitude difference of 0.3). Thus Antares is 6.309 573 8 × 1.318 256 7 = 8.317 637 9 times as bright as Ras Algethi, or 8.3 times as bright, approximately. Just remember when you are working to find a difference in magnitudes, add and subtract magnitudes but multiply and divide intensities.

Adding magnitudes involves the use of the simple formula (from *Astronomical Formulae for Calculators* by Jean Meeus (Willmann-Bell, Inc., Richmond, VA, 1979)):

$$m_c = m_2 - 2.5\log(10^x + 1)$$

where m_c is the combined magnitude of the system,

$\quad$ $x = 0.4\,(m_2 - m_1)$, and,

$\quad$ m_1 and m_2 are the magnitudes of the stars.

A distinction must now be made. When we talk about the brightness of a celestial object as seen from Earth, we use apparent magnitude (designated m). This is a measure of how bright an object appears to us. But there is also a standardized magnitude that allows objects to be directly compared in terms of their real brightnesses. This is called absolute magnitude (designated m). The absolute magnitude of a celestial object (not associated with the solar system) is the brightness that object would have if it were at a distance of 10 parsecs (32.6 light years). Put another way, the absolute magnitude is a measure of the star's luminosity – the total amount of visible energy radiated by the star. As you might guess, the absolute magnitude tells astronomers much more about a star than its apparent magnitude.

Note: For comets and asteroids a totally different "absolute magnitude" system is used. In such cases, absolute magnitude is defined as the brightness of a comet or asteroid as it would appear to a theoretical observer standing on the Sun if the object were one astronomical unit away.

Within the two divisions of apparent and absolute magnitude there are numerous magnitudes related to how bright objects appear at different wavelengths. Visual magnitude, centered around the yellow and green areas in the spectrum, gives a good approximation of the brightness of a star or other object as seen with the eye. Blue magnitude, a remnant of blue-sensitive photographic emulsions of the past, is yet another magnitude, as is red. Even invisible light from stars can be given its own magnitude. For example, ultraviolet and infrared magnitudes are commonly measured for stars. A photometer with standard color filters is used to determine these magnitudes.

By comparing two different color magnitudes of a star,

astronomers can obtain what is called a color index. The most widely used color index is the difference obtained when one takes a star's blue magnitude and subtracts the visual magnitude. You may see this written as B–V. A quick thought exercise shows that when this value is greater, the star is redder and when it is smaller or negative the star is bluer. The working range of color indices for stars is approximately −0.5 to 2.5.

The brightness of the night sky

How bright is the sky at night? Sky brightness is measured in terms of magnitudes per square arcsecond. Back in the 1980s a study was undertaken at Cerro Tololo Inter-American Observatory (CTIO), part of the National Optical Astronomical Observatory (NOAO) group. The study was published in NOAO Newsletter #10. Astronomers measured the brightness of the night sky throughout the lunar cycle. The values in the table below are from CTIO but should serve as reasonable approximations for most dark sites. The letters B, V, and R represent the filter "colors" of Blue, Visual, and Red. Each number given is in magnitudes per square arcsecond. Remember, the larger the number, the darker the sky.

Notice that the difference is greatest when the light is bluest. This is because our atmosphere scatters blue light the most, so the sky background will be brighter in that color when there is lots of light to scatter. In the table opposite this occurs at Full Moon when the sky is 3.2 magnitudes brighter than at New Moon viewed through a standard blue filter but only one

Orion. (Kodak Elite Chrome 200, pushed 1 stop. Nikon F2 with a 105mm Nikkor lens. Photo by Ulrich Beinert of Kronberg, Germany)

Brightness of the night sky

Lunar Age (days)	B	V	R
0 (New Moon)	22.7	21.8	20.9
3	22.4	21.7	20.8
7 (approx. First Qtr)	21.6	21.4	20.6
10	20.7	20.7	20.3
14 (approx. Full Moon)	19.5	20.0	19.9

magnitude brighter when measured through a standard red filter. A visual filter (or the eye) would see an approximate brightening of the overall sky background equal to 1.8 magnitudes.

Estimating limiting visual magnitude

During an observing session, it is always a good idea to make an estimate of limiting visual magnitude (also called faintest star detection). Not only will this help you determine how good (or bad) the sky is at that particular time, it will also allow you to judge the quality of your recorded observations months or years from that session. Also, making repeated estimates over a long time span cannot help but make you a better observer, more conscious of little details.

Some observers estimate limiting magnitude by eye while others use a telescope. Generally, estimates of telescopic limiting magnitude are done by those whose observing involves very faint objects at the limit of detection.

Observing Tip: If you perform a telescopic limiting magnitude estimate, note in your observing log the telescope aperture and the eyepiece (magnification) through which the estimate was made.

Roger N. Clark, author of the fabulous *Visual Astronomy of the Deep Sky* (Cambridge University Press, 1990) states:

Faintest star detection is a significant function of magnification, both in observational experience and in models of the eye. The magnification dependence is more so for brighter skies. The darker the skies, the less magnification required before you reach the ultimate limit. For example, as morning twilight begins and faint stars disappear, use higher magnification and you can get them back (at least for a while). This magnification dependence probably also accounts for some of the variability in faintest star reports by different observers and the same aperture telescope (of course experience is a factor too).

Most limiting magnitude estimates are performed near the zenith, where sky conditions are usually the best. This is almost always true when estimating limiting visual magnitude by eye.

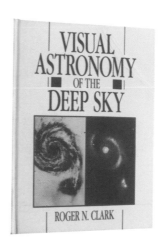

Roger Clark's excellent *Visual Astronomy of the Deep Sky.* (Photo by the author)

If, however, you are studying a particular nebula in depth or cluster of galaxies which lies far from the zenith, you may want to make your estimate near the object(s). Just be sure to do so at an equivalent zenith distance (altitude).

Measuring by eye the seeing and limiting magnitude are two of the exercises you can do to help train your eye for ordinary observing – you'll get more enjoyment from it because you'll see more than you would without doing such exercises.

CHAPTER TWO

Equipment

2.1 Telescopes

Before the telescope

Year	Event
3000 BC	Glass first appears, in Egypt
1500 BC	Oldest known glass vessels made
425 BC	Optical properties discussed
1000 AD	Atmospheric refraction explained
1278 AD	Glass mirror invented
1285 AD	Spectacles invented

Refracting telescopes

Refraction is the bending of light as a result of it passing from one medium (such as air) to another (such as glass). A refracting telescope makes use of this property by using a lens with curved surfaces. As light goes from air to glass and then back to air, its path is deviated toward the optical axis of the lens. If the surfaces of the lens are shaped properly, the light is brought to a focus.

The first telescope was constructed by the Dutch spectacle maker Hans Lippershey (1570–1619), who, on 2 October 1608, filed a patent application for "an instrument for seeing faraway things as though nearby." This was a tube with a convex lens at the front and a concave lens in the rear, which one would look through. The device magnified objects approximately 3×. At the time, there was some contention as to who first invented this device so a patent was never granted.

The Italian inventor Galileo Galilei (1564–1642) built his own telescopes beginning in 1609. Galileo was the first to use the new device to study celestial objects and what he saw revolutionized astronomy forever.

You probably know that the earliest telescopes had very poor optical quality. The lenses had many and varied aberrations. Telescope makers found that if they made their systems with a large focal ratio they could at least minimize the optical defects. The most famous of these inventors were the Dutch astronomer Christiaan Huygens (1629–1695) and the German astronomer Johannes Hevelius. Huygens constructed loosely connected telescopes with focal lengths of 3.6, 7, and 37.5 meters and made significant discoveries with them, including being the first to correctly identify the nature of Saturn's rings. He also reduced the color problems of the objective lens and used optical stops (in his smaller telescopes) to reduce light reflected by the telescope's walls.

Hevelius made telescopes with focal lengths of 18, 22, and 46 meters! These were much more massive units in wooden frames, coupled by pulleys and moved about with the help of a team of assistants. In addition to being difficult to point accurately, they were essentially useless if there was any wind.

One of the problems with early lenses was a defect known as chromatic aberration. White light is made up of all colors. Unfortunately, the colors, when passed through a simple lens, do not focus at the same point. Blue light is refracted more than red light.

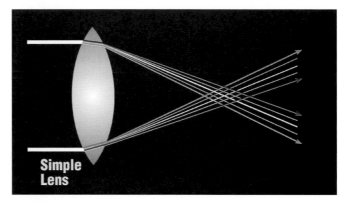

Chromatic aberration. A simple lens will not bring all colors to focus. (Illustration by Holley Y. Bakich)

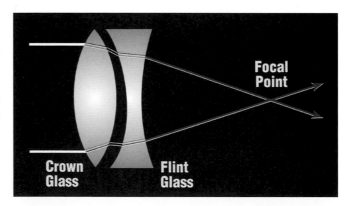

A two-element achromatic lens dramatically reduces chromatic aberration by bringing the red and blue wavelengths to the same focus. (Illustration by Holley Y. Bakich)

In 1729, Chester Moore Hall (1703–1771) devised a lens design which used crown and flint glasses and which gave a relatively color-free image. The word for this type of lens is achromat (not color dependent). At this time, telescope making was a huge business so Hall worked in secret. He actually had the crown and flint lenses made by different optical shops. Hall's achromatic lens was 2.5 inches in diameter and had a 20 inch focal length. This was a monumental achievement, not least because 60 years prior, no less a personage than Isaac Newton had stated that the construction of an achromatic lens was impossible.

The nineteenth century saw a tremendous increase in the technical quality of glass and achromatic lenses. This was highlighted, in 1819, by the production of the lens for the Dorpat Refractor by Joseph Fraunhofer (1787–1826). This was a lens 240 mm in diameter with a focal ratio of f/17.7, which compares favorably with achromats produced today. This is the instrument used by F. G. W. Struve in his discovery and measurement of double stars. When you look at a double star catalog today and see the symbol Σ, it is one of the double stars discovered by Struve with this telescope.

In America, in the middle of the nineteenth century, Alvin Clark (1804–1887) and his sons began making quality telescopes. Their firm made numerous quality refractors, large and small. Their crowning achievement was the largest refractor ever built, which went into operation in 1897, for the Yerkes Observatory of the University of Chicago. Its lens is 1.016 m in diameter. For a detailed history of the Clark firm, I highly recommend the book *Alvin Clark & Sons: Artists in Optics* by Deborah J. Warner and Robert B. Ariail (Willmann-Bell, Inc., 1995).

In the twentieth century, achromatic refractors continued to improve. In the 1920s two problems were tackled: light loss caused by reflections off air-to-glass surfaces and internal reflections within the lens system. Both of these problems were solved by the Clarks and by Carl Zeiss, Inc. of Germany. These firms introduced oil-spaced objectives to solve these problems. The oil eliminated internal reflections and increased transmission by over 2% at each surface. It also smoothed out errors caused by irregularities in the lens' surfaces. The cells which held these lenses had to be nearly perfect, however, or expansion and contraction due to temperature would cause the oil to leak out! Also, after a decade or so, early oils became cloudy and had to be replaced.

In the 1950s, coatings (most notably magnesium fluoride, MgF) were developed and these reduced light loss and internal reflections without the need for oil. A new type of glass, composed of calcium fluoride, CaF_2, was also invented. (The first fluorite objective in a telescope was offered in 1977, by Takahashi Ltd. of Japan.)

In 1951, United Trading Company began selling a high-quality line of refractors known as Unitron. These telescopes were heavily advertised from the 1950s through the 1970s (see any issue of *Sky & Telescope* during this time). All Unitron telescopes were supplied with well-corrected, air-spaced achromatic objectives.

The first refractor lens to be labeled "color-free" was a triple lens system offered by Roland Christen of Astro-Physics, Inc., in 1981. Only two apochromatic lenses were available at that time, both f/11 magnesium fluoride coated oil-spaced triplets. The smaller was a 150 mm and the larger a 200 mm. See Christen's article in *Sky & Telescope*, October 1981, p.376, *An Apochromatic Triplet Objective*. This was the beginning of the new age of apochromatic refractors.

> **Note:** Although apochromats are labeled "color-free," different wavelengths of light do not come to exactly the same focus, although a much better focus than in achromats. Today's apochromatic objectives have two to four lens elements. At least one is made with fluorite or ED (extra-low dispersion) glass, which provides even better color correction.

Thomas Back, designer of high-quality optical systems and owner of TMB Optical in Cleveland, Ohio, has given a wonderful definition of what it takes for a lens to be regarded as apochromatic:

> Any telescope objective that has a Strehl ratio of 0.95 or better at the peak photopic null in the green–yellow part of the visual color spectrum, centered at 555 nm, coma corrected over its full aperture, diffraction limited from the C (red) to F (blue) wavelengths with no more than 1/4 wave spherical optical path difference (OPD) and the violet g wavelength with 1/2 wave or less OPD P–V spherical, satisfies the modern definition of "Apochromatism." Lenses of this quality will be free from secondary color in focus and have extremely sharp and high contrast images.

Advantages of refractors

Good-quality achromatic or apochromatic refractors offer some advantages over reflectors. The first relates to the fact that refractors, by default, have a totally clear aperture. This means that there is no central obstruction causing light to be scattered from brighter to darker areas. Thus, the contrast is better in refractors. Refractors are often cited as the premier instruments for planetary or double star observing.

A second advantage of refractors is their low maintenance. Lenses do not require recoating. In addition, the optical tube assembly of a refractor does not generally require collimation. The lens is fixed into the tube and usually does not become misaligned, lacking some major trauma.

Disadvantages of refractors

Because the refractor is a closed-tube assembly, it can require a longer amount of time to cool to ambient temperature. Today's thin-walled aluminum tubes have reduced this period significantly but it should still be taken into account.

A second disadvantage, related to achromatic refractors, is some chromatic aberration in bright images. This most commonly manifests itself as faint fringes of color around objects like the Moon or Jupiter.

The primary disadvantage of refractors is the expense that is involved in producing a large apo/achromatic lens. The reason is that an apochromatic triplet lens has six surfaces which must be figured. The cost ratio between a 150 mm apochromatic lens and a high-quality 150 mm mirror (only one surface to be figured) is at least 10-to-1.

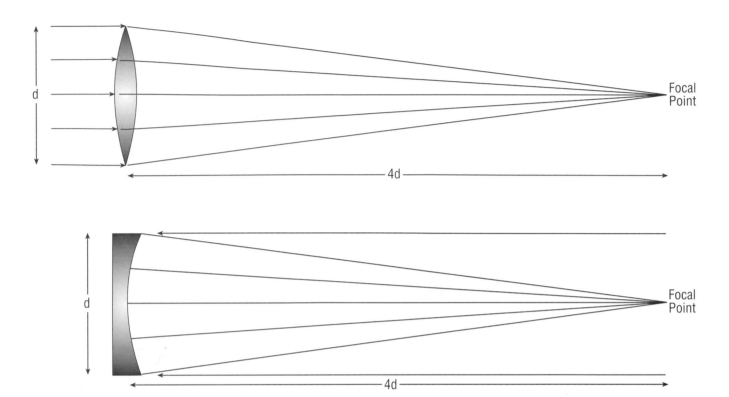

Two similar systems. *d* is the aperture. *4d* is the focal length. Because the focal length is four times the aperture, both are f/4 systems. (Illustration by Holley Y. Bakich)

Reflecting telescopes

The Scottish mathematician James Gregory (1638–1675) invented the first reflecting telescope. He published a description of the reflecting telescope in *Optica Promota*, which was published in 1663. He never actually made the telescope, which was to have used a parabolic and an ellipsoidal mirror.

The first working reflecting telescope was constructed by the great Isaac Newton in 1668. It had a spherical mirror with an aperture of 1 inch and a tube length of 6 inches. Not satisfied with his first effort, he completed an improved and somewhat larger reflector with an aperture of nearly 2 inches. The first "Newtonian" reflector was presented to the Royal Astronomical Society in 1671, and Newton was made a full member.

Early reflectors had mirrors made of speculum, an alloy comprising roughly 80% copper and 20% tin. Once figured and polished, this metal would begin to corrode after only a few months, whereupon it would need to be polished again. Care had to be taken to keep the same figure on the mirror with each polishing.

A Cassegrainian reflector is a type of reflecting telescope with a parabolic primary mirror and a hyperboloidal secondary mirror. Light is reflected through a center hole in the primary mirror, allowing the eyepiece or camera to be mounted at the back end of the tube. The Cassegrain reflecting telescope was developed in 1672 by the French

Kathy Machin, of Kansas City, Missouri, with her homemade 300 mm Dobsonian telescope at the Texas Star Party 2001. (Photo by the author)

sculptor Sieur Guillaume Cassegrain (1625–1712).

In the eighteenth century, Sir William Herschel constructed a number of reflecting telescopes with mirrors of various diameters and focal lengths. With his most famous, a "7-foot" (2.1 m focal length) reflector, he discovered the planet Uranus. This telescope had a mirror 165 mm in diameter. Speculum reflectors reached their height in the middle of the nineteenth century with the 1.8 m mirror in the telescope of William Parsons, Third Earl of Rosse, at Birr Castle, Parsonstown, Ireland.

Gil Machin, of Kansas City, Missouri, christens his 32 cm, f/13.5 classical Cassegrain at the 2001 Texas Star Party. Did I mention that Gil built, not only the telescope, but also the mount? Mechanical excellence and a good performing telescope as well. (Photo by the author)

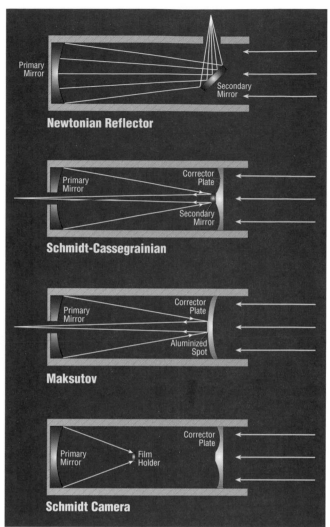

Telescopes using mirrors. Catadioptric telescopes also use a correcting lens. (Illustration by Holley Y. Bakich)

In 1835, a process for depositing a thick layer of silver on glass was developed by the German chemist Justus Leibig (1804–1873). This was a major step forward because when the silver tarnished it could be chemically removed and a new layer redeposited without altering the curvature of the mirror. Apart from tarnishing, silver is not the ideal reflective surface for a telescope mirror. Aluminum, for example, reflects 50% more light. John Donavan Strong, a young physicist at the California Institute of Technology, was one of the first to coat a mirror with aluminum. He did it by thermal vacuum evaporation. The first mirror he aluminized, in 1932, is the earliest known example of a telescope mirror coated by this technique.

Advantages of reflectors

Reflecting telescopes suffer no chromatic aberration. Mirrors have only one optical surface. An apochromatic lens has between four and eight. Mirrors are therefore much less expensive to produce. Telescopes over about 200 mm are all reflectors or catadioptrics (see later).

Disadvantages of reflectors

Because a secondary mirror is used, there is a central obstruction. This causes some scattering of light and loss of contrast. So-called planetary Newtonians have smaller central obstructions (some as small as 16% of the aperture).

All Newtonian reflectors suffer from coma. The smaller the focal ratio, the greater the coma. Also, the uncorrected diffraction limited field becomes smaller. The use of a coma corrector such as the Paracorr at f/5 and below significantly increases the diffraction limited area. This makes it easier to keep the object in that part of the field of view, particularly if your telescope doesn't have a drive.

Regarding maintenance, mirrors may require recoating after several years. And reflectors are quite sensitive to being bumped, jostled or transported. A reflector which is not set up as a permanent instrument should be collimated prior to each observing session. The shorter the focal length, the smaller the collimation tolerances for achieving diffraction limited performance, so accurate collimation becomes much more important.

Large reflectors with thick primary mirrors have a difficult time cooling to ambient temperature. Fans are sometimes employed to aid in the cooling process. Finally, very large Newtonian reflectors require a ladder for use when objects near the zenith are viewed.

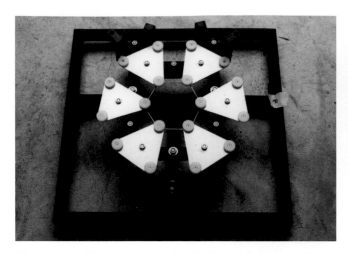

The mirror mount for a 600 mm mirror in a StarMaster telescope. Note the excellent design of the support points which distribute the weight. The open construction also aids in cooling the mirror. (Photo by the author)

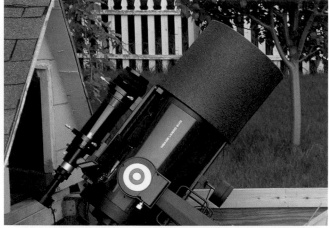

The Schmidt–Cassegrain telescope (SCT) at Everstar Observatory in Olathe, Kansas, is a Meade 250 mm LX200. (Photo by Mark Abraham)

Ritchey–Chretien telescopes

This telescope design was developed jointly by the American optician George Willis Ritchey (1864–1945) and the French optical designer Henri Chretien (1876–1956) in the first decade of the twentieth century. Ritchey, who built the 60 inch and 100 inch mirrors for Mt. Wilson observatory, was so upset by the refusal to use this design for the 100 inch that he publicly criticized it, and was fired. George Ellery Hale, who engaged Ritchey in some heated debates and was the person who fired him, refused to consider this design for the Mt. Palomar 200 inch, choosing instead a Cassegrain design. Personalities notwithstanding, the Ritchey–Chretien design has been used for major telescopes at Kitt Peak, Mauna Kea, Cerro Tololo, the VLT, and even the Hubble Space Telescope.

The lower the amplification factor of the secondary mirror, the flatter the field. The Ritchey–Chretien system has a secondary mirror which magnifies $2.7\times$, whereas the Schmidt–Cassegrain has a $5\times$ secondary. The Ritchey–Chretien design is coma-free, whereas the Schmidt–Cassegrain is not. Produc-tion-type Schmidt–Cassegrains use a spherical primary and secondary which do not correct for coma. Ritchey–Chretien telescopes have hyperbolic primaries and secondaries which correct for coma. Finally, the Ritchey–Chretien design has two optical surfaces. All Schmidt–Cassegrains have four. On the down side, some astigmatism and field curvature have to be compensated for.

Catadioptric telescopes

Catadioptric means pertaining or due to both reflection and refraction of light. They are also known as compound telescopes and are hybrids that have a mix of refractor and reflector elements in their design.

The first compound telescope was made by the German astronomer Bernhard Schmidt (1879–1935) in 1930. The Schmidt telescope had a spherical primary mirror at the back of the telescope, and a glass corrector plate in the front of the telescope to remove spherical aberration. The telescope (or Schmidt camera, as it is often called) is used for photography by placing photographic film at the prime focus.

The Schmidt design is the precursor of today's most popular telescope design, the Schmidt–Cassegrain. This combination of the Cassegrainian telescope with the Schmidt corrector plate was invented in the 1960s. Like the

M31. (8" Celestron Schmidt camera, f/1.5, 30 minutes on TP 2415 film, 13 Nov 1979, Naco, Arizona. Image by David Healy, Sierra Vista, Arizona)

Meade 300 mm SCT, showing the main panel. (Photo by the author)

Cassegrain reflector, a secondary mirror bounces light through a hole in the primary mirror to the eyepiece.

The second type of compound telescope was invented by a Russian astronomer, Dmitri Maksutov (1896–1964) in 1944. Details of a similar design were published by a Dutch astronomer, A. Bouwers, in 1940, who was experimenting with corrector plates in front of mirrors. The Maksutov telescope is similar to the Schmidt design, but uses a more spherical corrector lens. This new lens was used to produce a compact and rugged catadioptric telescope. The meniscus corrector used with a center-hole primary mirror Cassegrain configuration was to become known as the Maksutov–Cassegrain.

The Maksutov–Cassegrain was first popularized in the 1950s by Questar Telescopes. Questars provided views

The Meade 300 mm LX200 GPS Schmidt–Cassegrain telescope (SCT) set up in the backyard to cool. (Photo by the author)

The Meade 300 mm LX200 GPS Schmidt–Cassegrain telescope (SCT) getting ready for a night's observing. (Photo by the author)

similar to the finest apochromatic refractors of the same aperture, but were only one third of the physical length of the typical refractor.

In the early 1990s, the Maksutov corrector was coupled to a Newtonian reflector to create the Maksutov–Newtonian. The first popular Mak–Newt was introduced by Ceravolo Optical of Ottawa, Canada. This design is different in that the secondary mirror is a flat and imparts no power as do those of the Mak–Cass or Schmidt–Cass telescopes. Also, the optics path is not folded. The effective focal length of the telescope, and the overall physical length of Mak–Newt telescopes are about equal to the focal length. Finally, the primary mirror has no hole, so the focuser of a Mak–Newt is positioned as is the traditional Newtonian.

Unusual telescopes

Of course, the telescope designs described above are not the only ones ever constructed. For a look at some unusual designs, see David Stevick's "Weird Telescopes" page on the internet at

http://bhs.broo.k12.wv.us/homepage/alumni/dstevick/weird.htm

Active cooling

Some observers use one or more small fans to help bring large telescope mirrors more rapidly to ambient temperature. This can definitely help, especially during the early evening hours. The consensus seems to be that you should leave the fan on until the mirror reaches ambient temperature. Then, as the air temperature continues to fall, allow the mirror to radiate its heat away naturally.

Technical stuff: RMS, Strehl ratio and P–V

RMS refers to the root-mean-square of the wavefront deviation measured at best focus with light of wavelength 550 nm. Some regard this as a better measure of deviation than the P–V deviation (see later).

The Strehl ratio of a mirror is the light falling in the Airy disk as a percent of what a perfect mirror of the same dimensions would give. Disregarding complications by the secondary obstruction, a perfect aperture would have 83.7% of the light in the Airy disk and 16.3% in the rings surrounding the Airy disk. So if the Strehl ratio were 0.994, the light inside the Airy disk would be 83.7% × 0.994 = 83.2%, with 16.8% in the rings.

The P–V (Peak–Valley) deviation of the surface of a mirror is the maximum deviation minus the minimum deviation from the best-fit parabola. A convenient unit for the surface P–V is nanometers. It is also popularly given in terms of waves, in which case it is twice the surface P–V divided by the reference wavelength of light. The Rayleigh Criterion says that the P–V should be less than one-quarter wave, or surface P–V less than 68.8 nm.

A 250 mm, f/20 Maksutov–Cassegrain made by Yuri Petrunin of Telescope Engineering Company (TEC) of Golden Colorado. The central obstruction is less than 22%. (Photo by Robert Kuberek of Valencia, California)

Jeff Medkeff of Sierra Vista, Arizona, has originated an interesting thought problem related to all of the above indicators of mirror quality. He says . . .

The reason that P–V is not a very useful indicator of optical quality can be illustrated by a thought exercise. Imagine that you have one eight-inch objective mirror or lens that is optically perfect, except for a one-millimeter square patch that has a 300 wave error (a tower, or a pit, on the surface that is 300 wavelengths off optimal). This objective would have a very poor P–V rating; if a mirror, it would have a hugely bad 600 waves P–V wavefront.

Now imagine a second eight-inch objective. This one has a very slight linear trough running from opposite points on the edges, through the center of the mirror, but the bottom of the trough is only 1/8 wave lower than the optimal surface. If a mirror, this would result in a 1/4 wave P–V wavefront. Obviously, the second objective has better looking numbers. What will actually be seen if one looks through such telescopes?

The first telescope will render essentially perfect images. The one-millimeter square defective area will not be noticed by even the most critical observer, and the effect will be hard if not impossible to measure without dedicated optical testing. The second telescope will show noticeable astigmatism, where stars focus (or don't focus, rather) into lines rather than points, and planets will lose almost all their detail. Given the choice, everyone here would flock to the 600-wave P–V wavefront scope and abandon the 1/4 wave P–V wavefront scope to the trash heap, because the former would have a Strehl ratio hanging around 0.99 or so. As can be seen, Strehl ratio takes into account the amount of the surface that is affected by the error, and weights things accordingly.

This is of course exactly what optical quality measurements should do. They "should" empower the user to make rational decisions about what kind of telescope to use or what kind of optics they should be looking for. And they can and do succeed at that for some of us. But it can only happen if the user understands what the numbers mean, and to this extent knowing something about those measurements and how they are taken is essential.

The two sides to this coin are the ignorant masses who don't think anything worse than <insert number here> is any good on the one hand, and the optical cognoscenti who think that no optical quality measurements are ever useful to the end user on the other. Each position is clearly flawed, and the latter feeds off the discontent promulgated by the former. Education is the solution to bringing the former to an appreciation of where they err, and will reduce pressure and discontentedness in the latter. To that extent it's discouraging to see people hand waving and saying "good enough is good enough" and "just observe" and "who cares about optical quality." Obviously, a lot of folks already care about it if they are obsessed about their wavefront errors. Rather than harangue them, let's give them what they need to move on, and if they ultimately want to be amateur opticians rather than amateur astronomers, let them.

Telescope maintenance

The reason a telescope must be cleaned and aligned is to bring out its best performance. I equate it to tuning a musical instrument prior to playing it. Properly cleaning and aligning the telescope's components can make the difference between a good observing session and one where "firsts" or "bests" are recorded.

Collimation

Collimation is the alignment of the optical components of a telescope. Slight misalignments can cause or increase star image flaring, rob images of contrast, or prevent images from being uniformly in focus. Severe misalignments can reduce the light gathering capability of the telescope or make it impossible to bring objects into focus at all.

Meade 300 mm SCT collimation screws. Collimation is the bane of many an amateur astronomer. (Photo by the author)

A quick method of collimating a refractor

Most refractors hold their collimation very well. So, it isn't unusual to find them still correctly collimated after they have shipped to the customer. Unfortunately, many refractors are designed so that they are not easily adjusted for collimation by the owner. Those that can be adjusted usually have a three-pair, "push–pull" screw system built into the lens cell. Here's how to do it.

(1) Point the scope at a relatively dark wall or put the lens cap on.

(2) Slip a Cheshire tool in the focuser. (Note: The Cheshire is a development of the simple peephole tube, with an illuminated face (white or shiny) set at 45°. A hole in the side of the tube lets in light to make the reflection of the center spot visible against the bright face. There may be a "field stop" to better define the edge of the bright area. A Cheshire eyepiece or "tool" may be purchased from a variety of vendors.)

(3) Shine a bright light at the mirror in the Cheshire tool.

(4) Look through the Cheshire tool. You should see a bright round disc with a dark spot in the center, if the refractor is collimated. If it is out of collimation, you will see overlapping bright discs and black spots.

(5) Some refractors have three pairs of "push–pull" screws

in the front of the lens cell. You need to loosen and tighten them as needed to shift the image so that you see one round bright disc with a black spot in the center.

Tip: When doing any type of collimation, only turn the screws a tiny bit. Large motions are NOT required.

Collimating reflectors and catadioptrics

So much has been written about the processes and minutiae of collimating reflecting and catadioptric telescopes that to even summarize them here would be impossible. If you purchased your telescope new, follow the collimating instructions that came with it. One of the finest books ever written about evaluating and adjusting telescopes of all types is *Star Testing Astronomical Telescopes* by Harold Richard Suiter (Willmann-Bell, Inc., 1994, plus additional printings).

Star Testing Astronomical Telscopes by H.R. Suiter. (Photo by the author)

I do have a few quick tips for you to think about regarding collimation, based upon my own experiences:

- Collimation is much more critical as the focal ratio decreases (especially below about f/6). This is due to the increased curvature of the focal plane in these telescopes, resulting from a deeper mirror parabola.
- Be certain the primary mirror is centered in the rear of the telescope (aligned with the centerline of the telescope tube). If it isn't, there is an increased chance that the front edge of the telescope will vignette some of the incoming light.
- Although it is not critical for the light path to be reflected by exactly 90° from the primary off the secondary (two 45° angles), the most efficient design employs a 90° bend as this angle minimizes both the height of the focuser/eyepiece and the size of the diagonal.
- To aid in collimating an SCT, I highly recommend a set of Bob's Knobs, available via the internet at http://hometown.aol.com/rkmorrow/myhomepage/index.html This is such a great idea. I never liked using Allen wrenches in the dark.

- Laser collimators are wonderful tools once rough collimation has been done. There is, however, one danger in using collimation tools: they will indicate collimation based solely on the point that you've chosen as the center of the primary mirror. Choose this point carefully!
- Finally, after all your mechanical collimation is done, perform a star collimation. A star collimation uses light reflected from all of the mirror surfaces plus the eyepiece and so is the ultimate guide.

Star test

When a star is slightly de-focused on the inside or outside of focus, it will display a bright disk surrounded by a series of rings. If the optics are collimated, the disk and rings will be concentric. Practice making star collimations. They can be difficult to perform, as they require good seeing and high magnification (at least 2 × per millimeter). Keep the star centered in the field of view to reduce other effects such as field curvature.

When the image is a little inside or outside of focus, you should see rings of light and a darker center, and all should be round and well-centered. If not, note if the image improves if you move it a little off-center in any direction. If so, adjust the collimation screws to move the image toward the center of the field until it looks symmetric. And remember, very small motions of the screws (or knobs) are all that are required.

Cleaning optics

From time to time, optical surfaces get dirty. There's really only one way to avoid this and that's to keep them sealed in their original boxes unused. Certainly that is not what this book intends you to do!

Some tips from Meade Instruments will get us started on the road to proper telescope care. Meade points out, and I agree, that prevention is the best recommendation that a telescope owner can follow in keeping a telescope in top working order.

Dust and moisture are the two main enemies to your instrument. When observing, always use a dew shield. The dew shield not only helps prevent dew from forming and dust from settling on the lens, it also prevents stray light from reducing image contrast during observing.

Although dew shields go a long way to prevent moisture build-up, there can be times when the telescope optics will have a uniform coating of moist dew. This is not particularly harmful, as long as the dew is allowed to evaporate from the instrument, accomplished by setting up the telescope indoors with the dust covers removed. Never attempt to wipe down optics that are covered with dew. Dust and dirt may be trapped with the collected dew, and upon wiping the optics you may scratch them. After the dew has evaporated you will most likely find the optics in fine condition for the next observing session.

If you live in a very moist climate, you may find it necessary to use silica desiccant stored in the telescope's case to ward off moisture and the possibility of fungus growing on and within the coatings of the optics. Replace the silica desiccant as often as necessary. Packets of silica desiccant can be "restored" by baking in a kitchen oven set on the lowest setting for 15 minutes.

Those living in coastal areas or tropic zones should also cover the electronic ports on the optional power panel and the keypad with gaffer tape to reduce corrosion on the metal contacts. Apply a dab of a water displacement solution (such as WD-40) with a small brush on all interior metal contacts and the input cord metal contacts. The keypad and all separate accessories should be kept in sealable plastic bags with silica desiccant.

A thick layer of dust will attract and absorb moisture on all exposed surfaces. Left unattended, it can cause damaging corrosion. To keep dust at bay when observing, the telescope can be set up on a small section of indoor/outdoor carpet. If you are observing for more than one night in a row, the telescope can be left set up but covered with a Cosmic Storm Shield (see "Telescope accessories" chapter) or even a large plastic bag (such as the one supplied with the telescope).

Eyepieces, diagonals, and other accessories are best kept in plastic bags and stored in cases. All of the non-optical surfaces of the telescope should be cleaned routinely with a soft rag and alcohol to prevent corrosion. The cast metal surfaces and the individual exposed screws can also be kept looking new and corrosion-free by wiping them down with a water displacement solution (such as WD-40). Take care not to smear the solution onto any optical surface, and to wipe up any excess solution with a clean dry cloth. The painted tube can be polished with a liquid car polish and a soft rag.

In my long "career" as an amateur astronomer, I have cleaned few mirrors and fewer lenses. I have, therefore, called upon Leonard B. Abbey, an expert in the cleaning of optics. He has graciously allowed an article he wrote about this subject to appear here. For more articles about all aspects of amateur astronomy, I recommend his "Compleat Astronomer" website, on the internet at http://LAbbey.com

How to clean mirrors and lenses
Leonard B. Abbey, FRAS

The cleaning of optical surfaces, especially those of first-surface mirrors, is the most delicate and exacting task which the astronomer is called upon to perform. At the time of cleaning, a lens is most vulnerable to damage; damage which cannot be repaired. Yet if a telescope is to perform at its greatest potential, cleaning must be done time to time.

I have used the following method for over thirty years without adding a single scratch to the surface of a mirror or lens. It has the advantage of requiring only materials which are readily available at the neighborhood pharmacy or grocery store. The cost is less than twenty-five cents per cleaning.

First you must realize that usually the best advice on cleaning mirrors and lenses is…DON'T DO IT. Dirt and grease which are adhering to the surface of mirrors and lenses may degrade image quality slightly, but they will not damage the delicate optical surface until they are moved against it. The need to remove dirt without allowing it to move against the underlying optical surface is what makes cleaning such a tricky task. However, if your mirror or lens is so dirty that it must be cleaned, then this is the way to do it.

For mirrors

1. Blow all loose dirt off with "Dust Off" or another canned, filtered, clean air product. (Available in camera stores.) Take care not to shake the can while you are using it, and be sure to release a little air before using it on the optical surface. This will assure that no liquid is dispensed to make things worse! You can use a rubber bulb for this purpose, but it is not nearly as effective.
2. Prepare a VERY dilute solution of mild liquid detergent (e.g., Dawn). It is very important that the detergent does not contain any form of hand lotion or lanolin. This product usually comes in a plastic bottle with a dispenser spout. Dispense the tiniest amount possible into a clean cup. (One drop, if possible.) Fill the cup with water. Stir. Throw almost all of this water away, and refill the cup. Now you have a VERY dilute solution.
3. Rinse the mirror off under a moderate stream of lukewarm water for two or three minutes. Test the temperature of the water with your wrist, just as you would when warming a baby's bottle. Leave the water running.
4. Make a number of cotton balls from a newly opened package of Johnson & Johnson sterile surgical cotton, U.S.P. (The "U.S.P." is important. It means that you have REAL cotton instead of a polyester substitute.) Soak 2 or 3 balls in the detergent solution. The cotton balls should be fully saturated with the detergent solution. Do not squeeze any of the liquid out. Wipe the surface of the wet mirror with a circular motion, going first around the circumference, and then working your way towards the center. The only pressure on the cotton should be its own weight. For this first "wipe" you should use several fresh sets of cotton balls. As you move the cotton balls around the mirror's surface, rotate them slightly so that the dirt they pick up is moved away from the mirror's surface, and toward the top of the balls.
5. Throw the cotton balls away.
6. Repeat the process with new cotton balls, using a LITTLE more pressure.

7. Rinse mirror thoroughly under tap, which has been kept running for this step.

8. Rinse mirror with copious amounts of distilled water (do this no matter how clean or "hard" your tap water is).

9. Set mirror on edge to dry, using paper towels to absorb the water which will all run to bottom of mirror. Keep replacing the paper towels as the mirror dries.

10. If any beads of water do not run to bottom, blow them off with Dust Off, or the rubber bulb. Any stubborn drops which remain on the aluminum surface can be picked up with the corner of a paper towel. The paper towel doesn't even need to touch the mirror's surface.

11. Replace the mirror in its cell, being careful to keep all clips and supports so loose that the mirror can rattle in the cell if it is shook. (Perhaps 0.5 to 1mm clearance.)

12. Spend some time realigning your scope.

13. If you do anything more than this, you will risk damaging the coating. But remember, if you follow these instructions any damage will almost certainly be to the coating, not the glass. When the mirror is re-aluminized it will look new in all respects.

14. You should not have to clean an aluminized mirror more often than once per year. Do NOT over clean your optics.

For objective lenses

DO NOT UNDER ANY CIRCUMSTANCES REMOVE A LENS FROM ITS CELL, OR THE CELL FROM THE TELESCOPE.

This restriction means that the above procedure must be modified. Only the front surface of the objective can be cleaned. If you remove the cell from the telescope, you will be in big trouble. There are very few people who can collimate a refractor. If you are reading these instructions, you are not one of them!

1. Blow loose dirt off with "Dust Off" or a rubber bulb, using the above precautions.

2. Soak the cotton balls in a 50:50 solution of Windex (commercial glass cleaner containing ammonia) and distilled water. Squeeze slightly so that the balls are not dripping wet.

3. Wipe front lens surfaces with the wet cotton, using only the pressure of the weight of the cotton balls. Follow immediately with dry cotton, using little or no pressure.

4. Repeat procedure, using slightly more pressure.

5. If some cotton lint remains on surface, blow off with Dust Off or rubber bulb.

6. Repeat this procedure if the lens is not clean, but if one "repeat" does not do it give up and leave it as is.

7. Inspect the lens to make sure that no cleaning solution has found its way into the lens cell, or between the elements. If this has happened, leave the telescope with the lens uncovered in a warm room until it is dry.

For Schmidt–Cassegrain and Maksutov telescopes

The only optical surface you should attempt to clean is the front of the corrector plate. Use the instructions for cleaning refractor lenses. If your SCT needs more cleaning than this, send it back to the factory for cleaning.

For eyepieces and Barlows

Follow the procedure given for objective lenses, but use Q-Tips (U.S.P. cotton on plastic sticks) instead of cotton balls. You may, of course, clean both surfaces. The eyebrow juice on the eye lens of eyepieces may require repeated applications. I think that this is OK in this case.

SOME DONT'S

1. Do not use any aerosol spray product, no matter who sells it, or what their claims are.

2. Do not use lens tissue or paper. It DOES scratch.

3. Do not use pre-packaged cotton balls, they frequently are not cotton.

4. Do not use any kind of alcohol, especially on aluminized surfaces.

5. Do not use plain water for the final rinse.

6. Do not use any lens cleaning solution marketed by funny companies, like Focal, Jason, or Swift. Dawn and Windex (or their equivalents in other countries) are inexpensive and commonly available.

Re-aluminizing a mirror

Eventually, if you own a reflecting or catadioptric telescope for a long enough period of time, you will need to have the mirror re-aluminized. Owners of Newtonian reflectors will have to re-aluminize their mirrors more often than owners of Schmidt–Cassegrain telescopes, due to the closed tube design of the latter.

The number one factor which contributes to the need to re-aluminize a mirror is improper cleaning and/or handling. Other factors are excessive dust, which is abrasive, and condensation which may contain acidic atmospheric aerosols.

How can you tell if it is time to have your mirror re-aluminized? The deterioration of the coating is a gradual process and the recognition that it has degraded can be difficult. Obvious signs, such as numerous scratches, blotches or hazy spots, are rare. Look instead for halos around the brighter celestial objects when your mirror is dust-free. Also, if you suspect deterioration in the coating, try to compare the view through your telescope side-by-side with a telescope of equal aperture. Finally, if you feel that the coatings could be better, check the calendar. If it has been more than five years since your mirror has been re-aluminized, it is probably time.

Tip: If you are sending your primary mirror for re-aluminizing, send your secondary mirror as well.

Enhanced coatings

The same company which re-aluminizes your mirror can also add a coating to it. Some coatings are for the protection of the aluminum surface. Most overcoatings are silicon monoxide (SiO). Magnesium fluoride (MgF_2) is sometimes used, but it isn't as resistant to scratches as SiO.

Other mirror coatings, called enhanced coatings, raise the reflectivity of the surface of the mirror by 5–10%. Since an enhanced coating is considerably more expensive than a regular coating, sometimes costing double, we may well ask the question, "Is it worth it?" The general consensus is, "No, it is not worth it." Enhanced coatings are more susceptible to atmospheric conditions such as high humidity, airborne particles and acidic content than are regular coatings. Most amateur astronomers subscribe to the rule that enhanced coatings deteriorate much more rapidly than regular ones.

Rick Singmaster, owner of StarMaster Telescopes, has extensive experience with mirror coatings of all types. He provides some perspective on the "value" of enhanced coatings . . .

> Regarding enhanced coatings, we here at Starmaster do not recommend them on our primary mirrors, and we are not alone. Other major optical companies who do not advocate enhanced coatings include: Pegasus

> Optics of Brackettville, Texas, Spectrum Coatings of Deltona, Florida, Astro Systems, Inc., of LaSalle, Colorado, and Zambuto Optical Co., of Rainier, Washington. Swayze Optical, Inc., of Portland, Oregon, has only used them on approximately 5% of the primary mirrors they have produced.

> The failure/recoat rate on early Starmaster primary mirrors which were enhanced was more than double that of standard coated optics. Many primary mirrors wouldn't even come to focus because the enhanced coatings severely degraded the wave front.

> Another problem we encountered was that the contrast was lower with many of the enhanced coated primary mirrors. This was noticed when they were compared side-by-side with "standard" coated mirrors from the same coating company. Furthermore, this difference was also noted when comparing various coating companies' products so it wasn't limited to just one company.

> I personally made these comparisons when testing scopes side-by-side prior to shipping. Since I have done this for every scope we have made, this opinion is based on my own observation of a large sample.

> At Starmaster, our only concern is that our customers receive a fine performing scope. I don't have anything against trying to obtain more performance from a scope by whatever means, but when I see something which degrades performance time after time, I simply can't in good faith recommend it to my customers. Especially if it adds cost.

2.2 Mounts and drives

Altazimuth mounts

An altazimuth mount is the simplest type of telescope mount. The word is a combination of altitude and azimuth. Altitude is the distance between the horizon and the zenith. Azimuth is a measurement, in degrees, beginning at the north point of the horizon and proceeding through east. Thus, a telescope on this type of mount will move up and down, and left and right.

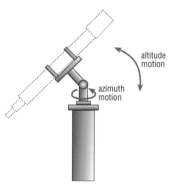

An altazimuth telescope mount. It moves up and down and left to right. (Illustration by Holley Y. Bakich)

Dobsonian mounts

Renowned telescope maker and amateur astronomer John Dobson built his first telescope in 1956. From that point on he was totally committed to helping everyone see the universe. The altazimuth mount that he invented has revolutionized amateur astronomy. It is a simple dual-pivot mount almost always combined with a Newtonian telescope

"Would you like GoTo with that 600 mm telescope?" There are few of us who would not love to own one of these large StarMaster telescopes, seen here at the factory awaiting their mirrors. (Photo by the author)

optical tube assembly. Just push the telescope up or down to change altitude and push it left or right to change azimuth. It is a terrific design and easy to use.

For a wonderful guide to building a Dobsonian mounted telescope, see the "Telescopes in Education" page on the internet at http://tie.jpl.nasa.gov/tie/dobson

Driven altazimuth mounts

A recent development in altazimuth mounts is the driven altazimuth mount. With motors attached to both the altitude and azimuth axes of the mount, the telescope can either (1) track an object across the sky once it is placed in the field of view or (2) interface to a computer to both find and track an object. Quality mounts employing either system are very accurate. Once found, an object can be followed without the observer continually moving the telescope.

Equatorial mounts

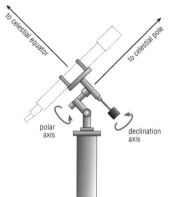

An equatorial mount. Polar alignment is very important. (Illustration by Holley Y. Bakich)

If the Earth did not move, an altazimuth mount would be all that any of us would ever need. But the Earth does spin and we must deal with it. The second type of mount is called the equatorial mount. It was designed to track the apparent motion of the stars. The way it does this is by aligning one of its axes parallel to the Earth's spin. Please note that there are two types of equatorial mounts – manual and motorized. If you have a choice, select the motorized version.

Polar alignment

The technique of properly setting up an equatorial mount is called polar alignment. The tripod is positioned so that an imaginary line extending from the telescope's polar axis

To North Celestial Pole

Earth's Axis

Polar Axis of Telescope

The polar axis of an equatorial telescope must be lined up with the Earth's axis. (Illustration by Holley Y. Bakich)

points to the north celestial pole. If you are observing visually, rather than imaging, you can align this axis with Polaris. Polaris is not at the exact north celestial pole, so the telescope will be off a little. Images, however, will remain in the field of view for quite a while.

To polar align your telescope mount to a greater precision, you must align the mount's polar axis with the actual north celestial pole (NCP). As of this writing, Polaris lies 0.77° from the NCP at right ascension (RA) 02^h33^m, approximately. To offset in the correct direction (by eye), find the star Kochab. Kochab is β UMi, the brightest star of the pair at the end of the handle of the Little Dipper. Kochab's RA is roughly $14^h 50^m$. Since its RA and that of Polaris differ by about 12^h, it lies halfway around the sky from Polaris. So offset Polaris by $3/4$ degree in the direction of Kochab. (Be certain to offset in the correct direction. The view through the eyepiece is often reversed from reality.)

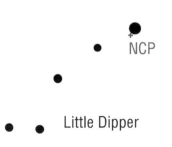

NCP

Little Dipper

The location of the north celestial pole as it appears in the sky. To scale. (Illustration by Holley Y. Bakich)

Tip: When polar aligning, be certain to move the entire mount, not just the optical tube assembly.

If you have a telescope on a permanent (non-portable) mount and you desire to do astrophotography, the most critical procedure you can perform is accurate polar alignment. One of the best descriptions of the technique I have found is Bruce Johnston's "Polar Alignment Made Simple." Find it on the internet at

http://members.aol.com/ccdastro/drift-align.htm

For southern hemisphere observers or visitors who desire a detailed procedure for finding the south celestial pole, see Dave Gordon's excellent and detailed description, complete with charts, at

http://www.aqua.co.za/assa_jhb/Canopus/Can2000/coobSCP.htm

Digital setting circles

Either altazimuth or equatorial mounts may be upgraded with a set of digital setting circles (DSCs), available from many astronomy vendors, for example, Lymax Astronomy, on the internet at

http://www.lymax.com/index.php

All commercial DSCs work on altazimuth telescopes as well as equatorials. There is an electronic rotary encoder which is placed on each axis, and a small processor to perform the transformation from the altitude and azimuth of the mount to the right ascension and declination of the sky.

DSCs appeal to many observers who like to understand the concepts of astronomical coordinate systems, but have the processor do the actual pointing. To set up DSCs, you must

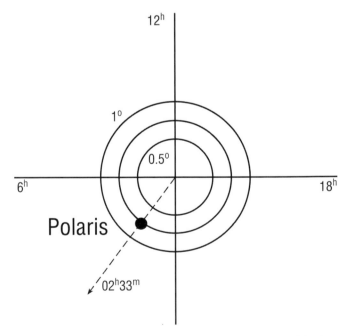

Polaris, the North Star, is not at the exact position of the north celestial pole (NCP). As of this writing, Polaris lies 0.77° from the NCP, and is moving closer. It will be at its nearest point in 2102 AD, when it will lie at a distance of 0.46°. (Illustration by Holley Y. Bakich)

first center either two or three bright stars (depending on the system) in the field of view. Then, after you enter an object's designation (for example M11 or NGC3242) you simply move your telescope until the readouts for right ascension and declination are at zero.

GoTo drives

A GoTo system is essentially nothing more than a DSC combined with a dual-axis motor driver. As with DSCs, GoTo drives require you to center one or two stars so that they can align. Most GoTo-driven telescopes manufactured today have a large database of objects from which you can select. In fact, some of the smaller telescopes equipped with a GoTo have objects in their databases which the telescopes have no hope of revealing.

Recently (as of this writing), telescope GoTo drives have been paired with Global Positioning System (GPS) electronics and electronic compasses. Such a system is the Meade 12-inch GPS. The result is an almost hands-free setup. With the Meade, for example, if you choose the alignment option "Automatic," the GoTo drive first acquires the GPS satellite constellation (as it is called). This allows it to figure out the position (longitude, latitude and elevation) of the telescope. Next, it finds north, then true north, which is offset from the compass' magnetic north. Then it electronically figures out whether the mount is level and compensates in both tip and tilt. After that, the first of two stars is placed into the field of view. You are asked to center it and press "Enter." This is repeated with a second star. Both stars, by the way, are selected by the telescope's computer, which knows whether or not they are above the horizon. That's it. After three button presses of the control paddle, you are ready to observe.

Personal comment: GoTo or not?

Many amateur astronomers with whom I have observed through the years believe that a person's first telescope should not have a GoTo drive, and their opinions have nothing to do with cost. Their contention is that all amateur astronomers should learn the sky, and star hopping is part of that basic education. In their minds, GoTo drive is an unnecessary crutch which should be avoided.

On the other hand, most celestial objects are faint, and locating them can be very difficult, especially for someone just starting out. Failure to learn how to find "things in the sky" is a common reason why people quit the hobby. Because of this, I favor GoTo mounts, and not just for beginners.

Experienced observers who know the sky can also benefit from GoTo mounts. I cannot tell you how many times I have heard an experienced amateur astronomer say, "Since I got my GoTo drive, I spend much more time observing!" And, really, isn't that what it's all about?

A set of Celestron anti-vibration pads. A highly recommended accessory to help combat the shakes encountered by portable tripods. (Photo by the author)

Mount stability

We call our instruments "telescopes," but the phrase "optical tube assembly on a mount" also works. In fact, it points out the fact that half of any "telescope" is the mount. Some would say the mount is the more important part. An unstable mount will render the finest apochromatic refractor unable to deliver quality images. If the mount is undersized, wind (the bane of most large telescopes,) will not be your only enemy. You will experience bouncing images even when you are focusing.

The test of a quality mount is sometimes given as "damp-down" factor. This is the time it takes an image to stabilize (for the vibrations to dampen) after the telescope is moved or refocused. Under no circumstances should the image take more than two seconds to damp down. Regarding wind, a spring Texas gale is going to move even the stoutest of mounts. A better test is how your mount responds to a slight breeze.

There is a trade-off to the "best possible mount," which would be affixed to a permanent pier sunk into several cubic meters of concrete in an observatory. Most observers want (and need) their telescopes to be portable. The degree of portability is a trade-off, to be sure. An amateur astronomer lucky enough to own a 600 mm StarMaster reflector realizes that the "quick-look" type of observing session is pretty much out of the question.

Backlash

The term backlash refers to unwanted slack (or "play") in a telescope's drive system. Some backlash is unavoidable, as inexpensive gear systems which eliminate all unwanted space cannot be made. To reduce backlash, see your mount's instruction manual and/or talk with others who have the same setup.

PEC

Precision error correction (PEC) corrects for mechanical irregularities in the RA motion of a telescope caused by imperfections in the drive. A basic problem with all telescope drives, including those in professional observatories, is that there are mechanical irregularities in the gears and worm which drive the worm wheel that points the optical tube. These irregularities occur with a period determined by the speed of rotation of the gears.

The idea is to correct the speed of the drive motor to compensate for the fact that the worm drives the worm wheel a bit faster or slower as it rotates. This is done by placing the drive into a "training mode" and manually training the computer to compensate for defects. This is a good solution which can remove as much as 90% of the irregularity.

The Hargreaves strut

In the September 2001 issue of Sky & Telescope (p. 114) a stabilizing arm called the Hargreaves strut is described with building instructions. I was intrigued and built one for my

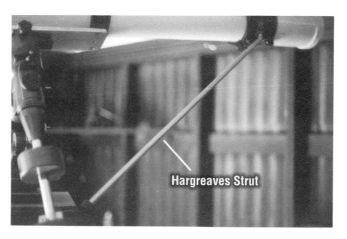

If you own a long focal length refractor, the Hargreaves strut may help stabilize your view. (Photo by the author)

100 mm f/15 Unitron refractor. What a difference! The damp-down time is much shorter now after focusing. Wind is also not as great a problem. Total construction time once all the parts were received was approximately one hour. I heartily recommend this apparatus to owners of large focal ratio refractors.

2.3 Eyepieces

Eyepieces are like stereo equipment. With stereo equipment, a high-quality item is recognized by all. Workmanship is admired. Faithfulness in reproducing the sound in as original a form as possible is valued. And yet, while listening to a familiar piece of music, each of us perceives something a little bit different about it. There may be some nuance meaningful to you that I have not identified. The end result is that we don't all end up with the same stereo equipment, or camera equipment, or automobiles, or . . . eyepieces.

To be fair, this is often due to financial considerations. The "recognized" best eyepieces are not cheap. It is tough for some of us to justify spending as much on our eyepieces as we did on our telescope. The investment, however, has to be looked at over the long term. If you upgrade your telescope (a flaw in the human spirit sometimes called "aperture fever"), you don't need to change your eyepieces. Even if your new telescope uses 2 inch eyepieces and your old one used 1.25 inches, there are adapters which can be purchased.

When thinking about adding an eyepiece to your set, good advice comes from Susan Carroll, an avid amateur astronomer living in Fort Scott, Kansas.

> When one is contemplating the purchase of a new eyepiece, you should, if possible, try it first in your own scope, with your own eyeballs, before you buy. Perhaps you could borrow one from somebody at a star party for an hour, or there may be a vendor that will let you try it out that night before you buy it.
>
> There are too many personal variables to take into consideration to go just on someone else's word. Are you near-sighted? far-sighted? wear glasses when observing? max. dilation of your pupils is 7 mm? or like us old types, closer to 5 mm? and the list goes on. Look on an eyepiece as an investment for your future observing. Don't be in a rush to buy the newest and latest "toys" – you may be sorely disappointed if you do.

Another consideration is the weight of some of the new eyepieces. The Tele-Vue Nagler 31 mm weighs one kilogram! This is as much as some binoculars. If you have a small to medium-sized telescope, forget about using this eyepiece and a number of others as well. If you purchase and use these eyepieces, remember to tighten your set screws just a little bit more if you are going to be moving your scope or if the eyepiece will encounter a position from which it might fall from the diagonal. If, on the other hand, your telescope is in the medium-to-large size range, save your money for high-quality 2 inch eyepieces. Purchase them one at a time, always staying alert for good deals in the used equipment market.

Eyepiece designs

A number of different eyepiece designs continue to be sold today. The main types are pretty well recognized by advanced amateurs. A couple of definitions are in order. The magnification of an eyepiece is

Magnification = focal length of telescope/focal length of eyepiece

> **Example:** I have been using Meade's excellent 305 mm LX200 GPS telescope for some time now. It has a focal length of 3048 mm. If I insert a 22 mm Tele-Vue Nagler, this will provide a magnification of 138.54, or approximately 139.

The field of view (abbreviated fov) is what you actually see when you look through the eyepiece. The apparent field of view (afov) of an eyepiece is the angular size of the light cone which is able to enter the eyepiece. Eyepiece apparent fields can range from 25–84°. Apparent field can be contrasted with true field, the actual part of the sky being viewed through the telescope, by the equation:

True field = apparent field/magnification

> **Example:** Find the true field for the Nagler 22 mm eyepiece when used with the LX200 GPS. The magnification was calculated above. Since this eyepiece has an afov of 84°, the true field is 84/139, or 0.604°. Now find the true field for a 25 mm Vixen Lanthanum eyepiece. The magnification will be 3048/25 = 122, approximately. This eyepiece has an afov of 50°. So the true field will be 50°/122 = 0.41°. This example may begin to show why amateurs love wide-field eyepiece designs. The Nagler in this example provides more magnification *and* a wider field than the Vixen Lanthanum, albeit at three times the cost.

When comparing different eyepieces, or similar designs by different manufacturers, always note the contrast between what you're viewing and the (hopefully) black sky. Observing a bright planet or the lunar terminator is generally a good test. How much light is scattered into the black area?

Another thing to be aware of is edge sharpness. For this test, nothing beats star images moved, or left to drift, from one edge of the fov to the other.

One final point before you head to the table of eyepiece designs. Just about all of today's eyepieces that are worth considering are coated on every air-to-glass surface. This reduces the light lost by reflection, scatter, etc., from around 4% to less than 0.2%. Therefore, no matter how many surfaces there are in an eyepiece, light transmission is no longer an issue. There are plenty of other differences you might notice, just not this one.

Summary of eyepiece characteristics

Huygens	Uses two simple lenses to reduce chromatic aberration (two elements). Has high field curvature causing limited afov of approximately 25°. Very short eye relief. Stay away from this design.
Ramsden	Uses two lenses of equal focal length (two elements). Has some chromatic aberration but field curvature is much less than the Huygens. Usable afov is 25–30°. Stay away from this design.
Kellner	Similar to the Ramsden design, but one of the lenses is replaced by a two-piece achromatic lens (three elements). Better correction for chromatic aberration and field curvature. Afov is 40–45° with good eye relief. Very low cost, and, with that as a consideration, not a bad eyepiece.
Erfle	Uses three achromatic pairs or two pairs and a single lens (five or six elements). Well corrected for chromatic aberration and field curvature. Low-quality Erfles may exhibit astigmatism at the edge of the fov. afov is a nice 60–70°. Good, low-cost eyepiece.
Orthoscopic	Uses an achromatic triplet and a single lens (four elements). Good correction for all optical aberrations and very uniform across the fov. Some internal reflections. afov is 45–50°. The Abbe orthoscopics are known for superb contrast. Very good, medium-priced eyepiece.
Plossl	Uses two achromatic pairs very close together. Some designs incorporate an additional simple lens (four or five elements). An inexpensive eyepiece to make. afov is 50–55°. Eye relief is generally 0.73 times the focal length of the eyepiece. More internal reflections than with the orthoscopic. A Plossl is in every amateur's set. Good, medium-priced eyepiece.
Tele-Vue Nagler	Six to eight elements, 82° afov! Introduced in 1982 by Tele-Vue, Inc. 10–12 mm eye relief. Contrast not as good as some eyepieces but nice, sharp images. Excellent, high-priced eyepieces valued by observers. Lots of different "types" (essentially, upgrades) of Naglers have come out since the originals.
Vixen Lanthanum	A five-element Plossl plus a one to three element Barlow; this provides 20 mm eye relief for all focal lengths. 45–50° afov. Pretty good contrast, tack sharp to the edge of the field. A more expensive wide-field version (65° afov) of these eyepieces is also made. Superb, mid-priced eyepieces.
Tele-Vue Radian	18, 14, 12, 10, 8 mm models are six-element. 6, 5, 4, 3 mm are seven-element. 60° afov. Weigh between 225–360 g. Slightly wider field than the Vixen Lanthanum, but at twice the cost. Excellent eyepieces.
Panoptic	Six elements, 68° afov. Eye relief equals 0.68 of focal length. Edge of field exhibits pincushion distortion which can be tolerated. Excellent but expensive eyepieces.
Meade SWA	Six elements. 67° afov. Some focal lengths have short eye relief. On the lower end of "expensive" but good eyepieces.
Meade UWA	Eight elements, 84° afov. Essentially a copy of the early Nagler design but slightly less expensive. Excellent.
Pentax XL	An Erfle design with a Barlow lens to provide long eye relief (20 mm) throughout the range of focal lengths (five to seven elements). 65° afov. Long eye relief of 20 mm throughout entire series. Excellent, though high-priced eyepieces.

Now, the caveat. Only by direct comparison between eyepieces of identical focal length on the same telescope can you be sure which performs better. I'm not saying throw out the numbers or ratings above but make sure, before you choose at least the first eyepiece of a certain type, that you have compared identical (or nearly identical) focal lengths of different types.

It is important, when comparing nearly identical eyepieces to choose suitable objects. Close double stars – those near the limit of the telescope's resolution – are excellent for such a test. Galaxies are not.

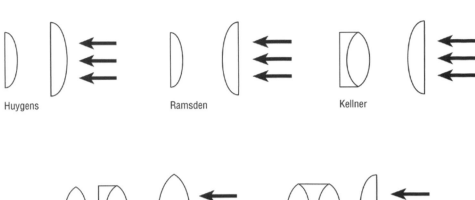

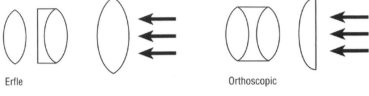

Various eyepiece designs.
Explanations are given in the text.
(Illustration by Holley Y. Bakich)

Parfocal eyepieces

The term parfocal refers to an eyepiece which is part of a set (two or more) that can be swapped with another of the set without adjusting the focus. This is a terrific convenience when it happens, but, admittedly, even those sets of eyepieces labeled parfocal are seldom that. To me, it doesn't seem that difficult but it would entail different barrel lengths within eyepiece sets.

Exit pupil

Rather than comparing magnification, some observers like to talk about the exit pupil. Exit pupil is defined as the diameter of the light shaft exiting an eyepiece. To put it another way, the exit pupil is the bright circle you would see on the surface of the eyepiece from about 30 cm away if you were to point the telescope toward a bright light. (This also works with binoculars. See that chapter for an illustration.)

When speaking of the exit pupil of an eyepiece, two formulas may be used:

Diameter of exit pupil = aperture of telescope/magnification

or

Diameter of exit pupil = focal length of eyepiece/focal ratio of scope

The subject of the proper exit pupil to be used for certain objects is hotly debated among advanced amateur astronomers. At the risk of drawing the ire of some, let me state that the importance of exit pupil size is overrated. Exit pupils are sometimes used to compare the views through

Making an eyepiece set parfocal

Ray Rochelle of Chico, California, offers the following steps so that you can make your eyepiece set parfocal.

It is really not hard at all. First you will need something you can wrap around the barrel of each eyepiece and it doesn't take much – thin plastic sheeting, stiff paper card, tape, etc. What you will be doing is cutting strips that will go around the barrels of the eyepieces and use a small piece of tape to secure them. To find the width of the strips follow these instructions.

(1) Set up your telescope to view stars, the Moon, something far away.
(2) Insert your highest power eyepiece e.g., 4 mm, 12 mm, whatever, and focus.
(3) From now on do not touch the focus knob.
(4) Insert the next eyepiece, and move it in and out manually until it focuses, then lock it down. Take a felt tip pen and draw a line around the (telescope end) of the eyepiece, where the eyepiece goes into the back or into the diagonal. There should be a space up to 1/2 inch from the base of the eyepiece to the line you just drew around the eyepiece.
(5) Do this to all your eyepieces before you go on.
(6) Take each eyepiece and cut a piece of plastic to cover the area you marked from the base of the eyepiece to the line you marked and secure with tape.
(7) Now you should be able to interchange each eyepiece and have to do very little if any refocusing, once you have focused one eyepiece.

This is just a way to bush each eyepiece the same. You can also buy, at no small figure, special bushings . . . ha ha!

A selection of eyepieces. From upper left, clockwise: 9 mm Vixen Lanthanum, 14 mm Meade Ultra Wide, Meade 55 mm Super Plossl, University Optics 28 mm Orthoscopic, Tele-Vue 22 mm Nagler Type IV, 20 mm Vixen Lanthanum. (Photo (and eyepieces) by Eugene Lawson, El Paso, Texas)

different telescopes. Observers obsessed with exit pupil size often try to "match" the exit pupil either to the type of object they are observing or to the pupil size of their observing eye.

Some amateurs will tell you that exit pupils less than a certain size (1 mm is the most frequently quoted number) cannot be used. This is simply not true. The image projected on the retina by exit pupils smaller than 1 mm have plenty of resolution for serious observing. What actually sets the lower limit to useful exit pupils is the seeing. Smaller exit pupil sizes have been used successfully during observations of the Moon, the planets and double stars.

Other (or perhaps they are the same?) amateurs argue that you can use exit pupils which are too large. Usually, this means anything over 7 mm. Why 7 mm? The reason stems from the fact that this number is often quoted as the fully dilated size of the dark-adapted human pupil. Any eyepiece exit pupil larger than this, it is said, is simply a waste of light.

Well, I concede that these amateurs are partially right, but not for the above reason. With any telescope other than a refractor (i.e., one which has a secondary obstruction) there comes a point where the exit pupil is so large that you begin to see the shadow of the secondary. Obviously, this can be a distraction. But matching the exit pupil of the eyepiece to the eye's pupil size has its problems too. First, unless you carefully measure your pupil size (there are gauges which allow you to do this, or you can make a trip to your optometrist) the "7 mm" standard can be wrong. There is, in fact, a wide range of fully dilated pupil sizes among observers (from 4 mm to 10 mm). In addition, that number decreases as we age. Last, but not least, is the fact that a perfectly matched exit pupil/eye pupil size is tough to observe with, as the eye must be held in exact alignment with the light exiting the eyepiece.

The key here, in my opinion, is to "take it to the limit."

Start with low magnifications (large exit pupils) and progress to higher magnifications until the image breaks down due to seeing, or at least until the detail you are studying cannot be improved. Many observers using non-driven scopes have their own "upper limit" to magnification. Without a drive, the image simply moves through the field too quickly and it is just easier to use lower powers.

Star diagonals

Newtonian reflectors have a built-in 90° bend in the optical path. Refractors and SCTs do not. Since the best views of celestial objects come when they are high in the sky, observing with a "straight-through" arrangement is often uncomfortable. The solution for this is a star diagonal, a machined piece which fits into the focusing tube and whose other side accepts an eyepiece. Star diagonals can be either 1.25 or 2 inches. Within the star diagonal is either a prism or a mirror.

With my long focal ratio refractor, if I had to view "straight through," I would probably confine my observing to objects within 30° of the horizon. Luckily, someone long ago invented the star diagonal, allowing comfortable viewing of objects even at the zenith.

Star diagonals come in two varieties – prism and mirror – referring, of course, to the optical element that does the bending of the light. The choice is simple: prism diagonals are less expensive; mirror diagonals are better.

A 2" Meade star diagonal with 1.25" adapter. (Photo by the author)

All star diagonals used in astronomy are of the 90° type. This means that the light is bent 90° from the telescope's optical axis. There are also 45° diagonals, used primarily for viewing objects right here on Earth. Don't buy a 45° diagonal.

The difference in mirrored star diagonals boils down to the amount of reflectivity the mirror provides. Ultimately, when you are pushing your equipment to its very limit and trying to see extremely faint objects, you want the star diagonal which will reflect the most light.

Mirror diagonals are superior to prism diagonals. The light throughput is a little bit better as a mirrored diagonal allows the light to contact only one optical surface. The main problems encountered with poorly manufactured star diagonals are astigmatism and miscollimation.

A different design has been developed by Astro-Physics, Inc., with their superb MaxBright star diagonal. Instead of a mirror, a series of coatings of film oxides are deposited, forming the reflective surface. For visual use, this diagonal reflects over 99% of the light striking it.

An occulting bar. The edge of the tape (blue here for illustrative purposes) is in focus and blocks half of the field of view. (Photo by the author)

A comparison of star diagonals

Diagonal	Size (inches)	Features
Lumicon	1.25 inch	96% reflectivity; enhanced aluminum mirror; threaded for 1.25 inch filters
Tele-Vue	2 inch	96% reflectivity; brass clamp ring on some models at extra cost; Everbrite version has dielectric coating, 99% reflectivity
Lumicon	2 inch	96% reflectivity; enhanced aluminum mirror; threaded for 48 mm filters
Astro-Physics MaxBright	2 inch	99+% reflectivity; very durable dielectric coating; internal brass locking ring; threaded for 48 mm filters

Occulting bar

Is one of your goals to see the ashen light of Venus? the faint companion of some bright star? the moons of Mars? the Merope nebulosity? To accomplish this, some observers have constructed an occulting bar. Some of these are as simple as a piece of black electrical tape. Others are as complex as, well, a small piece of aluminum foil.

An occulting bar is a piece of some type of thin, opaque material that you place in the focal plane of your eyepiece. Generally, you try to cover half the field stop. Then, when the eyepiece is rotated, the occulting bar acts to block the light from a particularly bright object (or bright part of an object) from reaching your eye.

The field stop is a ring (usually metal) which serves to sharply define the field of view. This ring is placed at the focal plane of the eyepiece. Note however, that some of the newer eyepiece designs have some of their lens elements in front of the field stop. Without disassembling the eyepiece (not a good idea) you cannot make an occulting bar.

For the ones where the field stop is accessible tape or use a small drop of rubber cement to affix the "bar" to the field stop. As mentioned above, material used for occulting bars includes black construction paper, aluminum foil, over-exposed film and black tape (plastic and paper).

Observing Tip: Don't wait for darkness to check your occulting bar. The best check is done in daylight. This will show any irregularities or places of greater transmission of light.

Barlow lenses

A Barlow lens is a negative lens which increases the effective focal length of an objective lens or mirror. Barlows are rated by magnification factors. One might magnify $2\times$, another $3\times$, etc.

A simple 2x Barlow lens. (Photo by the author)

Example: In a telescope with a 1500 mm focal length, a 15 mm eyepiece will render a magnification of 100×. If you insert a 2× Barlow, the magnification will be 200×. Thus the effective focal length has changed from 1500 mm to 3000 mm.

Forty or so years ago, when Barlows first appeared, they were simple units using single - element lenses. They worked, but they didn't do much for the image. Today's Barlows are multi-element, well-corrected, coated lenses with great light transmission. The number of amateur astronomers swearing never to have a Barlow has dropped markedly and continues to decline.

A Barlow lens can effectively double the number of eyepieces in your set, if the eyepiece focal lengths have been selected with this in mind. Also, with some eyepieces, using lower powers provides better eye relief. Stay away from adjustable Barlows. They have not yet arrived, in terms of image quality and workmanship. Note that you can "alter" the magnification your Barlow provides by placing it either before the star diagonal (usual setup) or between the star diagonal and the eyepiece. If you don't use a star diagonal, short extension tubes placed between the Barlow and the eyepiece can accomplish the same thing.

Some Barlows which claim $2 \times$ or $3 \times$ are not exactly those factors of magnification. To accurately calculate a Barlow's magnification, time the drift of a star across the field of view of an eyepiece. Then add the Barlow and use the same star, the same eyepiece and the same path of drift through the field of view. Divide the second number into the first and you will have your ratio.

Test a Barlow the same way you would test an eyepiece. Especially look for problems with contrast and vignetting of the image. If the Barlow scatters lots of light into the black areas of the fov, you can be sure that it will not reveal subtle detail on planets. Image contrast is a key factor with Barlows because a Barlow is generally employed to achieve high magnification. At high magnification, you're looking for detail. And nothing kills detail like lack of contrast. Compare Barlows of equivalent magnification to one another or compare a Barlow/eyepiece combination to the view through an eyepiece of half the focal length.

Paracorr

With the revolution in large, short focal length Dobsonian reflectors came a need for an optical component to correct a problem that plagues such scopes: coma. The trick was to do it without introducing other aberrations. Well, I can safely say that Al Nagler of Tele-Vue has accomplished this feat by introducing the Paracorr. Paracorr stands for "parabola corrector," and very much resembles a Barlow lens. In fact, it does introduce a slight magnification to the image ($1.15 \times$).

The Paracorr utilizes two multi-coated, achromatic doublets of high refractive index. Importantly, it introduces no false color into the images and no spherical aberration. The Paracorr increases the "sharp area" of fast Newtonians from (typically) $0.1°$ to more than $3°$. One feature I am thankful for is that the Paracorr is threaded to accept 48 mm filters. This is useful when making filtered observations at different magnifications – simply switch eyepieces.

An eyepiece projection setup. The set screw holds a 1.25" eyepiece securely in the housing. (Photo by the author)

Binoviewers

A binoviewer (short for "binocular viewer") is a device which replaces your one eyepiece with two. Just as in binoculars, the light is split by a prism, sending the beam to both eyes. This accessory is not inexpensive, with good-quality models costing anywhere between 500 and 1000 US$. Some observers swear by binoviewers. Others swear at them. Some of the relevant points follow.

Regarding light loss, you will lose surface brightness on objects because the light which previously went to your eyepiece now must be shared between two eyepieces. On faint objects, this is a problem. On brighter objects, or through a large telescope, the light loss is offset by a view which is much more comfortable. Eyestrain is reduced and

The Zeiss Binoviewer as it comes out of the box. (Photo by Robert Kuberek of Valencia, California)

The Zeiss Binoviewer, with a pair of 27 mm Tele-Vue Panoptic eyepieces, ready to attach to the telescope. This accessory belongs to Robert Kuberek of Valencia, California.

floaters are much less of a problem when both eyes are employed.

If you are considering the purchase of a binoviewer, there are two things you absolutely must check before you buy them. The first is whether or not they will bring an object to focus in your telescope. Many binoviewers require so much in-focus (the racking-in of your telescope's focus knob) that they cannot be used. In fact, there are binoviewer models which require nearly 12 cm of in-focus! To overcome this problem, many binoviewers incorporate a Barlow lens. While this takes care of the problem on most scopes, the Barlow/binoviewer/eyepieces combination often sends the magnification into the stratosphere. $3.5\times-4\times$ actual magnification is not uncommon with a $2\times$ Barlow. As I write this, manufacturers are working to solve the problem (and some like Tele-Vue, Takahashi and Zeiss, actually have).

The second check to make is whether or not you can actually use the binoviewer. Is the interpupillary distance of the unit too great or too small for your eyes? Can you merge the images from both eyepieces into one? Some observers find that, at very high or very low powers, they simply cannot get the images to join into a single image. This severely limits the number of eyepieces they can use. This is something that varies greatly from one individual to the next.

Oh, speaking of eyepieces, the realization often comes later rather than sooner that double the number of eyepieces will be necessary with a binoviewer. This is a serious deterrent to some (like me, for example). Apart from price, there may be other problems.

You have to be aware of the eye relief of your eyepieces. Eyepieces with very short eye relief are problematic, but so also are eyepieces with long eye relief. Holding an object continually in view becomes a struggle. Another consideration is the weight, especially of some of the new multi-element designs. For my f/15 refractor, even if I chose to add additional counterweights, I would worry about tube flexure. In addition to these considerations, you must match the eyepieces exactly. Trying to use one eyepiece that you bought new ten years ago and a pre-owned one that you bought yesterday may not work together. Tele-Vue Nagler eyepieces (Type VI Naglers have just been released as I write this!) are the prime example of this.

When you find a binoviewer/eyepiece combination that works in your telescope, you will enjoy uncounted hours of pleasurable observing.

Some binoviewer suppliers

Baader/Astro-Physics Binoviewer	http://www.astro-physics.com
Celestron, Inc.	http://www.celestron.com
Orion Binoviewer	http://www.telescope.com
Seibert Optics	http://www.SiebertOptics.bizland.com
Tele-Vue binoviewer	http://televue.com
University Optics	http://www.universityoptics.com

2.4 Filters

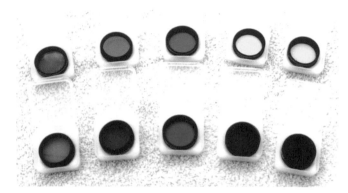

Serious planetary observers consider an eyepiece filter set to be a necessity. (Photo by the author)

Color filters

First, let me make a small point about filters. No filter makes any part of any astronomical object brighter. Because all filters subtract light of certain wavelengths from objects, filters make objects fainter. So when you hear someone exclaim, "Wow! This filter makes the nebula so much brighter!" remember that what they actually meant was, "Wow! This filter makes the nebula so much easier to see!"

From my experience, filters are feared by beginning amateur astronomers. They don't fully understand their use, they don't know exactly what the filter should show and the view to them is very unappealing. This probably stems from a desire to appreciate the image from an aesthetic viewpoint. Well, color filters have their purposes and I guarantee you that making an object "pretty" isn't one of them. Their purpose is to exaggerate differences in brightness. In fact, nearly all criticisms of color filters indicate that the observer was looking for changes in color rather than changes in brightness.

Color filters help overcome image deterioration caused by atmospheric scattering of light, they permit separation of light from different levels in a planetary atmosphere, and they increase contrast between areas of differing color. While using filters will not eliminate optical defects in the telescope they will help improve image definition even in a bad system.

Color filters are labeled along their circumference. In use, color filters are usually screwed into the eyepiece barrel. All eyepiece filters have threads which match the threaded inside barrels of eyepieces. Most filters fit 1.25 inch epieces and some fit the larger 2 inch models. In some cases, observers prefer to hand-hold the filter and move it back and forth between the eyepiece exit lens and their eye. The

purpose of such a maneuver is to be able to quickly compare the filtered view with the unfiltered one. Be aware that filters have been dropped and broken like this, especially in cold weather when the observer was wearing gloves. Certain manu-facturers also make filter holders. Such units allow up to four filters to be used simultaneously. Most of the amateurs that I know using such a device leave one filter slot empty so that they can also study the unfiltered view.

You would never stack filters of these colors for real observing. This shows how the filters screw into one another. (Photo by the author)

Filters work by "filtering," or blocking part of the spectrum (usually a very large part), leaving only the light that is transmitted to reach your eye. You can tell which part of the spectrum a color filter is transmitting simply by looking at it. A color filter is referred to by its Wratten number. The Wratten system was developed by Kodak in 1909 and has been the standard ever since. Filters used for photography, astronomy, and other applications all use this same standard.

Please be aware that all color filters work better in conjunction with larger telescopes. It is a simple rule of light throughput. On nights of excellent seeing, I have tried to use a violet filter on my 100 mm refractor to see cloud features on Venus. It just doesn't work. The filter only transmits 3% of the light hitting it. However, using a Meade 300 mm SCT, features are easily seen through the same filter.

We will discuss the effects of color filters in the chapters on individual objects. For a very good description of the use of color filters, see Susan Carroll's online article at
http://sciastro.net/portia/advice/filters.htm

The following table lists the most common color filters along with the percentage of light that they transmit.

Jeff Medkeff of Sierra Vista, Arizona, gives his recommen-dations: "If I could keep only two of my filters they would be No. 21 Orange and No. 82A (Light) Blue. If I could keep a third I would make it 25 A Red. If I could have a fourth it would be 80A (Dark) Blue. These would also be my recommendations for a first kit." See Jeff's excellent article about color filters online at
http://www.roboticobservatory.com/jeff/observing/colorfilter/index.htm

The most common color filters

	Color Filter	Transmission (%)
#8	Light Yellow	83
#11	Yellow-Green	78
#12	Yellow	74
#15	Deep Yellow	67
#21	Orange	46
#23A	Light Red	25
#25A	Red	14
#38A	Dark Blue	17
#47	Violet	3
#56	Light Green	53
#58	Green	24
#80A	Blue	30
#82A	Light Blue	73

Neutral density filters. Many observers use these for observing the Moon. (Photo by the author)

Color filters are often used by amateur astronomers to suppress the effects of our atmosphere. This does work, though to a limited extent. Using a #25A Red filter on objects low on the horizon and filters which are less red as objects are viewed progressively higher in the sky is said to improve the view.

Here is the rationale behind such thought. Scattering interposes a luminous veil between the observer and the object being observed. Scientists have shown that for particles in a planet's atmosphere of a given size, the scattering is inversely proportional to the fourth power of the wavelength of the light. Hence, violet light of wavelength 400 nm is scattered about 16 times more than deep red light of 800 nm. Our daytime sky is blue as a result of this property.

Unfortunately, although the science is sound, in practice I have found that the benefit from this procedure is small. Using a #25A Red filter to suppress the effects of atmospheric seeing, even if beneficial, seems to me to have extremely limited use. I almost never observe objects that close to the horizon.

Specialty filters

Neutral density filters

Another type of filter often used is the neutral density filter. These filters reduce the amount of light (by absorbing it) but do not filter any of the colors. Neutral density filters for visual use come in a wide range of transmissions, from 80% all the way down to 1%. Darker values can be obtained, of course, by stacking filters. In general, lighter neutral density filters are used for the planets and darker ones for the Moon. Please note that no neutral density filter ever tested was truly "neutral," but are close enough for visual use.

Polarizing filters

Polarizing filters reduce glare by only allowing light of a specific orientation to pass through. Single polarizing filters are often employed by double star observers when one of the stars is much brighter than the other. The hope is that the glare from the brighter primary star will be reduced, making the companion star easier to see.

Some observers use a polarizing filter to make deep-sky observations during Full Moon. They report significant contrast gain 60° from the Moon and astounding results at 80–90° although this is better seen at low power. On star-like objects, the polarizing filter helps a little at low power, but the limiting magnitude is the same as it would be if you used a higher power eyepiece.

A variable neutral density filter is the result of a cross-polarizing filter. These filters allow you to vary the brightnesses of images while you're observing. They consist of two polarizing layers mounted in a rotating cell. The light transmission may be varied from about 3% to 40%. Such filters are most frequently used for lunar observing.

Baader Moon/Skyglow Filter

Despite its name, this is not a filter with which to observe the Moon. It is a filter which cuts out skyglow caused by moonlight and other sources. As of this writing, I have not used this filter, but the reports I have gotten are heartening. Apparently, the Baader filter cuts the glare and haze from the Moon while preserving detail much better than a deep-sky filter. Stars were reported to be much brighter and more visible through the Baader filter than through the deep sky with the difference in transmission and contrast obvious.

Minus Violet Filter

Sirius Optics in Kirkland, Washington, manufactures a filter they call MV1 "Minus Violet." It reduces the false color fringing of blue–violet light most noticeable in relatively low-cost achromatic refractors. It does this without appreciably changing the color of the object. In use, it was found that the MV1 introduced the slightest amount of color and, in response to this, Sirius Optics has introduced the MV20.

The two claims made regarding the MV1 filter are that (1) it reduces the false color of the image and (2) that it makes focusing easier. The reasoning is that, since there is less unfocused light, it is easier to find the best point of focus. I can confirm both claims through my 100 mm achromatic refractor.

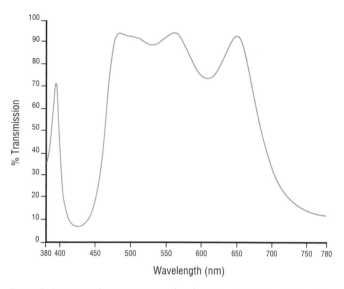

Transmission curve of the Minus Violet (MV1) filter made by Sirius Optics of Kirkland, Washington. (Retraced by Holley Y. Bakich)

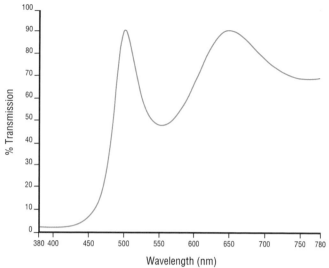

The Contrast Enhancement filter made by Sirius Optics of Kirkland, Washington, allows more light through than standard light pollution reduction (LPR) filters. (Retraced by Holley Y. Bakich)

Variable Planetary Filter

Another filter by Sirius Optics in Kirkland, Washington, is their variable color filter. This is a cell containing a color wheel which, when turned, changes the wavelength of the transmitted light. It works like a charm! In fact, your tendency may be to move the wheel too much, thus missing some of the details this filter may reveal. Plan to spend a lot of time (all of it enjoyable) observing the planets through this filter.

Sirius Optics' Variable Planetary Filter. (Photo by the author)

Light pollution reduction (LPR) filters

LPR filters come in two varieties: broadband and narrow band. The word "band" refers to the range of wavelengths that the filter will allow to pass through. Broadband filters are often termed "light pollution filters" by their manufacturers or sellers. Using these filters from mid-city, for example, will show some improvement in the images. They do not eliminate the effects of light pollution, however. On this subject, please note what Bob Haler of Lymax Astronomy in Kansas City, Missouri, has to say:

> From my personal experience, the only filter that seems to have miraculous anti-light pollution abilities is the OIII (oxygen 3) filter. I have used Lumicon® OIII filters in ridiculously light polluted environments (like

a parking lot full of mercury vapor lamps) and gotten excellent views of the Ring Nebula, the Dumbbell Nebula, and other bright nebulas. I am NOT saying that it is just like observing these objects from a dark location far from city lights. What I am saying is that you can get some good views of these objects and might be able to show them off in less than ideal conditions – like the grounds of your local grade school.

Broadband filters are many and varied. I have yet to be convinced that they are worth the money, but I won't go so far as to say that they do not filter out some unwanted light. In fact, they tend to work better from a dark site or one with only mild light pollution. Narrow band filters are a different story entirely and it is those we will consider here. First, let's take a look at Bob's favorite.

OIII

This filter is called this because it mainly allows light from doubly-ionized oxygen to pass through. Yes, I said doubly. OI is the designation of neutral oxygen, OII is singly-ionized oxygen and OIII is doubly-ionized. I note with pleasure the fact that astronomy is not the only science with unusual nomenclature and designations. Generally, the OIII filter has a bandwidth of around 10 nm, centered over a pair of spectral lines with wavelengths of 496 nm and 501 nm.

A Lumicon OIII filter. (Photo by the author)

Planetary nebulae especially benefit from this filter as do supernova remnants. Other nebulae should not be overlooked, especially if the nebula is bright and the light pollution is high. (Try M42, the Orion Nebula, for example.)

Observing Tip: Try your OIII filter on difficult double stars. Some observers have reported good results on pairs of very unequal brightness.

UHC

This filter has a larger bandpass (22–26 nm) than the other narrow band filters, but a much narrower one than any wideband filter. In use, it makes the background sky appear slightly darker and gives the stars a slight blue color. Objects that benefit from the UHC filter are nebulae of all kinds, especially bright and diffuse.

H-Beta

This filter has the narrowest bandpass of all, only 8 nm. The transmission is centered on the hydrogen beta line at 486 nm. Amateur astronomers purchase this filter for one purpose: to observe the elusive Horsehead Nebula in Orion. And it works. Using an h-beta filter, I have observed the Horsehead in telescopes ranging in aperture from 150 mm to 750 mm. There are a few other nebulae that look good through this filter (including the California Nebula in Perseus), but the number is small.

These narrow band filters may be obtained from many telescope dealers.

One more thought from Bob Haler before we move on: You will sometimes see "cosmetic blems" available at star parties and special sales for greatly reduced prices. Looking through them they work fine, it's just that looking at them, they look bad. You still get great views and you can save serious money.

Filter performance comparisons for some common nebulae

David Knisely, Prairie Astronomy Club, Lincoln, Nebraska Reprinted by permission

The following is a summary report of visual observations of emission nebulae comparing the performance of various filters intended for such objects. The instrument used is a 10 inch f/5.6 Newtonian, working at 59×, 70×, and 141×, as well as a few unaided-eye observations using the filters hand-held and looking up at the sky (for Rosette, North America, California, and Barnard's Loop). The filters used were Lumicon's DEEP-SKY, UHC, OIII, and H-BETA, and were usually all mounted in a modified Lumicon multi-filter adapter. This allowed rapid comparisons between filters, thus avoiding some of the judgement problems caused by the time needed to change filters or reports from inaccurate single-observation anecdotal accounts. Observing was done from a dark-sky site (visual naked-eye limit 6.5 to 7.0).

For detailed descriptions of the objects, see any of the various observing handbooks. Two methods were used for rating filter performance. In the first method, each filter was given a 0–5 point "Score" performance ranking behind it for each object observed. Example: OIII (4) means the OIII gave a large improvement in the view over non-filter use and contributes 4 points to its overall score total. Items such as overall surface brightness, area of nebulosity observed, and contrast of detail were used to judge how well a filter improved the view. However, since this judgement contains some of the personal preferences of the observer, the exact results may be somewhat subjective in the long run. Different observers might have slightly different ratings of various filters on various objects, so small differences in judgements are to be expected. Still, the scoring does on average give a reasonable idea of overall filter performance.

Scoring legend

(5): Very large improvement over no filter

(4): Large improvement over no filter

(3): Moderate improvement over no filter

(2): Slight improvement over no filter

(1): No improvement or fainter than no filter

(0): Much worse than no filter (object marginal or not visible)

Scoring totals for nebulae so far surveyed

(81 objects as of 8 June 2000)

UHC	284 points,	average rating 3.51
OIII	257 points,	average rating 3.17
DEEP-SKY	178 points,	average rating 2.20
H-BETA	112 points,	average rating 1.38

The second method is a somewhat subjective recommendation for the best filter to use on the given object, based mainly on a personal judgement, and thus is more a matter of opinion and taste. The nebular brightness, total area shown, contrast of details, and overall view are all weighted to give an opinion of which filter will work well for which object. Other observers would doubtless have somewhat differing views on recommendations for the specific objects which were observed. When objects were best seen in two filters (i.e., nearly equal or beneficial performance), both filters would be given the recommendation for the object, with the one yielding the better overall view being listed first and the "close second" best listed next to it.

Recommendations ranking summary

UHC best on 34 nebulae, close second best on 39 nebulae.

Total 1st and 2nd recommendations for UHC: 73 objects.

OIII best on 32 nebulae (biased by the inclusion of some planetary nebulae), close second best on 19 nebulae. NOT recommended on 7 nebulae.

Total 1st and 2nd for OIII: 51 objects.

DEEP-SKY best on 7 nebulae, second best on 2 nebulae.

Total 1st and 2nd recommendations for DEEP-SKY: 9 objects.

H-BETA best on 10 nebulae, second best on 2 nebulae. NOT recommended on 36 nebulae!

Total 1st and 2nd recommendations for H-BETA: 12 objects.

General trends in results

So far (with a few notable exceptions), the numbers show the UHC and OIII are the filters of choice for viewing nebulae, and to some degree supports the general recommendation that if only one filter can be purchased, it should be the UHC. In performance characteristics, the UHC filter tends to reveal a slightly larger and/or brighter area of nebulosity with many emission nebulae than the OIII does, while the OIII filter will often yield somewhat more contrast and dark detail on a given object. The OIII also tended to be a bit better for "blinking" small planetary nebulae than the UHC was, while the H-BETA often hurt the view of many planetaries. The inclusion of some planetary nebulae may have slightly inflated the score of the OIII filter, since in general, the OIII often does a bit better on those objects. The H-BETA tended to be most useful on a more limited number of objects than either the UHC or the OIII filters. This may be due at least in part to the fact that many of the so-called "H-BETA objects", are low-excitation very faint nebulae, and thus are near or beyond the visual limits of my ten inch. The DEEP-SKY filter almost always produced at least some gain in contrast for nearly every object observed (especially when some skyglow was present), but rarely produced a spectacular improvement of the view. Filter comparison results for each of the objects observed are shown below.

Specific object results

Each object is listed with the various filters. Each filter's numerical score (0, 1, 2, 3, 4, 5; see earlier comments) is given, followed by the overall filter recommendation for the specific object. In the RECOMMENDATIONS portion, a "/" between the two filters named indicates that both filters will work well on the object, with the one on the left side of the slash being a slightly better choice: i.e. "UHC/OIII" means the UHC is slightly better overall, but the OIII will be quite useful as well.

M1 Crab Nebula (SNR in Taurus)

DEEP-SKY: (3) Improves the contrast and brings out the wispy arc-like cusp on the eastern end.

UHC: (4) Darkens the background and reveals little hints of tattered detail on the edges with the eastern "cusp" now more visible.

OIII: (3) Much darker than in UHC, and appears slightly smaller and somewhat rounder, but with hints of filamentary detail on the edges and across the nebula at moderate powers.

H-BETA: (0) Barely visible.

RECOMMENDATION FOR M1: UHC/DEEP-SKY (H-BETA NOT recommended).

M8 Lagoon Nebula (diffuse nebula in Sagittarius)

DEEP-SKY: (3) Some increase in contrast, with a bit more nebulosity visible than without a filter.

UHC: (5) Large boost in contrast and visibility of outer nebulosity. Nebula appears much larger (nearly a degree wide) with some detail enhancement, especially in the outer regions.

OIII: (5) Slightly fainter than in the UHC, but shows slightly more contrast and dark detail than UHC does. Some of the outermost nebulosity fades, but detail in inner regions is remarkable. Faint red color can be noted in the brighter areas. May be the better filter under light polluted conditions.

H-BETA: (2) Dims the nebula considerably, with only the circular ball of haze around the Hourglass nebula and the external arc being easy to see.

RECOMMENDATION FOR M8: UHC/OIII

M16 Eagle Nebula (diffuse nebula in Serpens)

DEEP-SKY: (2) Faint diffuse nebulosity is slightly easier to see than without a filter. Not a great deal of detail visible in nebula.

UHC: (4) Large increase in visible nebulosity, showing wide diffuse fan of light in the shape of a broad "T" shape. Small darker inclusion becomes visible along the northern side.

OIII: (4) Slightly fainter than with UHC, with slightly less faint outer nebulosity, but shows more contrast and dark detail in the interior, including faint narrow "fingers" from south side into the center of the nebula with averted vision.

H-BETA: (2) Dims the nebula significantly, but "T" shape still vaguely visible.

RECOMMENDATION FOR M16: UHC/OIII, but H-BETA hurts the view.

M17 Swan (Omega) Nebula (diffuse nebula in Sagittarius)

DEEP-SKY: (3) Some improvement in contrast and detail, with the fainter loop of nebulosity to the northeast just becoming visible to form the omega shape.

UHC: (4) Noticeable improvement in contrast and detail, with much of the faint nebulosity on the outer regions and along the "omega" loop becoming quite easy to see.

OIII: (5) Slightly fainter than UHC, but contrast is also higher, with a rather striking dark area becoming noticeable along the west side of the swan's neck. Dark detail in interior of main bar is better defined than with UHC.

H-BETA: (1) Object is noticeably dimmed compared to the other filters, making the filter a poor choice for use on M17.

RECOMMENDATION FOR M17: OIII/UHC (H-BETA NOT recommended).

M20 Trifid Nebula (diffuse emission/reflection nebula in Sagittarius)

DEEP-SKY: (2) Small difference between filtered and unfiltered views with a slight gain in contrast with the filter, but with any light pollution, the filter may be of greater use.

UHC: (4) Nebula is slightly fainter than with DEEP-SKY filter, with a slight gain in contrast over the DEEP-SKY and more contrast gain over unfiltered views.

OIII: (3) Nebula is fainter than with UHC or DEEP-SKY, and main Trifid section appears slightly smaller (hurts the northern reflection nebulosity), but dark detail in the inner "lanes" shows up slightly better.

H-BETA: (4) Nebula is somewhat fainter than in UHC, but Trifid section shows a bit larger area of nebulosity than the UHC does. It kills the reflection nebula and reduces the brightness of the detail right around the central star.

RECOMMENDATION FOR M20: UHC/H-BETA.

M27 Dumbbell Nebula (planetary nebula in Vulpecula)

DEEP-SKY: (3) Some improvement in visibility of outer haze off the sides of the dumbbell, but the object is also slightly fainter.

UHC: (5) Large improvement in contrast and outer detail, with large "wings" of light off the Dumbbell's sides becoming easy to see. Interior seems brighter and bigger, with interesting greenish glow.

OIII: (4) Dimmer than with UHC, but interior shows more dark detail and contrast. "Wings" off the sides remain fairly easy to see.

H-BETA: (1) Nebula is dimmed greatly by the filter, extinguishing the fine outer detail and only showing the inner dumbbell-shape.

RECOMMENDATION FOR M27: UHC (OIII also useful in showing some inner detail, but H-BETA is NOT recommended).

M42 Great Orion Nebula (diffuse nebula)

DEEP-SKY: (3) A moderate boost in contrast can be seen, and much more outlying nebulosity is visible. This is a good filter for the general public, since it still will show the stars while enhancing the nebula.

UHC: (5) Large boost in contrast over no filter is noted. Outer nebulosity is quite easy to see, with southward loop being easily seen with averted vision. Bluish and greenish colors are quite easy to note with direct vision.

OIII: (4) A few of the outermost nebulosity areas are dimmed, but there is more contrast, with considerable improvement in light and dark detail, especially in the inner regions. Bluish and, occasionally, reddish tints are noted with large apertures. M43 is somewhat fainter than in the UHC filter, but narrow bandwidth of OIII may make it the filter of choice with light pollution.

H-BETA: (3) Much of the fainter outer areas of the nebula vanish, but fan-like main portion and M43 remain, with interesting contrast and changes in detail visible, including a brighter linear arc in the western part of the fan. Some reddish hints are also visible in the H-BETA.

RECOMMENDATION FOR M42: UHC/OIII (near tie)

M43 (north part of Great Orion Nebula)

DEEP-SKY: (3) Higher contrast than without filter, but not much detail enhancement except when there is some light pollution.

UHC: (3) Somewhat more contrast than with the DEEP-SKY with overall "comma" shape now easily seen.

OIII: (2) Dims the nebula, but overall shape is still easily visible.

H-BETA: (4) Really makes M43 stand out, with high contrast and some irregular dark detail in the overall comma-shaped nebula.

RECOMMENDATION FOR M43: H-BETA (UHC and DEEP-SKY also help).

M57 Ring Nebula (planetary nebula in Lyra)

DEEP-SKY: (2) Darkens the background slightly, and brings out hints of very faint nebulosity off the ends of the oval, but otherwise doesn't help much.

UHC: (4) Really darkens the background and stars, but begins to show a more prominent glow in the interior of the ring, with hints of faint outer nebulosity around the outer edges of the ring.

OIII: (4) Darkens the nebula and the background still further, but slight increase in contrast noted. Outer shell just visible with averted vision.

H-BETA: (0) Really kills things, with the nebula now being very dim.

RECOMMENDATION FOR M57: UHC/OIII. Nebula is bright and small enough not to really benefit enormously from filter use, but UHC does improve it to a degree (H-BETA is NOT recommended!).

M76 "Mini-Dumbbell" or Butterfly Nebula (planetary nebula in Perseus)

DEEP-SKY: (2) Marginal improvement over no filter, with hints of nebulosity off the sides of the dumbbell.

UHC: (4) Much more nebulosity visible, including faint patches or "wings" off each side of the dumbbell, along with some interior detail.

OIII: (4) Nebula is slightly fainter, but shows more contrast, with some dark detail being seen near each lobe of the dumbbell. The patches off to the sides of the dumbbell look like partial loops.

H-BETA: (0) Dims the nebula almost to extinction at moderate powers.

RECOMMENDATION FOR M76: OIII/UHC (H-BETA NOT recommended!).

M97 Owl Nebula (planetary nebula in Ursa Major)

DEEP-SKY: (2) Slight improvement over non-filter use (hints of the "eyes").

UHC: (4) Much higher contrast than with DEEP-SKY filter. One eye and hints of the other are seen.

OIII: (5) Increase in contrast over UHC. Both eyes visible with hints of irregular outer edge structure.

H-BETA: (0) Nearly obliterates the nebula.

RECOMMENDATION FOR M97: OIII/UHC (H-BETA NOT recommended).

NGC 40 (planetary nebula in Cepheus)

DEEP-SKY: (3) Slight increase in contrast and detail (brighter opposing sides), but object does not really require a filter.

UHC: (3) Slightly fainter than in DEEP-SKY, but shows a bit more contrast.

OIII: (2) Somewhat fainter than in UHC, but disk still quite visible.

H-BETA: (2) Somewhat fainter than in UHC, but very slightly brighter than with the OIII filter (a "near" H-BETA object).

RECOMMENDATION FOR NGC 40: DEEP-SKY/UHC (near tie).

NGC 246 (planetary nebula in Cetus)

DEEP-SKY: (2) Defines it a bit better than without a filter, but still mainly a diffuse roughly circular glow around a few stars.

UHC: (3) Higher contrast, with nebula now a fairly well-defined moderate-sized dim disk with hints of brightness variations in the interior.

OIII: (5) Dramatic increase in contrast over the UHC! Shows several dark spots in the interior and hints of sharp filament-like outer edge of the disk.

H-BETA: (0) Really kills the nebula (barely visible).

RECOMMENDATION FOR NGC 246: OIII. (H-BETA NOT recommended).

NGC 281 (diffuse emission nebula in Cassiopeia)

DEEP-SKY: (3) Nebula is somewhat easier to see (barely visible without filters), with the edges being more defined.

UHC: (4) Noticeable improvement in contrast and detail, appearing larger than with DEEP-SKY filter, and containing some dark detail.

OIII: (4) Nebula is dimmer, but interior dark lane-like detail becomes more noticeable, and the overall nebula shape is better defined than in UHC.

H-BETA: (2) Dims the nebula much more than OIII, with no more detail than is seen with the DEEP-SKY filter (dim).

RECOMMENDATION FOR NGC 281: UHC/OIII.

NGC 604 (HII region in galaxy M33 in Triangulum)

DEEP-SKY: (2) Slight increase in contrast over unfiltered view, but easy to see without filters.

UHC: (3) Much easier to see than in DEEP-SKY, standing out well as an oval puff, with much of the detail in the galaxy remaining visible.

OIII: (4) Considerable increase in contrast, almost "blinking" over UHC and unfiltered views. Galaxy is much fainter, but nebula really stands out.

H-BETA: (2) Much dimmer than in the other filters, but nebula is still seen.

RECOMMENDATION FOR NGC 604: OIII/UHC.

NGC 896/IC 1795 (diffuse nebula in Cassiopeia)

DEEP-SKY: (3) Noticeable increase in visibility, with nebula being only a glow without the filter. Two areas of dim diffuse nebulosity seen, one large (IC 1795) and the other smaller (NGC 896).

UHC: (4) Much more prominent, with better definition and a little dark detail, along with a wispy outer arc curving around from south part of IC 1795.

OIII: (4) Dimmer than in UHC, but more dark detail visible with faint outer loop-like structure visible arcing south, almost connecting the two patches.

H-BETA: (1) Barely visible.

RECOMMENDATION FOR NGC 896/IC 1795: UHC/OIII (H-BETA NOT recommended).

NGC 1360 (large planetary nebula in Fornax)

DEEP-SKY: (2) Slight increase in contrast with nebula easier to see than without a filter (but still visible without a filter).

UHC: (4) Significant improvement in contrast, with nebula appearing larger and noticeably oval. Some irregular interior detail and central star noted.

OIII: (4) Even more contrast than UHC, with clear interior arc-like detail, but central star much fainter. Nice dark background.

H-BETA: (0) Kills the nebulosity with only the central star and a small hint of haze around the star being visible.

RECOMMENDATION FOR NGC 1360: OIII/UHC (H-BETA NOT recommended).

NGC 1499 California Nebula (diffuse nebula in Perseus). Without filters, the nebula is barely visible as a faint brightening of the field with no detail.

DEEP-SKY: (2) A slight increase in contrast was noted, but otherwise, the view was similar to that without a filter.

UHC: (2) Slight increase in contrast over the DEEP-SKY filter, making the edges of the nebula slightly easier to see, but nebula is still somewhat difficult. Hints of vague brightness variations across the object are noted.

OIII: (1) Nebulosity is quite dim in a very dark field.

H-BETA: (4) Dramatic increase in contrast noted, making the object fairly easy to notice, with well-defined borders. Some faint filamentary detail is also noted. California Nebula is visible to unaided eye when H-BETA is used.

RECOMMENDATION FOR NGC 1499: H-BETA.

NGC 1514 Crystal-Ball Nebula (planetary nebula in Taurus)

DEEP-SKY: (2) Nice faint round puff around a faint star, easier to see than without a filter.

UHC: (4) Significant improvement in contrast, well-defined hazy ball with hints of dark detail in the interior of the nebula.

OIII: (4) More contrast than in UHC, with dark detail and arc-like forms in the main shell. Dimmer than in UHC but a bit better overall.

H-BETA: (0) Almost wipes out the nebula.

RECOMMENDATION FOR NGC 1514: OIII/UHC (H-BETA NOT recommended).

NGC 1999 (diffuse nebula in Orion)

DEEP-SKY: (2) Slight enhancement over no filter, and easy without one.

UHC: (1) fainter than DEEP-SKY or no filter.

OIII: (1) fainter than UHC or DEEP-SKY.

H-Beta: (1) fainter than DEEP-SKY, UHC, or no filter.

RECOMMENDATION FOR NGC 1999: DEEP-SKY

NGC 2022 (planetary nebula in Orion)

DEEP-SKY: (3) visible without a filter, but stands out better with DEEP-SKY (small fuzzy disk).

UHC: (4) Noticeably improves the contrast, with an almost annular form visible at higher magnifications.

OIII: (5) Much higher contrast and darker background than in UHC, but UHC or no filter may be bit better for high power observations of details.

H-BETA: (0) Almost wipes it out (barely visible).

RECOMMENDATION FOR NGC 2022: OIII/UHC (H-BETA NOT recommended).

NGC 2024 Flame Nebula (diffuse emission/reflection nebula in Orion)

DEEP-SKY: (3) Noticeably improves the contrast with the dark lane-like detail visible.

UHC: (3) Darker than in DEEP-SKY but with only a slight increase in contrast.

OIII: (2) Darker than in UHC, with less detail than in UHC.

H-BETA: (1) Darkest of all three filters, but the nebula remains visible with detail similar to that of OIII.

RECOMMENDATION FOR NGC 2024: DEEP-SKY/UHC (near tie).

NGC 2174 (diffuse nebula in northern Orion)

DEEP-SKY: (2) Very faint glow around a single star with hints of detail (much easier to see than without a filter).

UHC: (4) Large increase in contrast over DEEP-SKY filter, showing a large circular area of haze with vague irregular interior dark detail.

OIII: (4) Dimmer than in UHC, but has more contrast, showing some dim lane-like structure.

H-BETA: (0) Dims the nebula almost to extinction, showing less than the DEEP-SKY.

RECOMMENDATION FOR NGC 2174: UHC/OIII (near tie) (H-BETA NOT recommended).

NGC 2327 (diffuse nebula in Monoceros)

DEEP-SKY: (2) Very faint diffuse roughly circular haze around 7th mag. star.

UHC: (3) Object is larger with slightly better definition than in DEEP-SKY.

OIII: (2) Nebula is now very faint, with only the area round the star visible.

H-BETA: (4) Object is not quite as bright as in UHC but is much better defined, showing a dark inclusion from the northeast and a brighter arc-like western edge.

RECOMMENDATION FOR NGC 2327: H-BETA/UHC

NGC 2237-9 Rosette Nebula (diffuse nebula in Monoceros)

DEEP-SKY: (2) Some increase in contrast, but nebula is still more of a diffuse haze around the central star cluster with hints of irregularity.

UHC: (5) Noticeable increase in contrast, with more outer nebulosity visible and some irregular light and dark structure being visible. Nebula was visible when UHC was held up to unaided eye!

OIII: (5) Higher contrast than with UHC, with more dark irregular detail throughout the region, but not quite as much nebulosity visible as in UHC.

H-BETA: (1) Very faint glow around the star cluster, not much better than without a filter (but much dimmer).

RECOMMENDATION FOR NGC 2237-9: OIII/UHC (near tie).

NGC 2264, Cone Nebula (near S Monocerotis)

DEEP-SKY: (2) Slight increase in contrast, with dim diffuse haze now visible and brightest spot WSW of S Mon.

UHC: (4) Faint nebulosity now visible over entire field, nearly a degree wide. Dark southern inclusion "Cone" faintly visible in southern part of nebula.

OIII: (3) Dimmer than in UHC, with only the area SW of S Mon being easy to see (Cone not visible).

H-BETA: (1) Only a hint of a glow SW of S Mon.

RECOMMENDATION FOR NGC 2264: UHC.

NGC 2346 (planetary nebula in Monoceros)

DEEP-SKY: (2) Slightly easier to see than without a filter.

UHC: (3) Some increase in contrast, with hints of annularity.

OIII: (3) Some increase in contrast, slightly fainter than in UHC.

H-BETA: (0) Nearly extinguished.

RECOMMENDATION FOR NGC 2346: UHC/OIII (near tie) (H-BETA NOT recommended).

NGC 2438 (planetary nebula in Puppis)

DEEP-SKY: (2) Noticeably easier to see than without a filter. Hints of annularity.

UHC: (3) Notable increase in contrast, easier to see than in DEEP-SKY with annular form more noticeable.

OIII: (4) Much higher in contrast with annular form now fairly obvious.

H-BETA: (0) Nearly kills it completely.

RECOMMENDATION FOR NGC 2438: OIII (H-BETA NOT recommended).

NGC 2467 Thor's Helmet (diffuse nebula in Puppis)

DEEP-SKY: (2) Sparse star cluster with a single star and a faint ball of haze around it which is brighter on the south side.

UHC: (4) Much more contrast and detail than in DEEP-SKY with brighter linear band or arc on south side.

OIII: (5) More contrast and detail than in UHC with the arc connected to a "loop" which runs through the central star. Faint hints of outer detail.

H-BETA: (1) Really dims it.

RECOMMENDATION FOR NGC 2467: OIII/UHC (H-BETA NOT recommended).

NGC 2359 (diffuse nebula in Canis Major)

DEEP-SKY: (2) Better defined than without a filter but still low in contrast.

UHC: (4) Higher contrast than in DEEP-SKY, with arc-like detail off of a central oval mass.

OIII: (5) Even more contrast than in UHC with oval mass now looking like a loop with tendrils off each end.

H-BETA: (0) Kills most of the nebulosity.

RECOMMENDATION FOR NGC 2359: OIII/UHC (H-BETA NOT recommended).

NGC 2371-2 (planetary nebula in Gemini)

DEEP-SKY: (2) Two adjacent faint spots, helped somewhat over non-filter use.

UHC: (4) Enhanced over DEEP-SKY, with the two lobes showing hints of contact.

OIII: (4) Slightly higher contrast than UHC. Hints of faint outer wings.

H-BETA: (0) Kills the nebulosity.

RECOMMENDATION FOR NGC 2371-2: OIII/UHC (near tie) (H-BETA NOT recommended).

NGC 2392 Eskimo Nebula (planetary nebula in Gemini)

DEEP-SKY: (2) Enhanced slightly over non-filter use (easier to see the outer of the two shells).

UHC: (4) Darkens the sky background and enhances the nebula, making both shells quite easy to see.

OIII: (4) Jet-black sky background with higher contrast than UHC, but the two shells almost seem to merge (tones down the central star).

H-BETA: (0) Only the inner shell is visible, much fainter than in UHC, OIII, or DEEP-SKY.

RECOMMENDATION FOR NGC 2392: OIII/UHC. (H-BETA NOT recommended).

NGC 3242 Ghost of Jupiter (planetary in Hydra)

DEEP-SKY: (2) Slightly enhanced over non-filter use (easy without filters).

UHC: (4): Much higher contrast with faint circular outer halo-like shell beyond the two inner shells now visible.

OIII: (4): Much darker background but the two inner shells really blaze out.

H-BETA: (1): Much fainter (only the innermost shell is easily seen).

RECOMMENDATION FOR NGC 3242: UHC/OIII (near tie) (H-BETA NOT recommended).

NGC 4361 (planetary nebula in Corvus)

DEEP-SKY: (2) Somewhat higher contrast than without a filter.

UHC: (4) Large increase in contrast with faint diffuse outer extensions seen.

OIII: (4) Higher contrast, a bit sharper than in UHC but nebula appears slightly smaller.

H-BETA: (0): Nearly killed by the filter.

RECOMMENDATION FOR NGC 4361: UHC/OIII (near tie) (H-BETA NOT recommended).

NGC 6210 (planetary nebula in Hercules)

DEEP-SKY: (2) Stands out a bit better, but filters are not needed.

UHC: (4) Increase in contrast with faint hints of close outer shell north and south of main disk.

OIII: (4) Darkens the background and also shows hints of the outer shell.

H-BETA: (1) Dims the nebula, showing only the brighter inner core.

RECOMMENDATION FOR NGC 6210: OIII/UHC (H-BETA NOT recommended).

NGC 6302 Bug Nebula (planetary nebula in Scorpius)

DEEP-SKY: (2) Somewhat more contrast than without a filter.

UHC: (3) Noticeable improvement in contrast with central core region now seeming much brighter and outer EW flarings much easier to see.

OIII: (3) Makes the core region really stand out, although the nebula is not quite as bright as in the UHC.

H-BETA: (0).

RECOMMENDATION FOR NGC 6302: OIII/UHC (H-BETA NOT recommended).

NGC 6334 (diffuse nebula in Scorpius)

DEEP-SKY: (2) Nebula is a large very faint glow which is brightest around one star near the south end.

UHC: (4) Two separated patches around two stars near the south end, plus fainter patches and dark spots visible in a dim diffuse haze to the north.

OIII: (3) Fainter than in UHC, but still visible.

H-BETA: (3) Similar to OIII view but slightly fainter.

RECOMMENDATION FOR NGC 6334: UHC (OIII and H-BETA also useful).

NGC 6445 (planetary nebula in Sagittarius)

DEEP-SKY: (2) Makes it stand out better.

UHC: (4) Noticeably improves the contrast over DEEP-SKY.

OIII: (3) Darker more contrasting field, but slightly fainter than UHC.

H-BETA: (0) Kills the nebula almost completely.

RECOMMENDATION FOR NGC 6445: UHC/OIII (H-BETA NOT recommended).

NGC 6537 (diffuse nebula in Scorpius)

DEEP-SKY: (2) Not easy to see without a filter, as DEEP-SKY just barely brings it out.

UHC: (3) Noticeable boost in contrast, showing some irregularity and a brighter portion around a tiny group of stars.

OIII: (4) More contrast than UHC, with the patch around the tiny star group greatly enhanced.

H-BETA: (1) Almost kills the nebulosity.

RECOMMENDATION FOR NGC 6537: OIII/UHC (H-BETA NOT recommended).

NGC 6543 Cat's Eye (planetary nebula in Draco)

DEEP-SKY: (2) Makes it stand out better.

UHC: (4) Noticeable contrast improvement with faint diffuse outer halo visible. Faint patch west of main nebula barely visible (IC 4677).

OIII: (4) Really darkens the background and boosts the visibility of the outer halo. IC 4677 now slightly easier to see.

H-BETA: (1) Really dims it but is still visible.

RECOMMENDATION FOR NGC 6543: OIII/UHC (H-BETA NOT recommended).

NGC 6559/IC 4685 (diffuse nebula in Sagittarius)

DEEP-SKY: (2) Not visible without filters, glow around one star with hints of extensions north and northwest.

UHC: (4) Noticeably enhanced with some light and dark structure.

OIII: (2) Still visible but much fainter than in UHC.

H-BETA: (2) Visible with slight structure.

RECOMMENDATION FOR NGC 6559: UHC.

NGC 6781 (planetary in Aquila)

DEEP-SKY: (3) Easy without a filter but shows more contrast with hints of annularity even at low power.

UHC: (4) Noticeable boost in contrast with strong annular form and glowing interior (notable brightening along south side).

OIII: (4) Really darkens field and enhances the annularity.

H-BETA: (0) Kills the nebula completely.

RECOMMENDATION FOR NGC 6781: OIII/UHC (H-BETA NOT recommended).

NGC 6804 (planetary nebula in Aquila)

DEEP-SKY: (2) Slightly easier to see than without a filter.

UHC: (3) Brings the nebula out well.

OIII: (4) Nice high contrast a bit better than UHC.

H-BETA: (0) Nearly wipes it out.

RECOMMENDATION FOR NGC 6804: OIII/UHC (H-BETA NOT recommended).

NGC 6888 CRESCENT NEBULA (diffuse nebula in Cygnus)

DEEP-SKY: (2) Slight improvement over no filter, with the brightest segment of the crescent going through a star along the northern end of the nebula fairly easy to see.

UHC: (4) Nebula is now much easier to see, appearing as a large nearly complete oval ring of dim nebulosity with brightness variations and a dimly glowing interior.

OIII: (5) Complete oval ring with glowing interior and slightly higher contrast than with UHC, but overall nebulosity is fainter than with UHC.

H-BETA: (1) Very dim, with only the brightest arc portion which was seen in the DEEP-SKY filter visible at all in a very dark field. Nebula almost gone.

RECOMMENDATION FOR NGC 6888: OIII/UHC (near tie) (H-BETA NOT recommended!)

NGC 6960-95 Veil Nebula (SNR in Cygnus)

DEEP-SKY: (3) Nebula is easier to see than without a filter, with both sides of the loop being visible, including the section through 52 Cygni.

UHC: (4) Large increase in detail and contrast! Nebula really stands out with some filamentary detail. Hints of other strands in the interior of the loop.

OIII: (5) ENORMOUS INCREASE IN CONTRAST AND DETAIL with wonderful fine filaments and strands visible even between the two main arcs, making the entire complex closely resemble its photograph. OIII is the filter of choice here.

H-BETA: (1) Very dim, but still visible (forget it!).

RECOMMENDATION FOR NGC 6960-95: OIII (UHC is helpful, but not quite as much as the OIII (H-BETA is NOT recommended).

NGC 7000 North American Nebula (diffuse nebula in Cygnus)

DEEP-SKY: (2) Nebular overall form is easier to see than without a filter, but only slightly.

UHC: (5) Very noticeable improvement in contrast over the DEEP-SKY filter, with both "Florida" and "Mexico" now quite easy to see.

OIII: (4) Some improvement in contrast and detail, with brighter "spine" on east side of "Mexico" and some faint dark detail being easy to see, but nebula is somewhat fainter than in UHC.

H-BETA: (3) Detail is similar to OIII, but nebulosity is fainter than OIII.

RECOMMENDATION FOR NGC 7000: UHC/OIII but both H-BETA/DEEP-SKY are useful on the object (UHC was brighter, but OIII shows more contrast).

NGC 7009 Saturn Nebula (planetary nebula in Aquarius)

DEEP-SKY: (2) Does show the anses on each end of the planetary a bit better, in the form of two small puffs.

UHC: (4) Anses become more spike-like, with noticeable increase in contrast.

OIII: (4) Nebula is dimmer, but contrast is a bit higher, especially in the interior, where inner shell detail can be seen.

H-BETA: (1) Nebula is noticeably dimmer, appearing as just a disk.

RECOMMENDATION FOR NGC 7009: Filters are not needed, but OIII/UHC will help bring out the fainter detail (H-BETA NOT recommended).

NGC 7023 (emission/reflection nebula in Cepheus)

DEEP-SKY: (3) Noticeable boost in contrast with nebular glow expanded over no filter. Darker areas noted on east and west sides.

UHC: (2) Dimmer than in DEEP-SKY and slightly smaller, but nebula still shows up better than without a filter.

OIII: (2) Dimmer than in UHC but still shows some hints of detail.

H-BETA: (1) Dimmer than UHC and OIII. Only central area around the star remains visible.

RECOMMENDATION FOR NGC 7023: DEEP-SKY.

NGC 7027 (planetary nebula in Cygnus)

DEEP-SKY: (2) Easy without filter, but DEEP-SKY makes it stand out a little better as a small bluish-green oval.

UHC: (4) Really makes the nebula almost blaze out and hints at large faint irregular outer shell. High power reveals off-center central star and an interior arc southeast of the central star.

OIII: (4) Core slightly dimmer than in UHC, but outer shell is easier to see with hints of detail in the outer shell.

H-BETA: (0) Really dims the nebula!

RECOMMENDATION FOR NGC 7027: OIII/UHC (near tie) (H-BETA NOT recommended!).

NGC 7129-33 (diffuse nebula in Cepheus)

DEEP-SKY: (2) Slight increase in contrast, with a faint haze visible around a central group of 4 to 6 stars.

UHC: (3) Haze now easier to see with more contrast, but still rather diffuse with some faint detail which is brightest in the northern portion. Two other faint patches visible slightly away from the north one.

OIII: (3) Brings out a little more detail (dark inclusion in one side?).

H-BETA: (1) Dims the nebula significantly, although it is still there.

RECOMMENDATION FOR NGC 7129-33: UHC/OIII.

NGC 7293 Helix (planetary nebula in Aquarius)

DEEP-SKY: (2) Large dim roughly circular fuzzy patch with slightly darker middle, easier to see than without a filter, but does not have a lot of contrast.

UHC: (4) Noticeable increase in contrast, showing a clear fat slightly diffuse ring with a glowing center and hints of structure. Nebula is now quite easy to see.

OIII: (5) Much more contrast than the UHC, with hints of helical nature and indications of outer filamentary nebulosity. Dimmer than in UHC, but stands out better than in the UHC. Best performance of all the filters.

H-BETA: (0) Barely visible in this filter (almost kills the nebulosity).

RECOMMENDATION FOR NGC 7293: OIII/UHC (H-BETA NOT recommended).

NGC 7538 (diffuse nebula in Cepheus)

DEEP-SKY: (3) Boosts the contrast making it easier to see than without a filter.

UHC (4) Darkens the background and brings out the nebulosity more than DEEP-SKY.

OIII: (4) Dimmer, but contrast is a bit higher.

H-BETA: (0) Dims it nearly to extinction.

RECOMMENDATION FOR NGC 7538: UHC/OIII (H-BETA NOT recommended).

NGC 7635 Bubble Nebula (diffuse nebula in Cassiopeia)

DEEP-SKY: (2) Vague diffuse oval fuzzy area around a bright star.

UHC: (3) Oval area of nebulosity noted around the star running roughly east–west with large very dim diffuse extensions noted to the northwest and southeast. A dim "Y"-shaped patch can also be seen just north of the central star.

OIII: (4) Higher contrast, with the "Y"-shaped patch now much more definite.

H-BETA: (1) Very dim, not as good as OIII, but nebula is still visible.

RECOMMENDATION FOR NGC 7635: OIII/UHC.

NGC 7662 Blue Snowball (planetary nebula in Andromeda)

DEEP-SKY: (2) DEEP-SKY filter does darken the background somewhat.

UHC: (3) Really darkens the background, but adds only a little nebulosity.

OIII: (3) Dims the nebula slightly, giving a jet-black sky background and a bit more interior contrast (but not much more detail).

H-BETA: (1) Significantly dims the nebula over the OIII.

RECOMMENDATIONS FOR NGC 7662: Filters are not really needed, but UHC/OIII may help with locating it at low power via "blinking" (H-BETA NOT recommended).

NGC 7822 (faint diffuse nebula in Cepheus)

DEEP-SKY: (2) Very faint large elongated (east-west) glow around a few stars.

UHC: (3) Glow is noticeably enhanced over DEEP-SKY, shows some irregularity.

OIII: (2) Fainter than UHC, but still visible.

H-BETA: (2) Fainter than the UHC, but shows about as much detail as UHC.

RECOMMENDATION FOR NGC 7822: UHC (H-BETA and OIII also useful).

IC 405 Flaming Star Nebula (diffuse emission/reflection nebula in Auriga)

DEEP-SKY: (3) Nebula visible as a very faint diffuse glow with irregularities around and to the east of AE Aur. Not clearly visible without filters.

UHC: (2) Slight increase in contrast showing faint arc-like filament north-east of AE. Faint background glow fainter than in DEEP-SKY filter.

OIII: (1) Only hints of nebulosity.

H-BETA: (2) Hints of arc and one other faint patch north of AE but nebula is fainter than in UHC or DEEP-SKY.

RECOMMENDATION FOR IC 405: DEEP-SKY/UHC (no filter helps a lot, and may be mostly a reflection nebula).

IC 410 (nebula associated with NGC 1893 in Auriga)

DEEP-SKY: (2) Faint glow running EW through the "Y"-shaped cluster NGC 1893 with southward extension off east end.

UHC: (4) Detailed arc-like irregular nebulosity running east-west and then curving south with darker inclusion along southwest side.

OIII: (4) Brings out more dark detail along the east and south sides, but nebula is dimmer. Really stands out, as nebula follows the form of the cluster.

H-BETA: (0) Nebula is almost wiped out.

RECOMMENDATION FOR IC 410: OIII/UHC (H-BETA NOT recommended).

IC 434 Horsehead Nebula (diffuse nebula in Orion)

DEEP-SKY: (2) Little change is seen from viewing without a filter. When visible, it appears as a weak dark gap in the dim north-south nebulosity, and the shape is hard to see. Nebula is difficult, unless viewed under very dark and clear conditions.

UHC: (3) Horsehead now stands out weakly, showing some of the horsehead shape with averted vision, a definite improvement over no filter or the DEEP-SKY.

OIII: (0) No horsehead seen. IC-434 nebulosity only hinted at.

H-BETA: (4) Nebula still dim, but horsehead shape now fairly easy to see, showing up with more contrast than with the UHC filter. East edge of IC 434 seems brighter than the rest of the nebula with the H-BETA.

RECOMMENDATION FOR IC 434: H-BETA (UHC also helps, but OIII NOT recommended).

IC 1318 Gamma Cygni Nebula (diffuse nebula in Cygnus)

DEEP-SKY: (2) Brings out a large faint diffuse nebulosity in two elongated segments with a darker area between them east of Gamma Cygni. Larger area well northwest of Gamma also visible.

UHC: (3) Increase in contrast noted over DEEP-SKY filter, with dark gap between the patches east of Gamma Cygni now much more notable.

OIII: (1) Filter almost extinguishes the nebulae (very faint).

H-BETA: (3) Nebula is fainter than in UHC, but has higher contrast with a very dark sky background in the areas around the nebulosity.

RECOMMENDATION FOR IC 1318. H-BETA/UHC (near tie) (OIII NOT recommended).

IC 1396 (nebula SW of Mu Cephei)

DEEP-SKY: (2) Diffuse haze around a weak open star cluster, quite large with some vague brightness irregularities and a possible dark inclusion in the south side (B161).

UHC: (3) Nebulosity more visible and dark inclusion is much more definite, but the glow is still faint. Some variations in brightness are noted, but the object is still rather diffuse.

OIII: (0) Kills the nebulosity.

H-BETA: (1) Nebulosity is visible, but is extremely dim in a dark background.

RECOMMENDATION FOR IC 1396: UHC/DEEP-SKY (OIII NOT recommended).

IC 1848 (diffuse nebula in Cassiopeia)

DEEP-SKY: (2) Some increase in contrast with nebula appearing as an elongated faint haze going through a sparse cluster.

UHC: (4) Much easier to see, with nebula now elongated EW, brighter on northern side.

OIII: (4) Noticeably darker than UHC, but a little higher contrast.

H-BETA: (1) Very dim.

RECOMMENDATION FOR IC 1848: UHC (H-BETA NOT recommended).

IC 2177 (diffuse nebula in Monoceros)

DEEP-SKY: (2) Long faint irregular diffuse band of haze not easily seen without filters. Extends southward from open cluster NGC 2235.

UHC: (3) Easier to see, with somewhat more contrast. Narrower slightly sinuous core filament imbedded in more diffuse haze visible for nearly 2 degrees.

OIII: (2) Nebula barely visible, with most of outlying nebulosity gone.

H-BETA: (3) Core filament is fainter than UHC, but considerably more contrast.
RECOMMENDATION FOR IC 2177: H-BETA/UHC.

IC 4628 (diffuse nebula in Scorpius)

DEEP-SKY: (2) Faint diffuse irregular glow not visible without filters.
UHC: (4) Noticeable improvement, with nebula now easy to see and rather detailed, with some irregular lane-like detail.
OIII: (2) Much fainter than in UHC, but still visible.
H-BETA: (3) Shows some interesting filamentary detail, but not as bright or as detailed as in UHC.
RECOMMENDATION FOR IC 4628: UHC.

IC 5067-70 Pelican Nebula (diffuse nebula in Cygnus)

DEEP-SKY: (2) Nebular overall form is easier to see than without filters, with some hints of detail and the overall form.
UHC: (4) Very noticeable improvement in contrast over the DEEP-SKY filter, with both the "beak" and the "body" now fairly easy to see.
OIII: (4) Improvement in contrast and detail, but nebula is dimmer than UHC.
H-BETA: (2) Nebulosity is visible but is very faint.
RECOMMENDATIONS FOR IC 5067: UHC/OIII; DEEP-SKY also useful on the object (UHC was brighter, but OIII shows more detail).

IC 5146 Cocoon Nebula (diffuse nebula in Cygnus)

DEEP-SKY: (2) A bit better than no filter, but object is still easily seen as a dim roughly circular irregular patch in some stars without a filter.
UHC: (3) Slightly higher contrast with more irregular interior dark detail.
OIII: (1) Fainter and slightly smaller than in UHC (OIII hurts it).
H-BETA: (3) Dimmer than UHC but shows larger area of outer nebulosity and slightly better defined dark detail in the form of irregular lane-like features.
RECOMMENDATIONS FOR IC 5146: H-BETA/UHC (near tie) (OIII NOT recommended).

PK205+14.1 Medusa Nebula (large planetary nebula in Gemini)

DEEP-SKY: (2) Slight increase in contrast but nebula is still just a very faint diffuse hazy area.
UHC: (3) Noticeable increase in contrast with vague "C"-shaped arc now visible.
OIII: (4) Dimmer than in UHC, but slightly more contrast, with hints of filaments in the dark part (looks almost annular).
H-BETA: (0) Completely kills the nebula.
RECOMMENDATION FOR PK205+14.1: OIII/UHC (near tie) (H-BETA NOT recommended).

PK164+31.1 Headphone Nebula (large planetary nebula in Lynx)

DEEP-SKY: (2) Very slight increase in contrast over non-filtered view.
UHC: (3) Noticeably easier to see as two spots connected by a vague annulus.
OIII: (3) Much easier to see the spots, but the annulus fades somewhat.
H-BETA: (0) Completely kills the nebula.
RECOMMENDATION FOR PK164+31.1: UHC/OIII (near tie) (H-BETA NOT recommended).

Sh-2-13 (diffuse nebula in Scorpius)

DEEP-SKY: (2) Dimly visible as a very faint glow but not without the filter.
UHC: (4) Boost in contrast, becoming very patchy but still dim.
OIII: (2) Fainter but still visible.
H-BETA: (2) Similar to OIII.
RECOMMENDATION FOR Sh-2-13: UHC.

Sh-2-54 (diffuse nebula in Serpens)

DEEP-SKY: (2) Dim diffuse glow not visible without filters.
UHC: (4) Noticeable contrast gain, with considerable light and dark detail.
OIII: (2) Much fainter than in UHC but still visible.
H-BETA: (3) Better than in OIII with a little detail.
RECOMMENDATION FOR Sh-2-54: UHC.

Sh-2-84 (diffuse nebula in Sagitta)

DEEP-SKY: (1) Only hint of nebula.
UHC: (3) Faint diffuse "L"-shaped patch with irregular edges.
OIII: (1) Dark field with just a hint of nebulosity.
H-BETA: (2) Fainter than in UHC.
RECOMMENDATION FOR Sh-2-84: UHC.

Sh-2-101 (diffuse nebula in Cygnus).

DEEP-SKY: (2) Very faint moderate-sized diffuse haze in two segments around 3 stars (2 stars on the west and one on the east).
UHC: (3) Higher contrast but still faint. Two definite patches visible with hazy arc-like extensions. One on the west appears larger.
OIII: (2) Very dim but still visible.
H-BETA: (3) Almost as much nebulosity visible as in UHC, but dimmer.
RECOMMENDATION FOR Sh-2-101: UHC/H-BETA.

Sh-2-112 (diffuse nebula in Cygnus NW of Deneb).

DEEP-SKY: (3) Faint star with very faint small diffuse patch of nebulosity to its immediate south. Much easier to see than without a filter.
UHC: (4) Almost fan-like diffuse patch extending from the faint star to its south. More nebulosity visible than in DEEP-SKY, but still somewhat small.
OIII: (4) Fainter than UHC, but the nebula now

envelopes the star in a diffuse faint haze. Darker inclusion from the northeast now visible.

H-BETA: (1) Really dims it!

RECOMMENDATION FOR Sh-2-112: OIII/UHC (H-BETA NOT recommended).

Sh-2-132 (diffuse nebula in Cepheus)

DEEP-SKY: (2) Better than without filters as without filters the object is only hinted at. Just a very faint diffuse irregular glow around several stars roughly elongated east–west.

UHC: (3) Makes a patch on south edge easier to see and hints of other detail.

OIII: (4) Increases contrast with an arc across the northern side and a patch in the middle. Higher contrast but dimmer than UHC.

H-BETA: (2) Dims it more than OIII, but nebula remains visible.

RECOMMENDATION FOR Sh-2-132: OIII/UHC.

Sh-2-155 (diffuse nebula in Cepheus)

DEEP-SKY: (2) Very faint diffuse area of haze around two widely-spaced stars (better contrast than without a filter).

UHC: (1) Only hint of nebulosity visible.

OIII: (1) Little if any nebulosity visible.

H-BETA: (0) No nebulosity visible.

RECOMMENDATION FOR Sh-2-155: DEEP-SKY (probable reflection nebula).

Sh-2-157 ("fingers" diffuse nebula in Cassiopeia)

DEEP-SKY: (2) Not easily seen except as a vague elongated brightening in a rich star background.

UHC: (3) Elongated large diffuse and dim oval feature with two dim northward-pointing arcs.

OIII: (3) Nebula is fainter than in UHC, but still visible with increased contrast, especially in the two "finger" patches.

H-BETA: (2) Fainter than in OIII, but nebula is still visible.

RECOMMENDATION FOR Sh-2-157: UHC/OIII.

Sh-2-170 (faint diffuse nebula in Cepheus)

DEEP-SKY: (2) Not easy to see without filters. Round very faint very diffuse patch of haze around a group of 6 or 7 faint stars.

UHC: (3) Somewhat easier to see than in DEEP-SKY, with a bit more contrast.

OIII: (2) Still visible, but fainter than in UHC.

H-BETA: (2) Still visible but fainter than in OIII or UHC.

RECOMMENDATION FOR Sh-2-170: UHC.

Sh-2-171 (very faint large diffuse nebula in Cepheus)

DEEP-SKY: (2) Plainly visible over non-filter, but still faint and diffuse.

UHC: (3) Slight enhancement over DEEP-SKY with some light and dark areas.

OIII: (2) Fainter but shows more enhancement in several dark lane-like structures.

H-BETA: (2) Nebula remains visible, but just a bit fainter than in the OIII.

RECOMMENDATION FOR Sh-2-171: UHC (DEEP-SKY and OIII filters also useful).

Sh-2-261 (diffuse nebula in Orion)

DEEP-SKY: (2) Slight increase in contrast making the nebula faintly visible and easier than without a filter.

UHC: (3) Nebula now clearly visible but still faint.

OIII: (3) Nebula visible, but fainter than UHC with a bit more contrast.

H-BETA: (2) Nebula is still visible, but not quite as good as in UHC/OIII.

RECOMMENDATION FOR Sh-2-261: UHC/OIII (near tie).

Sh-2-276 Barnard's Loop (diffuse nebula in Orion) (Naked-eye observations with filter over unaided eye)

DEEP-SKY: (1) Hint of a glow in telescope, but not visible naked eye.

UHC: (2) Faint arc-like glow visible under good conditions over Orion's Belt, continuing southward east of the belt.

OIII: (0) No nebulosity seen.

H-BETA: (3) Faint glow visible both over the belt and curving down southeast along Orion's southeastern side. Very faint, but noticeably easier to see than in UHC filter.

RECOMMENDATION FOR Sh-2-276: H-BETA/UHC (OIII NOT recommended).

Sh-2-235 (diffuse nebula in Auriga)

DEEP-SKY: (3) Diffuse oval faint fuzzy patch, slight southern extension.

UHC: (3) Slightly more contrast than DEEP-SKY, but fainter.

OIII: (2) Fainter than UHC or DEEP-SKY.

H-BETA: (4) Faint, but two patches are now seen with brighter one on the north. More contrast than DEEP-SKY or UHC.

RECOMMENDATION FOR Sh-2-235: H-BETA/DEEP-SKY (UHC also helps).

vdB93 (Gum-1) (diffuse nebula in Monoceros near IC 2177)

DEEP-SKY: (2) Slight boost in contrast, showing more nebulosity than without a filter.

UHC: (3) More contrast and nebulosity visible, but still faint.

OIII: (1) Fainter than in UHC, with only hint of a glow around the central star.

H-BETA: (4) Better defined than any of the other filters, with more light and dark detail. Fainter than in UHC, but shows better contrast and detail.

RECOMMENDATION FOR vdB93: H-BETA/UHC (OIII NOT recommended).

No.14 welder's glass. An inexpensive way to look at the Sun. (Photo by the author)

Solar filters

Visual

So many items have been used as solar filters over the years that it would be impossible to list them all. Smoked glass, overexposed film of every kind, food packaging, compact disks – all of these, and many more, have been employed as we've tried to get a better look at our daytime star. There are certain considerations regarding solar filters, and SAFETY is at the top of the list. A good solar filter will not transmit harmful ultraviolet or infrared radiation. It will also drop the brightness of the Sun to a comfortable level. Safe is great, but we don't want to be squinting the whole time we're looking at the Sun.

Solar filters for visual use in conjunction with telescopes are very popular and there are many different brands from which to choose. Visual filters may be made of either coated glass or optical Mylar. The solar image through Mylar is a pale blue; through the glass filters it ranges in color from white to

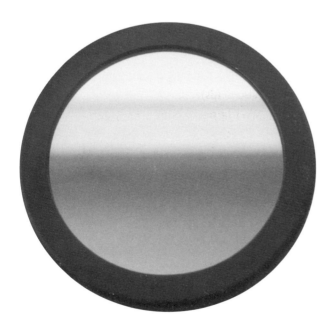

JMB Identiview solar filter. Made to fit my 100 mm Unitron refractor. (Photo by the author)

yellow or orange. The glass filters are more expensive, but more durable.

All solar filters fit over the objective end of the telescope. As such, they are sometimes called pre-telescopic filters. Some filters cover the entire objective and are known as full-aperture solar filters. Others have smaller openings offset from the center of the front aperture and are known as off-axis solar filters. This is done to eliminate the obstruction caused by the secondary mirror support. All solar filters should employ a round opening. Other shapes introduce pronounced diffraction patterns which are very distracting.

Warning: NEVER use a solar filter which fits into the eyepiece. Such filters have been known to crack due to heat buildup, with devastating results.

The most common pre-telescopic solar filters have a density of 5, which corresponds to a transmission of only one-thousandth of one percent (0.001%). This provides roughly a 12.5 magnitude extinction. Some solar filters are density 4 which have transmissions of one-hundredth of one percent (0.01%) and an extinction of about ten magnitudes. At least one model is density 4.5 which works out to a transmission of 0.003 and an extinction of 11.25 magnitudes. Photographic-only (absolutely NOT for visual use) solar filters with densities of 3 (transmission 0.1%, extinction 7.5 magnitudes) and 3.5 (transmission 0.03%, extinction 8.75 magnitudes) are available.

A number of manufacturers make these filters and almost all large astronomy dealers sell them. I do have a favorite. It is the Baader AstroSolar film, available directly from the source, Baader Planetarium, on the internet at http://www.baader-planetarium.de/com/sofifolie/sofi_start_e.htm

I first heard of the Baader AstroSolar film from Bob Kuberek, a friend and accomplished observer who lives in Los Angeles. He had ordered some and was impressed with what he'd seen. After I answered "no" to his question as to whether I had used Baader AstroSolar or not, Bob sent me some (which he had already mounted in a cell to fit my telescope, I might add!). Because I live in the desert southwest of the US, I use some form of solar filter nearly every day, so there was no lack of units to provide side-by-side comparison tests.

The increased brightness of the solar image through the Baader AstroSolar was the first surprise. The light transmission is much higher than one expects, but the image is not washed out. In fact, I found the contrast to be highest through the Baader filter. More brightness with higher contrast? It seemed too good to be true. Details only hinted at through other filters were visible in the AstroSolar. Faculae which could be seen on the solar limb through the other filters were resolved mid-disk through the Baader filter. Sunspots showed the same shapes as through other filters but more delicate details were visible. In a few instances, tiny sunspots not visible through any other filter were seen through the Baader AstroSolar.

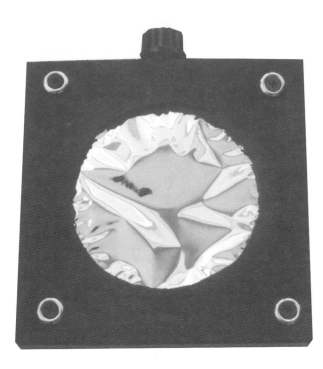

BaaderAstro Solar Filter. Thanks to Robert Kuberek for the filter. (Photo by the author)

It is therefore without hesitation that I recommend the Baader AstroSolar film as your primary solar filter. Some care must be exercised in the mounting of the film in the cell. You don't want to stretch it tight. To do so introduces stresses in the material. Also, although some pinholes are inevitable in any solar filter, you don't want to be making any of those yourself. Speaking of which, Roland Christen, of Astro-Physics, Inc., once commented on this "problem." With his permission, I quote his reply here.

I would like to discuss a subject that comes up quite often – how safe is the Baader film, and what about pinholes? The film has been tested for transmission of energy from UV into the far infrared, and does meet all criteria for a safe solar filter.

Nevertheless, we get film sent back to us by individuals who question the effectiveness of the filter. There are two reasons people send the film back. One is that the image of the Sun appears too bright, and the other that they can see pinholes when the film is held up to a strong light, or when they remove the eyepiece and look directly down the tube at the lens when the scope is pointed to the Sun.

The reason the image might appear too bright is that the customer is using a very low power – usually when the filter is used with a "shorty" refractor at 20× – 40×. There is no way to make the image dimmer at that power and also make the filter useable for medium to high power studies of disc detail. The answer is to use a neutral density filter of ND1 to ND1.8. These are

available at most photo supply stores. We can supply a 48 mm version from Baader that will thread into our Maxbright, or they can be threaded directly into our 2 inch to 1.25 inch adapter. By the way, using the Maxbright adds another measure of safety to the solar image because the multiple oxide layers do not transmit energy below 750 nm, so there is essentially no IR energy transmitted to the eye. This is on top of the reduction in IR that the Baader film produces.

Pinholes are the second reason people send their film back. Even though the utmost care is taken to provide a pinhole-free filter, it is impossible to screen out every pinhole. This is a problem with all filters, glass included.

How dangerous are pinholes? Recently a customer sent his film back because he had one eye with a detached retina, and did not dare to risk his other good eye on a possible defective filter. He could see pinholes when the scope was aimed at the Sun, and he looked directly into his diagonal with the eyepiece removed. His opthamologist examined the filter, and felt that the pinholes would be unsafe. In fact, if you ask any opthamologist if there is such a thing as a safe solar filter the vast majority would say absolutely NO! For this particular individual, we will advise that if he feels in the least unsure, then he should NOT look through this filter (or any other) at the Sun. However, let's explore what pinholes do, and how they may transmit solar energy to the eye.

Pinholes at the lens end by themselves do not transmit any sort of image to the eyepiece. Instead, they scatter light in all directions, and a very very tiny amount of this light reaches the eyepiece entrance aperture as a diffuse and even illumination. The vast majority of a pinhole's energy goes every which way to faintly illuminate the walls, baffles, focuser parts, etc. of the tube in an even, diffuse glow. If you were to somehow cut out all the light that the filter transmits directly to the eyepiece and retain only the light from pinholes, you would see NO solar image at the eyepiece, but only a faint background. There is NO focused image of the Sun to burn a hole in your retina, just a faint background light. This light is many times fainter than the actual image produced by the rest of the filter. The net effect is to reduce the contrast ever so slightly by filling in the dark areas of sunspots with faint illumination. Finally, pinholes can be dabbed out with a spot of paint, or with a black marker. To reduce the chance that your filter will develop pinholes, be sure not to handle the surface, and do not stretch it tightly over your filter cell.

Hydrogen alpha

Observing the Sun at the wavelength of hydrogen alpha light is one of my observing passions, and I've made not a few

The energy rejection filter for the author's DayStar hydrogen alpha filter. The filter (center, glass) is housed in a cell custom made for a 100 mm Unitron refractor. (Photo by the author)

filters have different bandpass widths. The widest of these can be nearly 2 Å and the narrowest 0.3Å. Generally, the bandpass is given in Ångströms (Å), one Ångström being one-tenth of a nanometer. I have looked through h-alpha filters with a bandpass of 1 Å. Prominences were easily seen but chromospheric detail was low. Through a filter with a bandpass of 0.5 Å there was great chromospheric detail but few prominences. For this reason, some h-alpha filters are tunable, that is, the center frequency of the bandpass may be shifted slightly to either side. Some filters accomplish this by tilting the filter stack while others use temperature control to vary the central wavelength.

You will often see h-alpha filters with the descriptor full-width, half-maximum (FWHM). This specifies the total bandpass (full-width) of a filter at 50% of its maximum transmission level (half-maximum). If a filter has a FWHM of one ångström, it means that the bundle of wavelengths allowed through the filter at 50% efficiency is 1 Å wide; below the 50% level the bandpass is wider, and above it the bandpass is narrower. The FWHM is the standard method of specifying narrowpass filter characteristics.

As of this writing, there are two main suppliers of h-alpha filters. DayStar (http://www. daystarfilters.com) and Coronado Instruments (http: //www.coronadofilters.com). In development is a relatively inexpensive, tunable bandwidth h-alpha filter from Sirius Optics, makers of a number of popular astronomical filters. As promising as this sounds, I have not yet observed with this filter.

The suppliers DayStar and Coronado receive high marks from me. I have used at least six DayStar and three Coronado h-alpha filters and have found them all to be superb. Please note: these filters are not inexpensive. The last dedicated h-alpha setup that I used cost nearly 10,000US$. But what a view!

converts. When we speak of hydrogen alpha or h-alpha, we are referring to a reddish-colored light with a wavelength of 656.28 nm. The Sun's chromosphere and solar prominences give off light at this wavelength. Normally, these features are only visible during total solar eclipses (and the chromosphere is only glimpsed momentarily just before or after totality). With a filter which blocks out all other light, however, these features can be observed on any clear day.

Solar h-alpha filters may comprise two parts. The filter stack is inserted into the eyepiece end of the telescope. Equally important is the energy rejection filter (ERF), which is a pre-telescopic filter. The ERF looks like a piece of glass (it may be mirrored or red), but it is very important in preventing ultraviolet and infrared radiation from reaching the main filter. The coating of h-alpha filters can be destroyed by ultraviolet light over a period of time, and heating of the filter by infrared radiation can cause the bandpass of the filter to widen to the point of uselessness.

Most pre-filters are smaller than the apertures of the telescopes they are used on. This is intentional as some h-alpha filters require a specific focal ratio (say, f/30). Although the reduced aperture limits the resolution of the system and the image is not as bright as with other telescope/filter combinations, working with an f/30 light cone definitely produces a high-quality image.

All h-alpha filters center on 656.3 nm. However, h-alpha

Daystar 0.7 Å hydrogen alpha filter. (Photo by the author)

2.5   Telescope accessories

After you have made a careful selection of a telescope, mount, and eyepieces, the choice of accessories is next on the list. A decade or so ago, I remember buying my first convertible. After some shopping around at various dealerships, I chose the make, model, and color and I was able to drive away in my new automobile that same day. However, when I turned on the radio, I realized my vehicle was incomplete. For such a great car, I had to have a quality sound system. So it is with telescopes. In this section I will briefly discuss a number of telescope accessories. Those accessories, such as eyepiece filters, which require more detailed discussion, are covered elsewhere. Some of these accessories I regard as "essential" while others fall into the category of, well, for lack of a better term, "nice to have."

- Essential
  - Single-power finders
  - Finder scopes
  - Observing chairs and stands/ladders
  - Lights and magnifiers
  - Batteries and power supplies
  - Storage cases
  - Protective scope covers
  - Dew removal systems
- Nice to have
  - Fans and SCT Coolers
  - Clocks
  - Under the telescope
  - Focused
  - Apodizing screens
  - Camera mounts
  - Counterweights
  - Illuminated reticules
  - Off-axis guiders
  - Tripod stabilizer

Single-power finders

I consider this the first of the absolutely essential accessories. Single- (or unity-) power finders are real time savers. They are also simplicity itself to use. Simply look through them at the sky and move your telescope. Lock down the telescope motions and look through the eyepiece. If you have taken the time to align your finder, the object you want to observe should be in the eyepiece's field of view.

There are two brands of single-power finders that I recommend: the Telrad and the Rigel QuikFinder. Neither of these is a telescopic finder. They are called "heads-up" displays. When you look through either of these, what you

The Rigel QuikFinder. Great for a small telescope. (Photo by the author)

see is a bull's-eye pattern superimposed upon the sky. The diameters of the circles in the Telrad bullseye are 4°, 2°, and 0.5°; those in the QuikFinder are 2° and 0.5°. To use either, you just sight through the finder and move your telescope until the object is centered in the smaller circle. If you have taken time to align your single-power finder to your telescope, the object you want to observe will be in the field of view of the eyepiece.

> **Observing Tip:** Start with a low-power eyepiece in the telescope to provide a wide starting field of view.

The Telrad was invented in the late 1970s by amateur astronomer Steve Kufeld (1942–1999), of Huntington Beach, California. Steve was a long-time member of the Los Angeles Astronomical Society and used club events to test out designs. Since 1982, Telrads have had an injection-molded plastic body. Telrads can be ordered from many reputable telescope dealers, such as Lymax Astronomy, on the internet at http://www.lymax.com (or use the toll-free number provided on the site). Rigel QuikFinders may be ordered from Rigel Systems, on the internet at http://members.home.net/rigelsys/rigelsys.html

Your choice of single-power finder depends a lot on your telescope. Telrads are much larger than QuikFinders and occupy a correspondingly large footprint on your telescope. This is not a problem for most Schmidt–Cassegrain telescopes and certainly not for large reflectors. I own a 100 mm refractor, however, and a Telrad would simply be too large for it. The QuikFinder's base is only 6 × 5.5 cm and it rises 13 cm off the tube. I find this perfect for my small

The Telrad, available from many astronomy retailers, is the most popular of the single-power finders. (Photo by the author)

requires more brainpower than they are willing to assign to the task. I humbly place myself in this category.

An option in some finder scopes is an illuminated reticule (more often than not using light from a red LED), which makes the crosshairs easier to see. Be absolutely certain that the reticule is dimmable.

Forget the small finders. Get one that has a front lens at least 50 mm or larger. Such a finder will let enough light in

An excellent finder scope. A magnification of 8 with a 50 mm objective lens. (Photo by the author)

refractor. Also, if this is a consideration, the Telrad weighs roughly 310 g while the QuikFinder weighs 95 g. One other, albeit minor, advantage to the QuikFinder is that you can put it in "pulse" mode. This is where the red circle display flashes at a rate which you set. The idea here is that you can more easily see the field or object if the display is not constantly on. This may fall more into the "convince yourself" category, but I admit that I use this feature. Personally, I set both the pulse and the display brightens to their lowest settings. At this writing, an add-on pulse feature was available for Telrad finders.

> **Observing Tip:** Carry a spare battery for whichever single-power finder you use. These units do not emit much light so it is easy to leave them on.

There is another class of single-power finder which projects a virtual red dot, rather than a bull's-eye pattern, onto the sky. Most of these are smaller and less expensive than either of the two already discussed. Personally, I far prefer the bull's-eye pattern to the dot.

Finder scopes

The best telescope in the world is relatively useless if you can't find anything with it. Therefore, I strongly suggest that you purchase a quality finder scope. Finder scopes come in two optical configurations. The first is the straight-through type and the other is the right angle finder.

The straight-through finder inverts (flips) the image, but it does have the advantage of allowing you to sight down the main telescope tube. This is more intuitive for most amateurs. Most right angle finders keep the field in correct orientation but some have complained that looking "at the telescope" trying to find objects is difficult. I offer a warning about certain right angle finders. If they do not employ an erecting prism in their design, a problem comes about because of the extra reflection introduced in the image. In such a case, the sky appears mirror-reversed when compared to charts. For some people, unscrambling the star patterns

so you will not become totally frustrated trying to locate fainter objects. As to the magnification of the finder, I will leave that up to you. Search around and observe with several until you can find a magnification, brightness, field of view and eyepiece design that you like. My choice is a 9 × 50 finder made by Celestron but there are many others from which to choose. This straight-through finder has crosshairs but they are not illuminated. It is compact and provides a bright, crisp image. In short, it is all that I need. You can purchase quality finder scopes from the same vendors who sell quality telescopes.

> **Tip:** Buy a finder scope with a mount that has two support rings and six adjustment screws, rather than a mount with one ring and three adjustment screws.

Once you have purchased your finder, it is very important that you take the time to align it with your telescope. If you are traveling to observe, and not simply setting your telescope up in the backyard, you'll probably want to align the finder prior to each session.

> **Tip:** Align your finder during the day.

To perform a finder alignment procedure, start with a low-power eyepiece in your telescope. During the day, or at least during early twilight, find and point your telescope at a small object 100 m or more distant. The flashing lights on the tops of transmission towers are excellent targets. Center your target in the field of view of the eyepiece. Lock the motions of your telescope. Now loosen (a little) the screws holding the finder scope to its mount. (Certain finders may have locking knurled nuts which also must be loosened.) Then adjust the screws until the crosshairs of the finder are centered on the object in the telescope eyepiece. Retighten the screws. Make them snug, but do not overtighten. Recheck your telescope and finder to be certain both are on the target. If you wish, insert a higher magnification eyepiece and recheck. Finally, lock the knurled nuts, if any. Remember, the closer the telescope and finder are aligned, the easier it will be to find what you are aiming at. Take a few minutes to do this correctly.

Observing chairs and stands/ladders

Comfort while observing is only overrated by beginners. If you are at all uncomfortable at the telescope you will do less observing and the observations you make will be less fulfilling and less accurate.

Having said that, nothing says comfort like a quality observing chair. My definition of "quality" chair is (1) sturdy construction;(2) padded seat; and (3) adjustable height. Personally, I don't like the "piano stool" type of chair because it has no back support, something that this observer craves after a long night at the telescope. The best chair I have used is the StarBound Observing Chair, available from many astronomy dealers. Plans exist for building much less expensive versions of this adjustable design. The Denver Observing Chair, developed by members of the Denver Astronomical Society, is one such chair. It can be constructed

Cosco ladder. An essential accessory. (Photo by the author)

An adjustable observing chair is a wonderful accessory to have. Mine is the StarBound Observing Chair, an anniversary gift from my wife. (Photo by the author)

for approximately one-fifth of the cost of the StarBound Observing Chair.

For large Dobsonian telescopes, observing chairs such as those discussed above simply will not do. They are too short and even a 300 mm Dobsonian pointed near the zenith places the eyepiece too high for the eye to reach when seated in an observing chair. In such cases, a step-stool or ladder must be employed.

Regarding ladders, there are ways to make using them more comfortable. If your ladder or step-stand has a metal frame, cut to length a piece of water pipe insulation and cover the top bar where your hands rest. Further, you could fasten carpeting to the steps of the ladder to make standing on them just a bit more comfy. Make certain the carpeting does not overhang the steps. Fasten it with very strong tape, a carpenter's adhesive, or small, round head bolts with washers. If using the latter, I suggest you place the bolts as near the left and right edges of each of the steps as possible, so as not to be standing on them for minutes at a time.

Lights and magnifiers

Please, for your own safety around experienced observers, do NOT take a flashlight and wrap a piece of red cellophane around the lighted end. This statement is meant to be humorous but a bright light, even if it is red, will draw the ire of those around you. Visit the Rigel Systems website at http://members.home.net/rigelsys/rigelsys.html and purchase one of their Starlite flashlights. This is the one with ONLY two red light emitting diodes (LEDs). My strong suggestion is that you do not purchase their Skylite model, which has two red LEDs and two white LEDs. I am simply not confident enough about my (or, to be perfectly honest, your) ability to select the correct switch at 4 a.m. when everyone's eyes are totally dark-adapted.

The advantage of the Starlite is simple: it is adjustable in brightness. As you progress further into amateur astronomy, you will find that the amount of light you use at star charts,

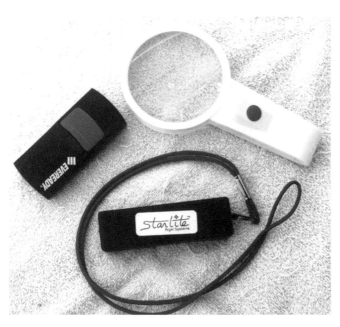

A selection of red LED flashlights including the best of the lot, the Starlite. (Photo by the author)

etc., diminishes greatly. Simply put, brighter light means you see less at the telescope, or at least that your eyes take longer to re-adapt to the darkness. In such cases, the color of the light is irrelevant. Even a pure red light, if too bright, will ruin your dark adaptation.

For those observers advanced in years (and I will place myself in this category), the small print on star charts can be difficult to read under low illumination. Inexperienced observers search out brighter lights. Experienced observers use a magnifier. This allows one to use their reading light at or near the lowest brightness level. I suggest using a non-illuminated magnifier. I purchased an illuminated magnifier from a reputable astronomy vendor. The light source within it was not very red, not at all adjustable and far too– bright – bright enough, in fact, to illuminate the white handle of the magnifier by light being transmitted through the plastic. Here's a free tip to any manufacturers of such magnifiers: use black plastic, not white!

Batteries and power supplies

If you only observe from one location, and if that location has alternating current power, consider yourself a lucky person. For the other 99% of us, some form of portable power is essential. There are several options, the most popular is hooking your equipment directly to your automobile battery. If you are doing this while observing alone, for your own sake start your car several times during the night to recharge its battery.

A portable power supply is an absolute essential if you are imaging or using a GoTo drive. (Photo by the author)

Mark Beuttemeier, of Sammamish, Washington, uses small 12-volt golf cart batteries in his StarMaster telescope. One usually lasts the night, but he keeps a spare charged up. For his LX200, he uses a 12-volt gelcell marine deep-cycle battery (could be backup batteries for security systems). Robert Kuberek, of Valencia, California, uses 2 DieHard deep-cycle marine batteries.

Storage cases

There is a well-known amateur astronomer who is a friend of mine (many of you would recognize his name) who keeps his very expensive eyepieces uncovered in a cardboard box in the back of his truck. Such abuse degrades the quality of whatever equipment is treated this way. Still, he is a superb observer who has produced many wonderful astronomical images. I don't, however, recommend his storage method to anyone.

As quality optical equipment, eyepieces and filters must be treated with at least some care. Not doing so invites dirt, scratches, dents, or worse. Most observers eventually acquire one or more of some type of foam-padded storage case. These cases come in many different designs and colors, but all have certain common characteristics.

A good storage case will be sturdy with quality hardware. It will be able to survive the occasional minor ding, drop or fall with minimum damage to the case itself and NO damage to the contents. Such cases are usually sealed to dust and

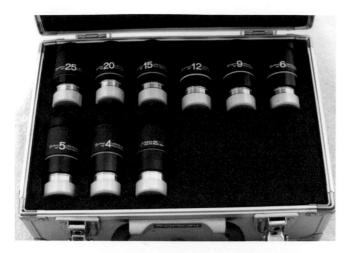

An accessory case protects its contents from dust and accidental drops. (Photo by the author)

some are even water-tight. If an observing trip which requires you to fly is in your future, you should purchase a case which is lockable. The inside of quality cases are all foam-filled. Most manufacturers use a type of foam that is in the form of pre-cut (though not quite all the way through) cubes. To use, simply figure out what size "hole" your equipment will need and remove just enough of the cubes.

> **Tip:** Most amateurs have found that, for larger objects at least, removing one fewer cube than necessary provides a tighter fit.

For some of your less delicate or expensive accessories, consider a large fishing tackle box. Usually, tackle boxes have several layers divided into lots of smaller compartments in addition to a storage area for larger items at the bottom.

> **Tip:** Whether or not your eyepieces are stored in a foam-lined case, always keep the plastic covers on them.

Protective scope covers

If you are going to be attending an extended star party and have chosen not to tear down and set up each and every day, you will need something to cover your telescope. Mother nature can be a harsh taskmaster. In the spring and summer,

Cosmic Storm Shields covering two medium-sized Dobsonian telescopes at the 2001 Great Plains Star Party in Parker, Kansas. (Photo by the author)

severe dust storms may occur in the US southwest desert and thunderstorms may happen at most other locations. Furthermore, the Sun itself can be an enemy, baking your telescope until it becomes nearly untouchable while heating the mirror so that stable images are not produced until hours after sunset, if at all.

A proper telescope cover – when secured with elastic or bungee-style cords – should keep out dust and rain, and also reflect sunlight to such a degree that the cool-down time when the scope is uncovered is short. I cannot overemphasize this last point enough. At one week-long star party, a friend with an 18-inch telescope covered it each morning with a reflective scope cover. When he uncovered it each evening, the scope was actually cool, despite the fact that the daytime temperatures were in the 90s and that bright sunlight was the norm. Purchase a scope cover specifically made for your telescope or have one custom made. Custom-made covers, such as "Cosmic Storm Shields" available at Lymax Astronomy (on the internet at http://www.lymax.com) do not cost significantly more than simple Mylar-coated covers.

Dew removal systems

I write this from the desert environment of western Texas where dew is extremely rare. I have lived and observed in other locations, however, but even here the memories of several observing sessions cut short by condensed moisture are fresh.

Dew forms on optics because the temperature of your telescope has fallen below the dew point (the temperature at which atmospheric water vapor will condense on an object). A very short – only somewhat technical – reason as to why this happens is because your telescope, under a clear sky, acts as a black-body radiator. As such, it is essentially radiating its heat to the sky, which is at a colder temperature than the atmosphere around your telescope. If the air temperature is near the dew point, this "extra" radiation will be enough to cool your telescope and cause dew to condense. I refer to dew, but be advised: under the right conditions frost may also form on telescope optics.

Of all telescope designs, probably the most plagued by dew are the Schmidt–Cassegrains, employing, as they do, a large glass corrector plate at the front end of the telescope tube. But refractors and reflectors are certainly not immune, not to mention eyepieces and finder scopes. To prevent the formation of dew, the telescope's optics must be brought back to a higher temperature than the dew point. This is somewhat of a balancing act, though. You don't want to overheat the optics. That would cause thermal currents which would degrade your observing.

If you notice dew forming on your scope, there are a number of ways to deal with it. There is also one way not to deal with it: under no circumstances should you wipe the dew off manually! You will scratch your optics. Safe ways to remove dew include the following. (1) Bringing the telescope indoors to warm up. I realize that this is often impractical or downright impossible. (2) In the field, you can turn on your automobile's heater and move the affected part inside to warm up. In each of these cases, leave all optics uncovered so the dew can evaporate directly to the air. (3) Use a portable hair dryer (set at the lowest air volume) to warm up your optics. Automotive supply stores often sell such units which operate from the direct current of a car battery. (4) Purchase a dew removal system which installs directly onto your telescope. Several systems are available but the two best known are the Kendrick Dew Removal System and the Orion Dew Zapper. The Orion system may be found on their website at http://www.telescope.com. The Kendrick system is available through many quality telescope supply dealers.

>**Observing Tip:** Dew shields, installed at the front of telescopes, are a good idea but will not stop dew from forming under harsh conditions.

Fans and SCT coolers

The glass used to make mirrors for reflecting telescopes can absorb a great deal of heat during the day. Heat buildup, and its subsequent release after sundown, will cause thermal air currents which can severely distort the image seen through the telescope. Some amateur astronomers have added cooling fans to their telescopes. The fans gently draw ambient air through the telescope to negate any thermal air current effects. Usually, the fans are of the low air flow, whisper type and are located underneath the primary mirror (the main source of heat absorption). Most of these fans operate on low-voltage direct current, which is easily transportable.

On 19 December 2000, I observed the companion to Sirius using Eugene Lawson's C11 SCT. He had left it at our house for a week, during which time I made full use of it. On that night, I allowed the telescope to cool for 5 hours before trying the difficult split. I knew that this would be necessary as I had observed with it quite a bit on the previous several

The CosmicOne SCT Cooler. A quick, clean way to cool down your Meade 300 mm or other SCT (see http://www.lymax.com. (Photo by Robert Haler)

nights. But what if a way could be found to speed up this process? What if a Schmidt–Cassegrain telescope could be cooled in 30 minutes, rather than 5 hours? These questions are rhetorical, of course, because such a device *has* been developed. It is called the CosmicOne SCT Cooler and it was invented by Robert Haler of Lymax Astronomy in Kansas City. I quote from the Lymax web site:

>The SCT Cooler is a device for ventilating the inside of catadioptric telescopes so that they may be brought quickly to ambient temperature. This allows image-ruining tube currents and heat pillars to be minimized in 10–20 minutes instead of the usual 45 minutes to 2 hours. The greater the temperature difference between the place your telescope is stored and the outside evening air, the more quality viewing time you can reclaim with a CosmicOne SCT Cooler. The SCT Cooler has a filtered air intake and other design features which minimize the possibility of dust infiltration and direct the air flow away from the internal optical surfaces.

To use the SCT Cooler, just unscrew the eyepiece holder from the back of your Meade or Celestron catadioptric telescope and carefully insert the cooler. Turn it on and in 20–30 minutes, you're observing with a cooled telescope. The most important consideration, at least to me, is the very fine filtration (at the micron level) that totally eliminates contaminants from the inside of the telescope. Catadioptric telescopes are notorious heat collectors, using, as they do, a closed tube. If you own one, you would do well to check out the CosmicOne SCT Cooler. Find it on the internet at http://www.lymax.com

Clocks

When recording an astronomical observation, it is simply imperative that the correct time is noted. A recently set wrist watch can do this just fine. Other options also exist.

If you have an observatory with alternating current, or if AC can be found (or made from DC) at your remote observing site, use a clock with large, red numbers. You can set it to either local time or Universal Time, depending on your preference, but be sure to note which when you're making your observing log entry. At this writing, I am still searching for a battery-powered LED clock with red numbers. Plenty of LCD clocks are out there, but no red LED ones powered by battery.

For critical observations where your eye must be continuously at the eyepiece, a short wave radio, tuned to one of the standard time frequencies, can provide a highly accurate time source. In America, the National Institute of Standards and Technology operates radio station WWV. Here is an informative quote from their website:

>WWV operates in the high frequency (HF) portion of the radio spectrum. The station radiates 10 000 W on 5, 10, and 15 MHz; and 2 500 W on 2.5 and 20 MHz. Each frequency is broadcast from a separate

transmitter. Although each frequency carries the same information, multiple frequencies are used because the quality of HF reception depends on many factors such as location, time of year, time of day, the frequency being used, and atmospheric and ionospheric propagation conditions. The variety of frequencies makes it likely that at least one frequency will be usable at all times.

In addition to short wave radios, a few observers use "talking clocks." The best of these may be found in stores that carry items specifically for the blind. These clocks have better speech quality than items in general stores and also come in various languages. Such clocks speak the time when a large button is touched. This is the only disadvantage. Unlike listening to a short wave broadcast of a time signal station, which is continuous, the observer must physically press a button on a talking clock for the time to be reported.

Under the telescope

Believe it or not, what you walk on while observing can significantly add to or subtract from your experience. An extreme example of this happened to me and some friends who attended the Enchanted Skies Star Party, near Socorro, New Mexico. Their dark sky site is quite nice, but the preponderance of lava rocks and cacti made for a dicey time that first night. After some manual labor the next day an area was cleared of all but the smallest rocks and all was fine thereafter.

At star parties, I have experienced many different types of "flooring" for telescopes. One observer likes to put a several centimeter layer of straw around his telescopes. This works fairly well when it is humid and not windy. After rain has fallen, there is an advantage in that the top layer of straw dries out quickly, leaving your footwear relatively dry. The straw also provides a good barrier between shoes and mud, if any. However, if there is wind, well, you can imagine how the straw does fly! Also, if it is very dry, a fine dust can be created by repeated walking on the straw.

Other options are available. Many observers use plastic tarps held in place by tent stakes or very large nails. When using tarps, I prefer ones with grommets. Such tarps generally exhibit better quality, as they have reinforced edges. I also prefer large nails rather than tent stakes, some of which, made from either plastic or thin metal, can be quite flimsy. I have also seen floor protectors, such as the ones used in offices under desk chairs, used as temporary flooring for observing.

Carpeting or plasticized turf has also made appearances at many star parties. If you are planning on using carpeting, it should be of a synthetic blend, as other types can absorb moisture and that can be a less than pleasant experience.

The finest "flooring" I have seen around any telescope was a plasticized turf with a very loose weave, approximately a centimeter thick. It allowed water to pass through it and did not absorb any. The best thing about it, however, was that it was slightly springy to walk on, providing somewhat of a cushion that really made a difference in the way my feet felt by the end of the night. At the end of the observing run, this floor is simply rolled up, but it is thick so leave enough storage and transportation space. Cleaning is also easy. Simply lay it out on a cement driveway or patio and rinse off with a garden hose. Those observers that use this type of flooring report that such an item may best be obtained in the camping section of large department stores. Note that care must be taken when choosing flooring material around or under a telescope – the material must not allow vibrations to be transferred to the telescopes.

Focusers

A chain is only as good as its weakest link. For some telescope systems, the weakest link is the focuser. Before we discuss features and usefulness, I want you to know that quality focusers are not cheap. Selecting one is also not trivial, and the installation, although simple for the most part, can be daunting to some. This having been said, once you use one of these elite focusers, you won't ever want to change. The two main types of focusers are the rack and pinion, in which a small gear drives the drawtube into which the eyepiece is placed, and the Crayford design, which allows the drawtube to ride on small bearings.

Of course, all focusers should hold your valuable eyepieces securely in place, irrelevant of the position of the telescope. This is accomplished by using knurled machine screws which apply tension to the eyepiece. This tensioning can be done directly by the point of the screw or, as in some designs, by a secondary mechanism upon which the screw tightens. This latter method is preferred by some who claim that direct contact by the screw may damage the eyepiece barrel. I have never had a problem with screws that make direct contact.

Meade motorfocuser. A nice accessory for fine tuning your focus. (Photo by the author)

Quality focusers also eliminate backlash. Backlash occurs when the direction of focus is changed if there is looseness or "play" in the focusing mechanism. It is not a serious problem but it is a very annoying one.

Focusers should be well-machined with no sharp edges. They should provide plenty of in-and-out travel to accommodate all your eyepieces. Many focusers provide a mechanism that allows you to adjust the "tightness" as you turn the focuser. A nice feature of some new focusers is a two-speed focusing mechanism. The "coarse" focusing knob is for quickly establishing a near-focus position, whereupon you would then switch to using the "fine" knob. I certainly recommend this feature.

New focuser designs may also incorporate motor control. Some even have a digital readout to allow you to return exactly to a previous focus setting. If you are contemplating the purchase of a top-of-the-line motor-driven focuser, make sure that it is geared correctly for your use. Some motorized focusers, intended for CCD work, are geared to such an extent that they are impossible to use visually.

The Feathertouch focuser by Starlight Instruments and the Next Generation Focuser by JMI are two examples of quality focusers.

> **Tip:** Purchase a focuser which will accommodate both 2 inch 1.25 inch eyepieces (with a slide-in adapter).

Apodizing screens

The subject of apodizing screens (sometimes referred to as apodizing "masks") brings to mind the old story, told about many different things, whereby some people swear by them while others swear at them. Apodizing screens are touted by a few vocal proponents as anti-diffraction devices which increase contrast, provide steadier images, produce a brighter Airy disk and more. I remain skeptical. A simple statement should suffice. If these claims were demonstrably true to large numbers of unbiased observers, I guarantee that every telescope manufacturer would provide apodizing screens for their telescopes. At this writing I note that the only apodizing screens available to the amateur astronomer are those that you build yourself.

An apodizing screen cuts down the light available to the telescope upon which it is installed. Because of this, even its supporters agree that it is not very useful for telescopes smaller than about 200 mm. Some screens introduce spikes around the images while others produce a rainbow-like pattern. Both of these can be very annoying. Also, apodizing screens are not at all useful for faint objects. In fact, the only objects I have seen observed with apodizing screens are the visible planets and double stars.

For those of you willing to experiment with this accessory, the apodizing screens I have seen are made from three layers of window screening. Each of the layers has a central hole cut into it. Usually, the holes measure 0.9, 0.78, and 0.55 of the diameter of the primary mirror. When sandwiched together, each layer is rotated a certain extent, usually 30°. If you are determined to construct one of these, please remember that I provide these numbers as starting points only, and not as absolutes.

2.6 Binoculars

Many amateur astronomers consider binoculars an accessory. I regard them as a necessity. In fact, for years I have suggested that beginning amateurs first purchase (or, better yet, borrow) binoculars to view the sky. Then, if their interest in astronomy continues, to upgrade to a telescope. If their interest wanes, at least the binoculars can be used for terrestrial viewing. Most of the advanced amateur astronomers I know own several binoculars. The following provides some facts and points to consider when choosing binoculars.

> **Note:** On the humorous side, you may hear someone referring to binoculars as a "pair of binoculars." This has become commonplace, but it is incorrect usage. They are simply binoculars. Technically speaking, I suppose you could refer to them as a "pair of monoculars," but you'll never actually hear that.

15 × 70 binoculars, about the highest magnification that can be hand-held for any length of time. (Photo by the author)

The numbers

Every binocular has a two-number designation, such as 7 × 50. The first number (in this case, 7) is the magnification (or, power). The second number (50) is the diameter (in millimetres) of each of the objective lenses. This example, 7 × 50, is the binocular I would recommend as a first unit for astronomical use. A magnification of 7 is in the "medium" range, which is high enough to bring out some detail. Too high a number here will: (1) overmagnify the motion of your hands, causing celestial objects to move around when the binoculars are hand-held; (2) limit the field of view, making objects more difficult to find for beginning amateurs. A note on point (1): most advanced amateur astronomers can easily hand-hold binoculars up to 16 power for short intervals of time. One other point is that binoculars of higher power (like telescopes with shorter focal length eyepieces) will reveal fainter stars by increasing the contrast between the star images and the sky background.

> **Observing Tip:** When hand-holding large binoculars, once you have focused them, move your hands more toward the front of the binoculars and they will be easier to use.

In our example, 50 mm is the aperture of each of the front lenses. This is a good size. The larger this number, the more light the binoculars will collect and the brighter the image will be. Binoculars with 50 mm front lenses collect more than twice as much light as those with 35 mm lenses. Astronomically speaking, this is a gain of nearly three-quarters of a magnitude in brightness. The disadvantages to larger front lenses are that the binoculars must be made (1) larger; (2) heavier; and, of course, (3) more expensive.

Optics

The components within a binocular which cause the image to appear "correct" are the prisms. Binocular prisms come in one of two basic designs: Roof prisms and Porro prisms. Roof prisms are more lightweight and smaller, but are not recommended for astronomy. Porro prisms are better, and are made of either BK-7 or BAK-4 glass. BAK-4 prisms (barium crown glass) are the highest quality available. BK-7 prisms (borosilicate glass) are also good quality, but the sharpness falls off slightly at the edge of the field compared to the view through prisms made of BAK-4.

Most of the better binoculars are multi-coated on all optical surfaces. We've talked about multi-coating in the "Telescopes" chapter, so suffice to say this is a desirable feature. Speaking of coatings, never buy a binocular with red lenses. Such "ruby" coatings are poor attempts to compensate for bad optics.

Mechanical considerations

While we're talking about what not to buy, steer clear of zoom binoculars, at least for astronomy. The idea of a variable-power binocular is a good one but it introduces too many compromises along the way. These binoculars function better when light is not a consideration, such as during the day when viewing terrestrial subjects.

Most binoculars focus both optical tubes simultaneously and are called center focus. Other models allow each optical

The way binoculars work. All components must be placed correctly so that no light will be blocked. (Illustration by Holley Y. Bakich)

tube to be focused independently. All else being equal, choose the ones with individual focus. Center-focus units add mechanical complication you simply do not need. Also, individual-focus binoculars tend to be more rugged and weatherproof. Besides, once you focus your binoculars on any object in the night sky, the focus will be good for any other object in the night sky, as all are "infinitely" distant to your binoculars. And since our eyes generally are not exactly the same, one of the eyepieces in center-focus binoculars focuses independently anyway.

Exit pupil

One of the most important terms when dealing with binoculars is the exit pupil. This is the diameter of the shaft of light coming from each side of the binoculars to your eyes. If you point the front of the binoculars at a bright surface, light or sky, you will see two small disks of light. These are images of the lenses. It is the same for both sides. The diameter of the exit pupil equals the aperture divided by the magnification. So for our 7 × 50 binoculars, the exit pupil diameter would be roughly 7 mm. For astronomical use, a large exit pupil is desirable. This is because the pupils in our eyes dilate in darkness. The wider the shaft of light, the brighter the image because the light is hitting more of our retina. This loose rule, however, is only true up to a point. The problem is that if the exit pupil of a binocular is too large to fit into your eye, you lose some of the instrument's incoming light.

Some people have dark-adapted pupils measuring nearly 9 mm in diameter. Others have small ones less than 5 mm. Our pupils are largest when we're young. From age 30 on, they start to contract, slowing in our later decades. Also, women of the same age tend to have larger pupils than men, on average. Unfortunately, there's no hard and fast rule that correlates pupil size with age.

You can measure your pupil size with a gauge available from certain telescope suppliers. There are a number of pupil gauges on the market (Holladay Pupil Gauge from ASICO, 26 Plaza Drive, Westmont, Illinois 60559, (AE-1573 – 35 US$)) and near vision cards with half or full pupils for matching the pupil size which can be obtained free from many ophthalmic or pharmaceutical companies.

Eye relief

Eye relief is the manufacturer's recommended distance the pupil should be from the eyepiece lens for best performance. Eye relief generally decreases as power increases. Eye relief less than 10 mm requires you to get very close to the eyepieces. This is no problem for advanced amateurs but for beginners, higher eye relief allows the head more freedom of movement. Also, those who choose to wear eyeglasses need a higher amount of eye relief.

A buyer's guide

7 × 50 Fujinon binoculars. The best the author has ever used. (Photo by the author)

Ready for a recommendation? Fujinon. That's it. They're the best. 7 × 50s, 16 × 70s, whatever. They're pricey, but not the most expensive binoculars by a long shot. If you purchase Fujinon binoculars based on this recommendation you will thank me for the rest of your life. Still, I can sense that some of you were probably hoping for something like the following list...

• Pick up the binoculars and shake them gently. Then twist them gently. Then move the focusing mechanisms several times. Then move the barrels together, then apart. What you're looking for here is quality of workmanship. If you hear loose parts or if there's any play when the binoculars are twisted or moved, don't buy them. Another thing to consider at this point is the weight of the binoculars. If you're going to be hand-holding them, try to imagine what they will feel like at the end of an observing session.

- Look into the front of the binoculars and check for dirt or other contaminants. Apart from perhaps some dust on the outside of the lenses, the inside should be immaculate. If not, don't buy them.
- Hold the binoculars in front of you with the eyepieces toward you. Point them at a bright area. You will see disks of light formed by the eyepieces. These are the exit pupils of the binoculars. They should be round. If they are not round, the optical alignment of the binoculars is bad and the prisms are not imaging all the light.
- Of course, you want to look through the binoculars. Try to do this outdoors and at night. Nothing will reveal flaws in the design of binoculars more than star images. If it is

The illuminated openings of these binoculars are the exit pupils. (Photo by the author)

impossible to test the unit at night or even outdoors, try to look through a door or window at distant objects. How well do the binoculars focus? Are objects clear? If there is any sign of a "double" image, this is a sign that the two barrels are not aligned. Do not buy them. If you are wearing glasses, can you get your eyes close enough to the binoculars to see the entire field of view? Move the binoculars visually across a straight line such as a phone wire or the horizon, if possible. Does the line look distorted near the edge of the field of view? (A tiny amount of distortion at the very edge of the field is not a huge problem.)

- Repeat the above tests with several different binoculars. Once you become more familiar with how binoculars compare, you will be well on your way to purchasing an excellent unit.

A further point. An optical flaw called curvature of field is present in all binoculars to some extent. This is due to the fact that a lens forms a sharp image on a surface that is curved. When the eyepieces are set to meet a part of the image that is in focus, say the center, their positions must be changed to make the edges of the image sharp. The quality of the binoculars is directly proportional to the extent that the unit minimizes this problem. An excellent instrument shows this only at the very edge of the field. When the entire field of the binoculars is in focus, it is termed a flat field. If you use a binocular that has a flat field you will not soon forget it. Such units are expensive, however, as a binocular that is flat across most of the field is very difficult to design.

Finally, a note about caring for your binoculars. Generally, binoculars come with lens caps, eyepiece caps and a case. Use

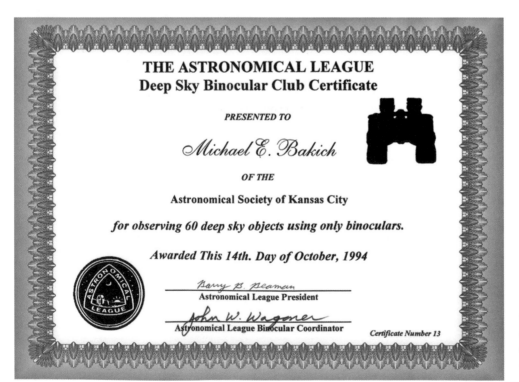

THE ASTRONOMICAL LEAGUE
Deep Sky Binocular Club Certificate

PRESENTED TO

Michael E. Bakich

OF THE

Astronomical Society of Kansas City

for observing 60 deep sky objects using only binoculars.

Awarded This 14th. Day of October, 1994

Barry B. Beaman
Astronomical League President

John W. Wagoner
Astronomical League Binocular Coordinator

Certificate Number 13

One of the many observing club awards given out by the Astronomical League is the Deep Sky Binocular Club Certificate. (Photo by the author)

them. These will help protect your binoculars from dust and moisture. Don't leave your binoculars exposed to direct sunlight, even if they're in their case. And because they are a piece of optical equipment, keep the vibrations (especially impacts) down to a minimum.

Binocular limiting magnitude

The formula for limiting magnitude gives a value of approximately 12.2 for 50 mm binoculars, 12.9 for 70 mm and 13.7 for 100 mm. These must be regarded as rough numbers only. Typical conditions of use for well-mounted binoculars under a dark sky with average seeing will decrease these values by about a magnitude. Naturally, experienced observers will see fainter objects than will beginners.

Image-stabilized binoculars

In the above discussion we stated that "too high a magnification will overmagnify the motion of your hands, causing celestial objects to move around when the binoculars are hand-held." But what if a solution could be found for this problem? After all, much of the allure of the night sky to amateur astronomers is in studying it under ever increasing magnification. To be able to enjoy both a wide field view (as binoculars provide) and moderately high magnification in hand-held binoculars is to have the best of both worlds. The solution is provided by image-stabilized (IS) binoculars.

IS binoculars use different methods to stabilize the image. Some have batteries which power a gyroscopic mechanism. Others use a non-powered design which relies on a gimbaled prism. In all designs, a button is depressed to engage the stabilization. The results can be quite dramatic. For example, I've never been able to hand-hold binoculars steady enough to obtain good, long looks at Jupiter's moons until I used IS binoculars.

The optics of IS binoculars vary in the same way as the optics of regular binoculars, and the same tests as listed above should be used if you are considering purchasing a unit. I will give you a head start, however. If you have the resources to consider IS binoculars, look at the Canon line first. I have found all models in their line to be easy to use, to have great optics and to be mechanically excellent. Of course, technology like this isn't free, or cheap. IS binoculars large enough to interest amateur astronomers cost, at this writing, between £500 and £1000 (1000–2000US$).

Binocular Mounts

Image-stabilized binoculars are a tremendous advance for amateur astronomy. But for the steadiest images possible, nothing beats mounting your binoculars to a tripod or custom binocular mount. Smaller, well-mounted binoculars with less magnification will, after a few short minutes of

"L" bracket for mounting binoculars to a tripod. (Photo by the author)

continuous use, beat hand-held binoculars of larger aperture and power.

The simplest binocular mount is an "L" bracket. This attaches to a mounting hole on the center post of the binoculars. The size of this mounting hole is designated 1/4 – 20, the same size as is found on the base of every 35 mm camera. The other end of the "L" attaches to a standard camera tripod. This is generally an adequate setup if the objects you're observing are not too high in the sky. For objects near the zenith, tripod mounted binoculars are uncomfortable – in some cases impossible – to use.

The other option is to use a custom-made binocular mount. Plans for binocular mounts that you can build are available. If you are mechanically inclined and have a well-stocked tool shop, I recommend that you build your own binocular mount. Here are several sites (out of many) with plans located on the internet:

http://home.wanadoo.nl/jhm.vangastel/Astronomy/binocs/binocs.htm
http://www.atmpage.com/binomnt.html
http://www.gcw.org.uk/bino/binonet.htm
http://www.home-dome.com/Astronomy/Projects/binocular_mount.htm

Most amateur astronomers, however, purchase commercially made mounts, and several are available. Most are based on a design which employs a movable parallelogram. Such an arrangement keeps the binoculars pointed at an object over a wide range of motion, thus allowing people of varying height to use them. This is ideal for observing sessions or star parties where a number of people will be viewing the same objects over the course of the session.

Tip: Choose a mount for your binoculars that is quite a bit sturdier than you require. This will allow you to easily upgrade at some point in the future.

Ruggedness in a binocular mount, as I define it, means that a few seconds after you have located an object in the field of view the image settles down and shows no vibration unless there is a strong wind blowing. If the image is not stable, your binocular mount may be at fault. Alternately, the fault could lie with the second half of the equipment needed.

The other necessity in a quality binocular mount is the tripod to which the mount attaches. Most camera tripods are inadequate for this purpose. They are simply not robust enough to handle the weight of the binoculars plus the weight of the mount, especially when extended. You will pretty much know immediately if your tripod is right for this task. A tripod can fail in more ways than by being flimsy. It may be that your sturdy tripod, even at full extension, is not high enough to allow you to stand under your binoculars and view objects near the zenith.

Parallelogram binomount (made of maple) and tripod (made of ash). With the binoculars mounted, the binomount measures 1.3 m in length. The tripod weighs 5.9 kg. (Photo by the author)

CHAPTER THREE

How to observe

3.1 Sketching what you observe

There are two things that I want to state up front. (1) You do not have to be an artist. (2) Sketching will make you a better observer. Sketching is also fun and an area of amateur astronomy where you can see definite improvement over time.

Not an artist?

Regarding point (1) above, be happy with what you sketch and know – with absolute conviction – that you will get better. As a first step, learn to draw a circle. Much of what you sketch will be circular (whole solar or lunar disk, planets), and all of the non-circular objects (or partial objects) will be located within the circular field of view of your telescope. You can use a compass to draw your circles. I used a graphic

computer program. The advantage to this is that I am able to print out as many circles as I need.

The size of the circle can be varied, but I have a suggestion. Do not make the circle too large. The first few times I sketched objects at the telescope, I used circles which barely fit on a full page of paper. The sketches took forever to make! After a few of these marathon sessions, I settled on a circle 10 cm in diameter; 7 or 8 cm would have probably been better but because I was just starting I didn't feel as confident with smaller circles which would have required a slightly finer touch to record detail. So I settled on 10 cm. Two of these will easily fit on a page with space left over for details. There are pre-drawn observing forms for the planets available from both the British Astronomical Society (BAA) and the Association of Lunar and Planetary Observers (ALPO).

Training your eye

Now, as for point (2), when you first start sketching celestial objects you will, by necessity, spend more time simply

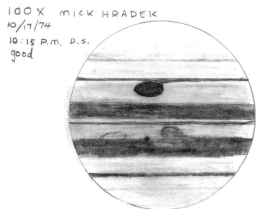

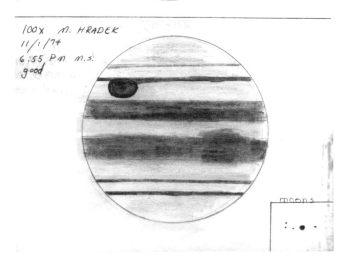

Mick Hradek of El Paso, Texas, made these two Jupiter sketches on 17 Oct 1974 and 1 Nov 1974.

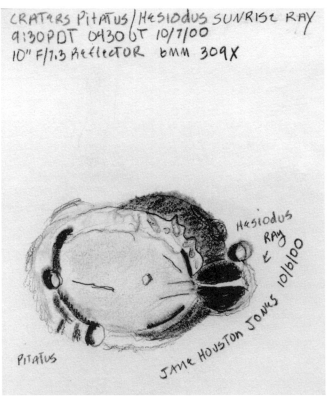

A very detailed lunar sketch by Jane Houston Jones of San Francisco, California. Her use of the term "sunrise ray" tells us that this crater was near the terminator when she sketched it.

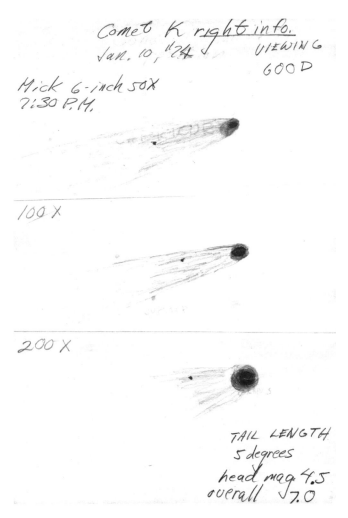

Comet K right info.
Jan. 10, "24   VIEWING
                  GOOD

Mick 6-inch 50X
7:30 P.M.

100 X

200 X

TAIL LENGTH
5 degrees
head mag 4.5
overall √7.0

These sketches of Comet Kohoutek were made on 10 Jan 1974 by Mick Hradek of El Paso, Texas.

looking at the object. This in itself is good. But to sketch, you must memorize small areas of detail before transferring what you are seeing to paper. The more objects you sketch, the more this will become part of your normal observing routine. You will start to notice things that you were unaware of before. In essence, your eyes (or eye, as most observers use one eye exclusively) are being trained.

This training of your eye is real and can be demonstrated. Here's how. Pick a celestial object. A planet, a star cluster, a nebula, it doesn't matter. As long as you can see it well through your telescope and make a few observations over a month's time, it will do. Observe and sketch the object just once a week for four weeks. That's it. Use the same telescope at the same location with the same eyepiece. Take 15 or 20 minutes each week to sketch the object. When you're done, you will have four sketches. Compare them. When you do, pretend that I'm directly behind you, looking over your shoulder. You will hear me say, "I told you so."

We have all observed an object and wished that we had a larger telescope with which to view it. Sketching helps us get the maximum detail possible out of the equipment we own right now.

Preparation and materials

In addition to your sketching forms, you will need a red flashlight, preferably one which can be dimmed. Ideally, a short table would be set up in a comfortable position near the eyepiece, but since this is almost always impossible, you will need a clipboard or some sort of sturdy board or notebook to back up your paper. If you can somehow fasten your red light to this, or suspend it above it, sketching will be a lot easier.

If possible, sit while sketching. I've said it before but it is worth repeating, comfort is everything while observing. This is especially true when you are looking for minute details and transferring them to paper.

Before you start, decide whether you want to make black-and-white or color sketches. My suggestion is to start with black-and-white. As you improve with details, then add color. Start with a common 2 pencil and a good-quality soft eraser. Practice lighter and darker strokes and fills. When you're ready, get some pencils of different shades, both lighter and darker. At least a dozen different shades exist for common art pencils, never mind the special ones. Any good art supply store will have all that you will need. Either buy several of each and keep them with your "sketching kit," or purchase a good pencil sharpener.

Tip: Don't hesitate to use your eraser!

White on black or black on white?

Most observers who sketch at the telescope use lead pencils on white paper. This is certainly a more intuitive approach, but some effort must be used to convert it to how it looked through the telescope, with the blacks and whites reversed. Of course, with today's graphics programs such as Adobe PhotoShop, you can scan your original sketch and almost instantly produce a negative image of it.

On the other hand, some observers sketch with white pencil on black paper. This method allows for more immediate direct comparison. Also, when examining the sketch later, it will remind you of what you saw in the telescope. The choice is yours.

Smudging

Some observers like to use dark pencils for nebulae and galaxies, drawing the outline of the object first. They then use a finger to smudge the outline into a blob representing the nebula. For this technique, coarse grade sketching paper seems to work much better, as does lots and lots of practice. Use a pencil with a dark, soft lead. An art supply store will provide some recommendations for paper and pencil when you tell them what you plan to do. Since this technique is obviously for deep-sky objects or comets, you have a choice whether to draw the outline of the nebula first or any stars in the field of view.

Color sketches

As with black-and-white pencils, colored pencils of all qualities may be found at your local art supply store. Start with a small set containing common colors.

To show different increments of the same color, press harder or more lightly with your pencil. If you want to make a darker shade, first color lightly with black. Then color over the black with a color of your choice. To make a lighter tint, first color firmly with white. Then color lightly over the white with a color of your choice. Experiment with how much black or white and color you need to get the value you want. This changes depending on the color you're using. For black, you can add more on top to make it darker. If black doesn't give you the look you want, try blending with other colors.

Note: Red light does not work well for color sketching. However, since most "colored" celestial objects are fairly bright (to trigger the color receptors in your eye), you may want to use low intensity white light when sketching in color.

A sketch of M81. Craig Molstad of Onamia, Minnesota, made this drawing 2 May 2002.

Other possibilities

The use of pencils, be they black-and-white or color, is only one possibility for sketching. Look around and see if a different technique appeals to you. Pen and ink drawings take a delicate touch but can be quite striking. Blendable chalk is another possibility. Some observers stick with charcoal and white chalk, blending them to obtain various shades of gray. Others use colored chalk.

At the telescope

Begin by drawing what you see. At first, notice details but don't dwell on them. Try to capture the essence of the object. If you've picked a non-comet solar system object, look at it

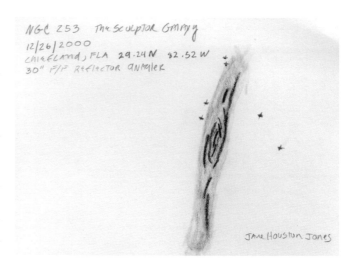

When the image is bright enough, you can even sketch deep-sky objects. Jane Houston Jones of San Francisco, California, using a 750 mm telescope, was able to capture details of shape and contrast in her sketch of NGC 253 in Sculptor.

as you would a clock face. For reference, note where features begin and where they end. Try to identify the exact middle of the object's circle.

If, on the other hand, you've chosen a deep-sky object as your subject, begin by comparing its size to the size of the field of view. Try to capture this ratio in your drawing. If individual stars are visible, sketch them in first and use them as reference points. Then (unless you've chosen a star cluster) begin to lightly shade in the outline of the object.

For either solar system or deep-sky object, compare and contrast features that you see. You can do this by choosing a pair of features and noting

• whether they are the same height and width
• whether they appear tilted at the same angle
• whether they are the same in terms of lightness and darkness
• how their borders compare
• whether they have the same degree of smoothness
• any differences in color
• where they are with respect to one another.

The big difference between solar system objects and deep-sky objects is that once you are satisfied with a sketch of a deep-sky object, you never have to draw it again. It will not change in your lifetime. (Supernovae in galaxies being an exception.) This is also true of the Moon, but there are enough details on the Moon for a lifetime's worth of sketching. With the Sun, planets and comets, however, you have objects which are in a continual state of change.

You never know what use you will have for your sketches. Let me give you an example. In the summer of 1999, near the time of solar maximum, I sketched the Sun's disk each day for 60 days. Because I used the same telescope/eyepiece combination and the same size form for each sketch, I was able to make an animation showing the Sun's rotation, and the effect it had on sunspot positions. You can view it on the internet at http://www.geocities.com/uni7777777777/gosungo.html

VALLIS SCHROTERI
SCHROTER'S VALLEY
12 DAY MOON 3/2 930PM
12.5 LB F5.75 202X

Jane Houston Jones of San Francisco, California, made this excellent sketch back in 2000. Note her technique, drawing dark features as more solid and following by shading in the lighter highlights.

Notice that the registration is not exact and the spot positions wander from day to day. You are viewing my skill level as an artist. Still, it was a valiant effort and you can do as well or better!

Tip: Light touch-up to fill in shaded areas or to add color after the sketching session is certainly permissible as long as you do not add any detail you did not see at the telescope.

Sketching log

You keep an observing log (after reading this book, how could you not?), so I suggest that you keep a sketching log as well. This may consist of nothing more than filling in the blanks on your sketching form. Fine. Just make certain that you fill them in. When you have filled a page, file it someplace safe.

Internet group

There is a group hosted by Yahoo! whose members are interested in sketching celestial objects. They are a friendly group willing to share techniques. Many have interesting websites. On the internet, go to http://groups.yahoo.com and join AstroSketch. You will need a Yahoo! registration ID to join. Follow the instructions on the site.

A sketching technique

Amateur astronomer and accomplished sketcher of celestial objects, Jere Kahanpaa, of Jyväskylä, Finland, has provided an illustrated explanation of his sketching technique.

Tools of the trade
The basic tools are very simple: a pencil and some thick white paper. I use an HB (medium soft) pencil for most objects. H2 is good for small detail. I think B2 or softer are not as suitable as they tend to smear all over the paper. An eraser can be used for thin absorption effects and small dust clouds but a very clean and sharp eraser is needed for a good effect.

The paper used should be quite thick (even thin cardboard) as thin sheets will soon be waterlogged and unusable (unless your observing site is in Sahara or any similar very dry place). A thin hard-cover book or a rigid sheet of any material is needed for a sturdy background.

Last but not least a faint red light is needed. Red is used because it seems to affect the night vision less than any other color. But even red light will affect your night vision so the light source should be as faint as possible. The light should be quite diffuse as sharp changes in the intensity will make evaluating the quality of the drawing much more difficult.

Example: NGC 7013
(The numbers in the following list correspond to the numbers on the sketches.)
1. Plot the brightest stars in the field. This is the most critical step as the rough net of bright stars will serve as a guide when drawing everything else.
2. Add fainter field stars. I usually also plot a small and faint point where the center of the object will be. It makes plotting the faint stars near the object much easier.
3. Add the rough out-line of the object. Compare the shape and dimensions with the "general grid" of field stars.

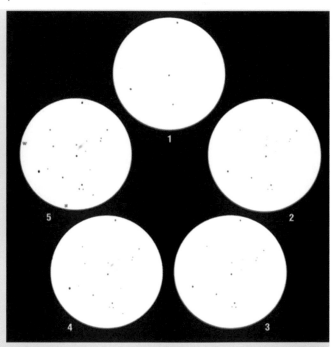

4. Add lead to the object until the contrast and details are correct.
5. Add faint stars around the object and check details. The drawing is now almost ready. Note the magnification, the name of the target, etc.

Also
Clean up the mess. This step can be done indoors. Use a black pen for nice round stars, carefully rub the drawing with your fingertip or a piece of paper to get a nice diffuse outline and finally spray a layer of fixative on the drawing. Add a detailed description and information about sky conditions, etc. The sketches shown were made through a 205 mm f/5 Newtonian on 9 Sep 1993, at 23:50 UT. A magnification of 133× was used.

3.2 Astrophotography

If you listen to many advanced amateurs (and vendors), it seems that these days you can't take an excellent astrophoto without a high-dollar CCD camera. This is simply not true. For a century and a half, astronomers – amateur and professional alike – used photographic emulsions to produce some outstanding images. And it still happens. If you are looking to capture the beauty of the heavens, don't discount film. Every image in this chapter was captured on a 35 mm camera.

David Healy of Sierra Vista, Arizona, with his venerable Celestron 14 set up to test films with a piggybacked 300 mm lens.

Cameras

If you are planning to do film astrophotography, selection of a camera is really quite simple. A suitable camera will have the capability to hold open the shutter for extended periods of time. This necessitates that the shutter control be of the manual type rather than the electronic or fully automatic variety. Most new 35 mm cameras have electronic shutter controls which make them unsuitable for long-exposure photography. This is because the shutter, while being held open, imparts a constant drain on the camera's battery. The small batteries in these cameras cannot withstand very long periods of holding open the shutter and will quickly go dead, closing the shutter. So an acceptable camera for astrophotography is one with a manual shutter control. With a remote shutter cable (with locking thumbscrew) installed and the camera's shutter speed set to the "B" (bulb) position, the shutter can be opened and held open indefinitely.

A second important feature to look for in a camera is mirror lockup. When the shutter button on a camera is depressed, two things happen quickly. First, the mirror flips upward to allow the light entering the lens to get to the film and second, the shutter opens. The motion of the mirror and its rapid sudden stop sets up a vibration in the camera that often causes blurred images on the photograph. The best camera for astrophotography is one that has the capability to "lock up" the mirror before the shutter is opened. This allows the vibrations to settle out before the film is exposed.

A third consideration is to obtain a camera with interchangeable focus screens and the availability of screens that are clear or bright. Many budding astrophotographers have tried to focus on a star with a normal focus screen (sometimes referred to as a diamond or split image screen). It is a very difficult feat. For astrophotography, a clear focusing screen is something you simply must have.

My vote for the best film camera of all time for astrophotography goes to the Olympus OM-1. It is a fully manual camera with mirror lockup and many different types of focusing screens are available. Well, I should say were available. The OM-1 has not been manufactured in many years. Still, a large number of these cameras were made so your chances of finding one are good. Camera stores that sell used equipment or pawn shops are worth checking out. As of this writing, over a dozen OM-1 bodies were available on the popular internet auction site eBay at http://www.ebay.com

Field of view

Knowing a lens' field of view is important in order to gauge how much sky that lens will cover. Using this information along with a star chart, one can determine what constellations or objects will fit within the field of view.

The formula for determining field of view is:

$$X = (57.3/f) \times d$$

where X = field size in degrees, f = focal length in millimeters, and d = film dimension in millimeters.

Remembering that the frame dimensions for 35 mm film are 24 mm vertically and 36 mm horizontally, we arrive at the table overleaf. The total area is given in square degrees.

Film

Most films used in astrophotography are professional films, which means you can't buy them in any but the largest cities

Lens field of view in degrees

Focal length	Vertical	Horizontal	Diagonal	Total area
28	49	74	88	3614
35	39	59	70	2315
50	27.5	41	49	1136
85	16	24	29	384
135	10	15	18	156
200	6.7	10.3	12	69
300	4.6	6.9	8	32

and they are more expensive than common film. It is also difficult to find a camera store that is willing to break open a box and sell you a single roll. So let me, up front, give you the best recommendation for a source of 35 mm film that I have found: B & H Photo, on the internet at http://www.bhphoto.com

When you get to their home page, click on "Film" on the left side. B & H stocks all current films and their prices are great.

Before we start considering individual films, let's take a look at a problem encountered in long-exposure astrophotography. Take two exposures through the same camera and with the same film, the first a 1-second exposure at f/1.4 and the second a 128-second exposure at f/16. In a perfect world, both exposures would look the same. The images would have the same densities and the same brightnesses. Unfortunately, we must deal with a property of film called reciprocity failure.

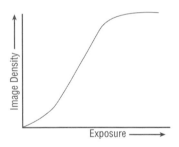

Reciprocity failure is represented at the top of this graph where increasing exposure does not add to the image density as quickly as during short exposures. (Illustration by Holley Y. Bakich)

Because of reciprocity failure, a 400 speed film which begins a long exposure will, after several minutes, act like a 100 speed film, or one even slower, depending on the film. The popular black-and-white film Tri-X is a good example. For short exposures, this film has a speed of 400. However, if you expose it for an hour (not an uncommon occurrence for astrophotography) its effective speed drops to 10! Because a great deal of astrophotography involves lengthy time exposures, this is a real problem for amateur astronomers. We either have to live with the effects of reciprocity failure or try to get around it. One way to minimize reciprocity failure is by selecting the correct film.

Another is to employ a technique known as hyper-sensitization.

Hypersensitization is a film treatment process that reduces reciprocity failure. This process replaces moisture and impurities in the film with a gas by pressure cooking it into the film, making it more resistant to reciprocity failure. The gas, by the way, is a combination of 92% nitrogen and 8% hydrogen and is called forming gas. When hypersensitization was discovered, the gas used was pure hydrogen, which is quite dangerous. Forming gas is inflammable, although some care must still be used in handling the pressurized containers.

The technique of gas hypering involves putting the film in an airtight canister, pumping the chamber interior to a very deep vacuum, purging the interior with forming gas and revacuuming for two or three cycles. The chamber is then pressurized with forming gas at about 3–6 psi and left to "bake" at about 50°C for periods of 4 hours up to 100 hours depending on the baking pressure, temperature, and film type used.

Commercially manufactured hypering kits are available from Lumicon http://www.lumicon.com. Some mechanically adept amateurs have even constructed units. However, hypersensitized film has one important drawback you may wish to consider.

M11 and star clouds. (8" Celestron Schmidt camera, f/1.5, 12 minutes on SO-115 film, 13 Nov 1979, Naco, Arizona. Image by David Healy, Sierra Vista, Arizona)

Hypered film has a very short period where it is usable. Once the film comes out of the hypering tank it is best to use it immediately and then develop it immediately. Some amateurs have had success sealing hypered film in airtight containers and freezing it until they are ready to use it, but even then its shelf life is less than a week. The main enemies of hypersensitized film are heat and moisture. Because of the short period of usability, ordering hypered film through the mail is unacceptable, in my opinion. By the way, developing of any hypered film can be done commercially at photo labs based on the film type. No special processing is required.

Black-and-white films

To be useful for astrophotography, a black-and-white film must have low reciprocity failure and a good degree of red sensitivity. As a film undergoes a long exposure, it loses sensitivity in certain ranges. Unfortunately, this is often in the red region of the spectrum. This region is best for recording nebulosity and if a film is not responsive in this area, your exposure will not be as good as one taken with a film with high red response. In technical terms, a black-and-white film should be sensitive to the emission line of hydrogen alpha, which lies at 656.3 nm. Because of this requirement, we immediately run into problems with most black-and-white emulsions.

The best film for black-and-white astrophotography is Kodak Technical Pan (TP 2415), preferably hypersensitized. This film is available in various formats, including 35 mm. It has very good response to red light (in fact, its spectral sensitivity is very uniform out to 690 nm), has high resolution and is inexpensive. Regarding its resolution, Technical Pan film can resolve an amazing 320 lines/mm. Compare this to the resolution of TRI-X at 80 lines/mm.

Another feature of Technical Pan film is that a wide range of image contrasts may be attained by using different developers. For those who understand the numbers, the highest contrast of 2.50 is attained by developing in DEKTOL and the lowest contrast of 0.50 is attained by developing in TECHNIDOL. Amateur astrophotographers are always looking for higher contrast so this lower value is of little use to them. It does show the film's remarkable range, however. Due to the low speed of TP 2415 (somewhere between ISO 15 and 25), many amateurs prefer to hypersensitize this film. This, of course, is not necessary for shooting bright objects such as those found in the solar system. I found Kodak Technical Pan film at the internet site http://www.fotoclubinc.com

Color films
Slides or prints?

To be brief, it is your choice. I've seen terrific images made on both slide and print film. I prefer slides as I do a lot of

M16 and the "Pillars of Creation." (Celestron 14 at f/6 with Lumicon guider, hypered 2415 film, 40 minutes, 7–8 Oct 83, Naco, Arizona. Image by David Healy, Sierra Vista, Arizona)

M8. (Celestron 14 w/Lumicon telecompressor (f/7), TP 2415 film hypered by Edgar Everhart, 24-minute exposure, 27 Sep 1981, Naco, Arizona. Image by David Healy, Sierra Vista, Arizona)

M46. (Celestron 14 at f/11, 45 minutes on hypered Kodak PPF 400 film, 16-17 Dec 1998. Image by David Healy, Sierra Vista, Arizona)

Film processing

I'll make this short. Process black-and-white film at home. Send color film to a lab. If you are going to shoot black-and-white film after reading this chapter, it will probably be Kodak Technical Pan 2415. This is the best film ever for black-and-white astrophotography and few photo labs are equipped to handle it to the specifications you are going to want. Read about processing Technical Pan film at the Kodak website, specifically at

http://www.kodak.com/global/en/professional/support/techPubs/p255/p255.jhtml

If you cannot, or do not wish to, set up a home darkroom, film can be loaded into a developing tank using a photographic changing bag.

Regarding color film (negatives or slides), I can make one important suggestion: have slides processed and returned to you uncut. Many amateur astrophotographers learn this lesson the hard way. Automatic slide mounting machines are poor judges as to where an astrophoto (especially of a faint object) starts and ends, and many a good shot has been ruined by the slide being cut where it should not have been cut. Once you have the uncut film in hand, you can cut and mount the slides in either plastic or glass slide mounts.

illustrated talks. Being able to drop slides in a tray is a real help. I suppose I could use a slide copier and duplicate the negatives on negative film, producing a positive, but that seems like a lot of extra effort. For displaying your images on a computer, only a scanner which will accommodate negatives or slides is required. (It is preferable to scan the original negative rather than the first generation print.)

Unlike the one great black-and-white film choice astrophotographers have, there are many terrific color films out there. Just about all color films are well-balanced over the entire spectrum and have good sensitivity in the red (remember the h-alpha line at 656.3 nm?). The new color films are fast, too. One problem is that new emulsions are being introduced all the time so that when this book is published there will probably be several good new color films which are not available now. If you see an emulsion you haven't tried, my suggestion is to purchase a roll and shoot. The following are my suggestions, as of the time of this writing.

- Slide films: Kodak Ektachrome Elite II 100 (nice fine grain, low reciprocity failure and the ability to be push processed); Scotchchrome 800 or 3200; Kodak Ektachrome Professional 1600. (This film is normally ISO 400 but is designed to be pushed two f stops to ISO 1600. Amateur astronomers have even gotten good results pushing this film to ISO 3200.)
- Print films: Kodak Royal Gold 1000; Fuji Super G800; Kodak Ektapress Multispeed; and many others.

Dr Alin Tolea, a member of the Bucharest AstroClub in Bucharest, Romania, took this wide-angle shot of the region around M8 and M20 in Sagittarius.

A word about traveling with film

Pack film in a carry-on bag (hand luggage) instead of in your normal luggage. X-ray equipment used at airport security checkpoints to screen carry-on luggage is less likely to damage undeveloped film than the more powerful units which X-ray checked luggage. Some have suggested placing your film in the checked baggage in lead-lined film bags to prevent potential fogging of film. It may go through just fine but the more likely scenario is that somebody will open your luggage and open the film bag. The choice is yours. At the walk-through checkpoint, you may be able to have the film hand inspected, but don't count on it. Security worldwide has been intensely tightened recently. But do bring some film with you on the plane, just in case your checked bags arrive late.

Another suggestion I will make is to try to choose a processing lab which has some flexibility to allow push processing your film, if you desire it. Some "one-hour" labs are locked into processing at a film's standard ISO speed.

Image processing

Whether you shoot slide or negative film, chances are that at some point you will want to transfer those images onto your computer and, probably, onto the internet as well. Within this transfer process is an opportunity to enhance – often dramatically enhance – your images. The processing end of astrophotography can make the difference between an OK shot and an excellent one.

The first step is to scan your images into your computer. As I mentioned above, it is preferable to use a negative scanner for print and slide film. Quality of scanners varies quite a bit, I'm sorry to say. Check around and read some reviews before purchasing one.

Once you have scanned your images in, software exists to enhance it. The best of the lot is Adobe PhotoShop, but it is not cheap. Another choice is JASC's popular software Paint Shop Pro. I will go into greater detail about image processing in the chapter on "The CCD".

Using a slide copier

Once we have processed negatives or slides, what can be done to produce positives or enhance weak images? If an enlarger is unavailable, a slide copier can produce beautiful astronomical pictures by creating reversal images from negative originals. Copying black-and-white negatives with unhypered Technical Pan film will yield black-and-white positive slides of excellent quality. The contrast of the resulting slides can be varied and adjusted by the proper selection of developer and processing times.

The same copy process will work with color negatives by using Type 5072 reversal film. This film is also processed in C-

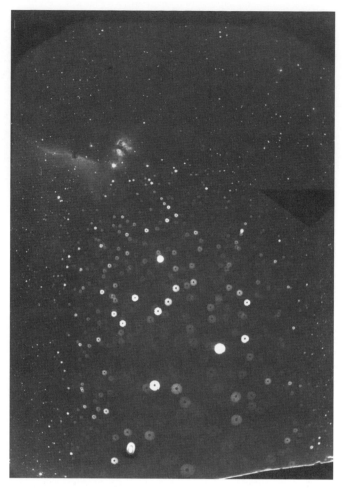

What can go wrong during an astrophoto? This example, by David Healy, Sierra Vista, Arizona, shows Orion's Belt – Murphy's Law Version (8" Celestron Schmidt camera at f/1.5, 103a G film, 5 minutes, 30 Oct 1978, Naco, Arizona). Among the mistakes: (a) film not flat in holder; (b) film trimming stuck on film; (c) film trimmed too small; (d) one of the "stars" is a processing flaw; (e) negative scratched; (f) negative dusty; (g) RGM-jet backing not fully wiped off; (h) filter filmholder used without filter.

41 chemistry and will yield high-contrast slides from negative originals. Color can be corrected with the use of CC filters and density adjustments made through exposure changes.

The contrast and color of original slides can also be adjusted through the copy process by using filters and exposure corrections on either Kodachrome 25 or Ektachrome Elite II 100 film. Remember, to bring out detail in dim originals, overexpose during the copy process.

Hard copy options

Few amateur astronomers have photographic enlargers at home. Most of us, then, must rely on our local photo lab to produce our prints from negatives or Cibachromes from slides. The slide provides the guide for what the final image should look like, but an example from a magazine will be useful to show the photo lab what you desire from an astronomical negative.

Rick Thurmond of Mayhill, New Mexico, photographed the Horsehead Nebula on medium format Ektachrome 200 through a 180 mm AstroPhysics refractor.

Some one-hour photo locations and some shopping centers now have the Kodak "Create-a-Print" machine. This digital scanning and dye sublimation printing device allows amateurs to create an 8×10 print from a negative. Basic digital image processing options such as color, density, and contrast controls allow a high degree of "fine tuning" before printing.

One-hour photo labs can also make prints from your astronomical negatives. The automated photographic printers these labs use can produce some surprisingly good results if you provide some guidelines. It is always a good idea to include a normal daylight or flash exposure of a more familiar subject at the beginning of the roll so the machine operator has a color correction reference. If the option exists, instruct that the image be printed for a dark sky and, if feasible, show the machine operator a finished example to demonstrate the expected sky density.

The home computer has also become a powerful tool which allows amateurs hard copy options for their astrophotography. Film scanners which produce high-resolution digital scans of slides and negatives are still rather expensive but very acceptable economical digital scans from 35 mm film can be obtained by using a Kodak Photo CD. Kodak will scan up to 100 35 mm images onto a CD at resolutions up 2048 × 3072 pixels at a cost of under 2 US$ per image. Specify a flat scan with no image processing. All enhancements can be made later with image processing software. With the addition of a color ink-jet printer, superb prints can be produced.

Tripod-mounted photographs

The equipment necessary for a tripod-mounted astrophotograph is minimal: a camera, lens, tripod, cable release (preferably with lock), and a watch with a second hand is all you need. Focus the camera on infinity and lock open the shutter. Many tripod mounted shots show an object in the foreground like a tree or mountain. It's really up to you. It also looks good to show the horizon, if you have a choice. If star trails are what you want try whatever exposure you feel like. Remember, any long-exposure astrophotography requires a dark sky.

A word about tripods, if I may. Buy a good one. This is one of the most important of all photographic accessories, especially when you consider the long exposures needed in astrophotography and the possibility that, at some point, you may wish to mount a pair of binoculars on your tripod. It is amazing how many people try to save money here and end up with a shaky, inadequate tripod that they eventually have to replace. A good tripod will work with any camera, so buy the best and strongest tripod you can afford.

Make sure the tripod you select has easy-to-operate convenient controls that can be worked with gloves on in the winter. Insure the tilt and pan movement locks are strong and will hold a front-heavy telephoto. A sturdy tripod will also allow you to mount two (or even more) cameras onto it at the same time. Use a bar or a plate for this setup.

Tip: Ascertain that the tripod you are considering will point to the zenith.

Lightning under Perseus. While doing tripod mounted astrophotography, the author captured this lightning strike on the Rincon Mountains, east of Tucson, Arizona.

The next necessity for tripod-mounted shots is a quality cable release. This item insures that exposures aren't ruined by vibrations caused by using your hand to trigger the camera shutter. A cable release should be at least 18 inches in length to isolate the camera from the operator's motions. There are several types of cable release locking mechanisms. I own both the set screw and the friction (or disk lock) variety and have used them a lot. Reliability is not a factor. Neither works well with gloves on. The friction type is slightly easier to operate one-handed.

Perseus. This piggyback image shows the Double Cluster (left) and the Pleiades (right). (50 mm lens, f/2.5. Photo by Steve Coe, Phoenix, Arizona)

Exposure times

Well, you've set up your camera on a tripod and pointed it at the sky. But you don't want to see "trailed" stars in your finished picture. Is there a way to figure out how long an exposure you can take without trailing? Yes, there is, and it can be determined with the formula:

$$X = 1000/f \times (\cos d)$$

where X = exposure time in seconds, f = focal length in

Circumpolar star trails. (Two-hour exposure on Kodak Elite Chrome 200, pushed 1 stop. Nikon F2 with a 50 mm Nikkor lens. Photo by Ulrich Beinert of Kronberg, Germany)

Maximum unguided exposure in seconds

Declination N or S	Focal length of camera lens (mm)			
	28	35	50	135
75°	138	110.5	77	28.5
60°	71.5	57	40	15
45°	50.5	40.5	28	10.5
30°	41	33	23	8.5
15°	40	29.5	21	7.5
0°	36	28.5	20	7.5

millimeters, and d = declination of object, in degrees.

The declination d is a factor because the closer you get to either celestial pole, the smaller the star trails per unit time. Star trails are greatest (as you would imagine) when your camera is pointed at the equator. Using the formula the table above gives some maximum exposure times.

Now, a word about setting the f stop of your camera lens. Well, two words: wide open. Many astrophotographers stop down the lens one or two clicks because the star images at the edges of the photograph are not pinpoints (due to lens irregularities). Rather than lose that light, which you will need if there are extended objects in the field (stars are pinpoint objects and are not affected by f stop), shoot wide open and then either electronically crop the image or mask the slide with black photographic masking tape. Such an item is sold by the better photo supply shops.

Piggyback astrophotography

With this setup, you are still shooting through the lens of your camera, however, it is attached or "piggybacked" to your telescope. Sometimes this is referred to as guided wide-

field astrophotography. When I shoot piggyback, it is a misnomer. I remove the entire optical tube assembly and simply use the mount and drive. (So, technically, I'm not "piggybacked" to the telescope.) This method allows one to take much longer exposures of the stars without having to

The heart of the Milky Way. Sagittarius is to the lower left. Red Antares, the heart of Scorpius, is to the right of the picture. (Four-minute exposure on Kodak Elite Chrome 200, pushed 1 stop. Nikon F2 with a 50 mm Nikkor lens. Photo by Ulrich Beinert of Kronberg, Germany)

worry about the stars making "trails" on the film.

To do basic piggyback astrophotography, you'll need everything listed for tripod mounted astrophotography except the tripod. In its place, you will need an equatorially driven telescope (or, as in my case, at least the drive). One more item you will need is a way to attach the camera to the telescope. This is usually accomplished with what is called a

Piggyback camera assembly with 135 mm lens connected to a 13 inch Newtonian via a ball socket camera mount. (Photo by Steve Coe, Phoenix, Arizona)

A hydrogen alpha filter will be attached to this fabulous Tamron lens which will be used for piggyback photography of nebulae. (Image by Robert Kuberek)

camera mounting bracket. For many telescopes, camera mounts are commercially available. Many amateurs have also made their own custom mounts. The requirement is that the mount be able to hold the camera securely so it does not move with respect to the telescope's tube assembly.

One important point about doing piggyback astrophotography: make certain that you balance your telescope after you have attached the camera and aimed it at the part of the sky you are going to shoot. The reason the telescope should be balanced at this time is that its balance can be made optimum only for certain positions (because of all the extra weight hanging off the telescope at strange angles). Some astrophotographers slightly off-balance their scopes so that the telescope tube moves slightly to the west. They do this because it ensures a good mesh of the telescope RA axis drive gears. If the balance is "perfect", backlash in

T-adapter and T-ring. The T-adapter is a standard piece. The T-ring is specific to the brand of 35 mm camera you are using. (Photo by the author)

Another way to mount a camera for piggyback astrophotography. Here, a telephoto lens is held by the finder scope rings. (Photo by Steve Coe, Phoenix, Arizona)

or similar arrangement), preferably illuminated and a drive corrector for the telescope's motor drive. The T-adapter is a machined tube, one end of which fits into the eyepiece holder. The other end is threaded to accept the T-ring. The other end of the T-ring has flanges exactly similar to any lens which couples to your 35 mm camera.

the RA drive gears can become a problem.

Exposure times for piggyback astrophotography vary greatly. Sky conditions, speed of film, accuracy of polar alignment, and drive all come into play regarding exposure times. In addition, you have to take into consideration the focal length of the camera lens you are using. If it is a "short" lens (50 mm or less), you can probably expose for 15–20 minutes. Longer lenses will magnify the errors in your alignment and your telescope's drive. Shooting a test roll of your chosen film is the best way to gauge exposure times. Start with an exposure of 1 minute and work up to 30 minutes or so.

Astrophotography through the telescope

Prime focus photography is accomplished by mounting the camera on the telescope and making the telescope the lens of the camera. For this type of photography you will need (in addition to the equipment listed above) a T-adapter and T-ring for your camera, a guide scope or off-axis guider, a reticule (or, reticle) eyepiece (one with a crosshair pattern

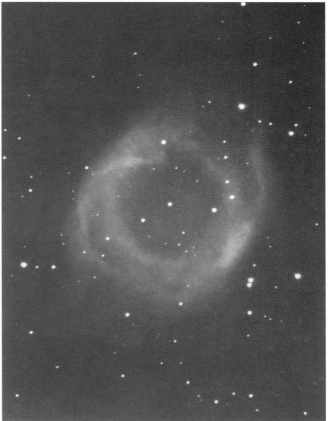

Helix Nebula (NGC 7293). (Celestron 14 at f/7, 1 hour on hypered Technical Pan 2415, 22 Oct 1984, Naco, Arizona. Image by David Healy, Sierra Vista, Arizona)

The Rosette Nebula. (22-minute exposure on Kodak Elite Chrome 200. 125 mm Borg apochromatic refractor at f/4 with Super Reducer (500 mm). Photo by Ulrich Beinert of Kronberg, Germany)

The most common method of guiding is by using a separate telescope attached to the main optical tube assembly. Since both scopes are driven on the same mounting, any variance seen in the tracking of one will also be seen by the other. One advantage to using a separate guide scope is that a suitable guide star can more easily be found since the guide scope can be independently aimed (within limits) of the imaging scope. Another advantage is that the images of any guide stars will be better than those provided with an off-axis guider.

The guide scope should have a primary focal length of at least half that of the main tube. Smaller and lighter guide scopes can be made to work by increasing their focal lengths with the use of a Barlow lens. Also, the mounting of the guide scope to the main telescope should be of high mechanical quality. This is so that the guide scope is kept in perfect alignment with the imaging scope. Even the tiniest bit of flexing in either the guide scope optical tube or in the mounting to the imaging scope will cause a loss of alignment of the two. So, even if you believed your guiding to be perfect, your final image may display trailed stars. This effect is called differential flexure.

Into the eyepiece end of the guide scope goes a reticule eyepiece. The two basic types of reticule eyepieces are illuminated and non-illuminated. Almost every astrophotographer I have met uses an illuminated reticule eyepiece. The illumination for the reticule pattern comes from a light emitting diode built into the eyepiece. This is powered either by an internal battery or, more often, by external power from the telescope.

Features to look for in a quality reticule eyepiece are variable brightness, blinking capability, diopter adjustment and reticule pattern. In my opinion, variable brightness is not really an option. An illuminated reticule eyepiece which

NGC 281, the so-called Pacman Nebula. (Celestron 14 on a Mountain Instruments MI-250 with a Graflex film holder/Lumicon giant easy guider. Shot on Fuji NPH film. Photo by Rick Thurmond of Mayhill, New Mexico)

has no dimming capability will limit your guiding choices to only the brighter stars. The ability of the illuminated reticule to blink is important for some astrophotographers who find staring at an immobile crosshair pattern tiring to their eyes. Some reticule eyepieces have a diopter adjustment which allows focusing of the reticule pattern. This can be an important consideration for those who choose to wear eyeglasses at the telescope. Finally, choose a reticule pattern that works for you. Bull's-eye patterns or crosshairs are the most popular, especially the dual crosshair pattern which forms a small box in the center of the field of view. Movement of guide stars when framed in such a box are generally easy to detect.

If an off-axis guider is used, it takes the place of the T-adapter. The camera is attached to the threaded side of the off-axis guider body with the T-ring just as it would be if you were using the T-adapter. The off-axis guider uses a "pick-off prism" that is located midway in the projection tube part of the off-axis guider body. An eyepiece barrel sticks out of the body of the off-axis guider at an angle perpendicular to the axis of the telescope. This eyepiece barrel is located over the pick-off prism so that a small fraction of the light coming through the scope can be diverted (picked off) into the eyepiece barrel. The pick-off prism extends down into the light path just enough to intercept some of the light coming through the scope's optics. This has no effect on the image since this light is so far off-axis that it would never get to the film plane of the camera anyway. A reticule eyepiece is inserted into the eyepiece barrel of the off-axis guider and through this you monitor the position of a guide star and make appropriate corrections to the scope drive if tracking errors are detected.

Unlike tripod mounted and piggyback astrophotography you cannot just focus a telescope on infinity, as you could with your camera. You must focus through the scope. The best way to do this is with the camera mounted on the telescope and a clear focusing screen in the camera. Pick a second magnitude star and make it as much a pinpoint as possible. The star may appear as a pinpoint through a range of the focusing knob's rotation. Target the middle of that range.

Once you center the object in the camera's viewfinder, choose a nearby guide star. If you have an off-axis guider, rotate it while looking through the reticule until a suitable star comes into view. If no suitable guide star is near the object you may have to move the object away from the center of the field of view of the camera to help you locate one. Guiding is crucial, and the key to a good shot. All great prime focus shots are guided either manually using a reticule eyepiece or the guiding is done automatically with a CCD autoguider.

Eyepiece projection photography

With eyepiece projection the image is projected onto the film plane through an eyepiece in the telescope. It is used mainly for high powered shots of the Moon and planets. The crucial piece of equipment is the eyepiece projection holder.

An eyepiece projection unit. The eyepiece slips in and is held securely by the thumb screw. One end screws into a T-ring and the other slips into the focuser, just like an eyepiece. (Photo by the author)

Because of the high magnifications used, eyepiece projection is the most challenging of all types of astrophotography. This is less true when a CCD camera is used, as very short exposures can be made and then electronically "stacked." The optics must be of high quality and the telescope must be vibration free and in perfect focus. Excellent seeing is the final requirement. Because of the possibility of error, only short exposures (no more than 30 seconds) should be taken.

All-sky astrophotography

If you can afford it, an 8 mm or 6 mm camera lens will provide some truly spectacular images of the entire sky. You can use such a setup either tripod mounted or piggybacked on a telescope. Unfortunately, such lenses cost a lot of money.

NGC 891. (In b/w, Celestron 14 at f/7, 1 hour 30 minute exposure on hypered TP 2415, guided by ST-4 Sky Tracker, 21 Oct 1995, Sierra Vista, Arizona. Image by David Healy)

A mirror such as this (from the Sandia Fireball Network) will allow you to capture all-sky images with your 35 mm camera. (Photo by Jim Gamble of El Paso, Texas)

The alternative is to use a spherical mirror above which a camera is suspended. A spherical mirror with a diameter 1.4 times that of its radius of curvature will reflect the entire sky. If the sphere segment is any smaller, less than 180° of sky will be reflected toward the camera.

All-sky camera setups have been built by many amateur astrophotographers. The most common method is to attach the camera to a mounting plate on an assembly having three or four support struts. This, then, is suspended over the mirror. A single strut holding the camera/mounting plate is sometimes used. In such cases, the strut is made of thicker metal.

Considerations in all-sky setups are the correct distance between camera and mirror, mounting hardware painted black, and focus. Most amateurs make the support plate round so that the occulted spot in the center of the image matches the shape of the entire image (round).

Focusing an all-sky camera is made much easier if it is done in the daytime. A sky full of broken clouds (rather than a totally clear one) is best for focusing. Once you have the camera focused, secure the focusing ring so that it does not move (generally with tape). You will have to make a compromise in focus, as points on the spherical mirror's surface are at different distances from the lens. Settle on a focus where most of the image is reasonably sharp. Additional sharpness can be obtained by stopping the camera lens down one or two f/stops.

Tip: Use a very fast film for all-sky photography. To use this system, shoot a test roll of film. Aim the camera assembly straight up and take many different exposures at several different f/stops. Make certain to record the exposure time and f/stop for each image.

An all-sky camera apparatus can also be adapted to an equatorial mount for guided photography. The easiest method is to simply secure it to the front of an upward-pointing Schmidt–Cassegrain telescope. One assembly I have seen has the circular base fitted over the capped end of a Celestron 8. For exposures up to half an hour in length, balance is no problem because the assembly is always pointing straight up. The exposure begins with the telescope aimed slightly east of zenith and ends slightly west of zenith. Because of the extreme wide-angle view, manual guiding is not needed with a reasonably well polar-aligned mount.

Cassiopeia. Not quite all-sky, but at least an entire constellation. (Kodak Elite Chrome 200, pushed 1 stop. Nikon F2 with a 105 mm Nikkor lens. Photo by Ulrich Beinert of Kronberg, Germany)

Two things to keep in mind about all-sky astrophotography are the coverage of the lens/mirror (you are in the field of view) and that the final image on film is backward. This is no problem if you are shooting slides but if print film is being used, be sure to print (or, if a photo lab is used, ask that the negative be printed) backward.

3.3 Digital and video cameras

The ultimate digital camera? A Canon D60 with a 300 mm f/1.4 IS (image stabilization) lens + 2x teleconverter. (Image by Charles Manske of Watsonville, California)

How about this as an example of a digital image? The diamond ring. Total solar eclipse of 21 Jun 2001. (Imaged from Chisamba, Zambia, with a Canon D30 by Charles Manske of Watsonville, California)

Digital astrophotography

Note: This chapter deals with consumer-type digital cameras and not with astro-specific CCD setups. Those are covered under "The CCD" chapter.

As a note about taking and displaying your images, allow me to use an example. Charley Manske is a friend of mine in California. In addition to being an amateur astronomer, he is also one of the finest digital photographers and digital videographers I know. We could all learn a lot from Charley's techniques. One of the things I've always marveled at is that Charley shoots a lot of images. Then he selects the few really, really great pictures and shares those. The others are seen by nobody but him. I guess this is proof of the old adage, "film is cheap (and digital images are cheaper)." As a beginning digital astrophotographer, it may take many tries before you have a really excellent shot. That's OK. Just think of Charley. You are not alone.

Charley also has a justification of sorts regarding the expensive model of digital camera that he chose. He says that he used to shoot a lot of film. One day, he sat down and added up the cost of just 100 rolls of film. It was quite a number! Add to that the processing (almost always equal to or a little greater than the film cost) plus the scanning technology needed to get the images into the computer and you can see how moving to digital is a good investment. Once you have the initial expensive equipment, digital media is "free."

The first (and as of this writing, the current) generation of digital cameras had many limitations for astrophotographers. Foremost among them was the fact that exposures were limited to only a few seconds at most. Recently, some models were featuring longer exposure times and even what would correspond to the old "bulb" setting. Another serious drawback was that early digital cameras did not have interchangeable lenses. Newer cameras can be adapted for telescope use in the same manner as standard 35 mm film cameras, by using a T-adapter or by a special adapter that attaches via threads right to the front of the camera's lens. Camera noise can be yet another problem, causing hot pixels and stray optical "noise" on the image. This can be easily dealt with during image processing.

If the object you are imaging requires an exposure longer than about 1/2 second, you will want to take a dark frame shot. This is used later to subtract out the hot pixels and noise produced by your camera during long exposures. Any dark frame must be taken at the same exposure and camera settings that you used for imaging. Take one before and after imaging each object, or if you change exposure time. A dark frame may be taken by covering the objective of the telescope or by removing the camera from the telescope and covering the camera lens (or opening, if no lens was used). In either case, make certain no stray light can get to the chip. That way, the only "image" you will get will be camera noise.

Apart from the disadvantages, a digital camera offers numerous advantages. The ultimate advantage over film is what might be called "instant gratification." You can see your images right after you shoot them. This is a huge advantage. Problems with focus, framing, etc., can be instantly dealt with. Digital also offers the advantage over film of being able to more easily combine images which can help bring out

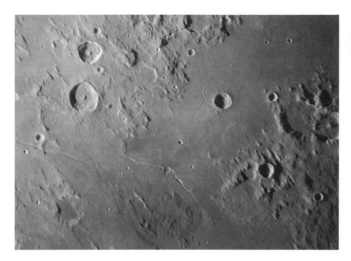

Hyginus-Rupes Recta Region. (Imaged by Arpad Kovacsy with a Nikon CoolPix 950 through an AP 155 EDT refractor)

finer detail. You can actually do this with film, but the negatives or slides must first be scanned into your computer. Finally, as noted above, if you do a lot of imaging, the cost savings over time place digital imaging far ahead of film.

Video astrophotography

Amateur astronomers are making great progress with the use of video cameras. Each day it seems that fainter and fainter objects are being imaged on video. At this writing, however, it must be stated that video astronomy is confined to the brighter celestial objects – the Sun, the Moon, brighter planets, and certain double stars. Some images of nebulae have been produced but not many.

The best video work that I have seen has resulted from this technology being used to record eclipses of the Sun and Moon. For a truly superb example, actually combining still digital images with digital video, see Charley Manske's Rendezvous, on the internet at

http://www.jivamedia.com/SolarEclipse2001/solar-eclipse-2001.html

It's the second image down from the top. Just select the transfer speed that best suits your internet connection.

Three types of video astrophotography are possible. The first type is "non-telescope" video. This is popular with amateurs who have long focal length lenses and who are interested in wider angle shots. Eclipses are a perfect target for this type of video astrophotography. Generally, after polar aligning, the telescope's optical tube assembly is removed from the mount and replaced with the camera. The system is rebalanced and you're ready to shoot! A couple of notes. (1) If you are shooting a solar eclipse, purchase (or make) a solar filter for your camera lens to use prior to totality. The Baader astrofilm is excellent for this. (2) It is critical to re-balance the assembly carefully, as the mount will have been balanced for the telescope. If this is not done, guiding errors will occur due to the mount's drive mechanism.

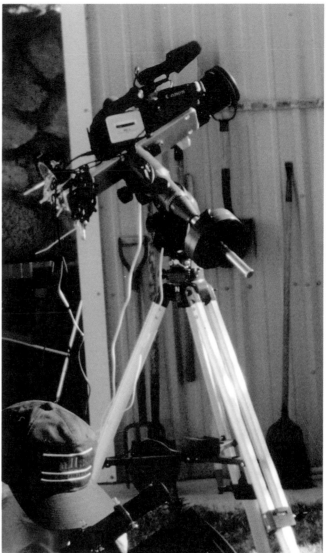

Video camera setup used to record the transit of Mercury, 15 Nov 1999. (Photo by the author)

The second type of video astrophotography involves the afocal method: taking a video camera and holding it up to the eyepiece of the telescope. Sound too simple? I have seen some very good results from this technique. If at all possible, use a camera which allows you to change the exposure manually. Experiment with all exposures on all objects you shoot. Also try different eyepiece/zoom combinations. Almost all video cameras have a zoom feature. Try a low-power eyepiece with the camera lens zoomed to maximum, a high-power eyepiece with the camera lens zoomed to minimum, and everything in-between. If you are shooting video in this fashion for any length of time, the best technique is simply to hand-hold the camera. Trying to position the camera on a tripod is problematic because the Earth is rotating. Whether the telescope is driven or you are moving it manually, as with a Dobsonian mount, continual repositioning of the camera will be necessary. Hand-holding the camera is just easier.

The third type of video astrophotography uses a small camera directly connected to your telescope. I have such a setup for use with my 100 mm refractor. You can imagine that on such a small telescope the camera must be pretty small. I use a PC-23C black-and-white security camera which I purchased from Supercircuits. Find them on the internet at http://www.supercircuits.com

For this setup I had to use a C-to-T connector. That is, I had to adapt the C-connector of the camera to a T-adapter which allowed me to use the camera in place of an eyepiece. I have obtained good video of the Sun, Moon, Jupiter, Saturn, and a number of bright double stars. I have gone so far as to have run video and telescope control lines from my backyard observatory to my entertainment center. Now, after I set up the video and center an object (for example, the Moon) outside, I can go indoors and switch back and forth between a television program and the Moon. The remote telescope controls allow me to view different areas of the lunar surface, as this setup has a smaller field of view than the entire Moon.

Supercircuits PC-23C monochrome video camera. Small package, high resolution. (Photo by the author)

Please be aware that this camera, while a good initial test bed, is not the "be all and end all" of video. Some day I will upgrade to a much more expensive megapixel unit which will definitely improve my video images.

It is worth noting that I called this section "Video astrophotography," because that's how these images are captured. It is the practice of some amateurs, however, to use video to produce better still images. The theory is that, within the large number of frames captured on video, some small number will be exposed during moments of excellent seeing. This technique works. I must add, however, that it does take time to scan the video for the best images, so be prepared for a lot of "away from the telescope" review. Once the best quality frames of video are identified, they are combined, or stacked, electronically using software such as PhotoShop.

Video capture of images may be done by hooking up the video camera directly to the computer. This is the best way. You can manipulate images, stack them, etc., with very little problem. If, however, you want to record on a tape medium

The double star α Herculis, 24 Jul 2001. (Imaged by Arpad Kovacsy with a Nikon CoolPix 950 through a Celestron CR-150 HD 6" refractor)

first, you will need a video capture board such as Play, Inc.'s Snappy, Dazzle Multimedia's Hollywood DV-Bridge, or ATI Tech's TV Wonder (and probably quite a few more by the time you read this).

However you choose to capture video images of celestial objects, a few common-sense rules apply. Since your telescope functions as your "lens," use the finest quality telescope you can buy or borrow. Next, use the most sensitive video camera you can buy or borrow. The more sensitive the camera, the higher (fainter) your limiting magnitude will be. Record onto a high-resolution media. This will set the final resolution. 8 mm and VHS should be your last choices. Hi-8 and S-VHS are much better and digital-8 is even better yet. Digital capture straight to a computer is best. If you don't have the resources to use a high-end system, all is not lost. In fact, valuable experience can be gained until you can upgrade. A number of video imagers report good results even using VHS tape. Make sure to use the fastest tape speed.

If you are imaging in black-and-white, try a yellow, orange or even a red color filter to help reduce the effects of the Earth's atmosphere. For most video cameras using CCD chips, such filtering will improve the image without significant light loss. This is because most chips of this type have very good sensitivity to red light. To obtain a bigger image, use a long focal ratio. Employ a good-quality Barlow lens if necessary. Many video astrophotographers work at f/30 or even longer.

Nikon Coolpix 950 with adapters necessary for astrophotography. (Photo by Arpad Kovacsy, Mt. Vernon, Virginia)

3.4 The CCD

The revolution is here. Truly this device, the CCD, has transformed amateur astronomy as it had been done until a decade or so ago. These devices, and the software associated with them, have enabled amateurs to image celestial objects from areas that would have never permitted photography. So much information exists about CCDs that I could fill this book several times over. This chapter, therefore, will only be a short overview.

How a CCD works

CCD stands for charge-coupled device, a rectangular, solid-state electronic chip made primarily of silicon. During manufacture, the chip has been divided into many small arrays of light-sensitive cells which individually form one small part of the entire picture. These "picture elements" are called pixels.

A Santa Barbara Instrument Group ST-10XME CCD camera (left side) is the newest addition to the Everstar Observatory in Olathe, Kansas. Note also the CCD on the guide scope, used for autoguiding. (Photo by Mark Abraham)

Delving a bit into the technical side, each pixel works because of the photoelectric effect. This occurs in certain materials, causing them to release electrons when struck by a photon of light. The electrons released by the pixel accumulate as light continues to strike the pixel during the exposure time. Once the exposure is made, the pixels of the CCD are dumped, or unloaded, to a computer with software able to turn the images into a picture (for viewing or storing as a file) or into a raw count (for photometry).

To obtain the highest resolution, astronomical CCD cameras up to this point have been monochrome. (However, change seems to be coming.) The reason monochrome chips are better is that color CCDs actually use groups of three pixels (in a triangular arrangement next to one another), one with a red filter, another with a green filter, and a third with a blue filter (a system known as RGB) to produce the color information for each point. Data from these three pixels are then merged electronically when the chip is dumped.

CCD cameras used by amateurs for astronomy are able to produce color images, although not directly. To obtain a color image, three images of a celestial object must be taken (through red, green, blue filters), which allows the CCD chip as a whole to mimic the action of a color chip. Once the three images are obtained, they are combined by a computer with the appropriate software.

Many cameras other than astronomical ones now use CCDs. Video camcorders and digital still cameras are good examples. Most camcorders have CCDs with arrays a few hundred pixels on a side. The best digital cameras may have chips with a thousand or so on a side. (Most are in a rectangular format, such as 640 × 480, etc.) Recently, digital camera

IC 2118, the Witch Head Nebula in Eridanus. (300 mm Nikon Lens attached to a Santa Barbara Instrument Group ST10 CCD camera. L(B)RGB = 120:10:10:20 minutes. Image by Robert Gendler, Avon, Connecticut)

NGC 1850 in Dorado. (LRGB: L=R+G+B, 600 s R, 600 s G, 900 s B through an ST-7E on a Celestron 11. Image by Steven Juchnowski, Balliang East in the State of Victoria, Australia)

NGC 3324 in Carina. (RGB: 600 s R, 600 s G, 900 s B through an SBIG ST-7E CCD on a Celestron 11. Image by Steven Juchnowski, Balliang East in the State of Victoria, Australia)

manufacturers have started advertising megapixel cameras. The highest as of this writing was a six megapixel camera. A CCD chip measuring 3000 pixels by 2000 pixels would accomplish this. CCD chips are not large. This means that each pixel measures from about 6–30 microns in diameter.

Types of chips

There are two types of chips which form the heart of the CCD: frontside illuminated and backside illuminated. These are sometimes referred to as thick and thin chips. Light striking the pixels of frontside illuminated (thick) chips must pass through a number of layers of the CCD first. Not so with backside illuminated chips. With these, the chip has been "thinned" so that the light strikes the pixels directly. Backside illuminated chips are usually more sensitive to light of all wavelengths, especially blue light.

Quantum efficiency

This is a measure of the sensitivity of the CCD chip. Unfortunately, every photon of light falling onto a CCD will not be registered. The percentage of light which is actually recorded by the chip is called its quantum efficiency (QE). Thin chips generally have a higher QE than thick chips.

Dark frame

Because CCD chips are electronic in nature, they generate heat. Unfortunately, heat (as well as light) will cause electrons to be generated by the pixels. Most CCDs are cooled in ways which vary with the manufacturer, but no matter how much they are cooled, some of this heat has an effect. This phenomenon is known as dark current. Fortunately, there is a solution, and that is to take a dark frame.

A dark frame (or dark subtraction) is taken when there is no light falling on the CCD chip. Covering the CCD camera or capping the objective of the telescope will usually make it dark enough. The dark frame must be of the same duration (exposure time) as the image, and should be made immediately after any image, while the CCD is at the same temperature as when the image was taken. This will duplicate the production of electrons due to heat in the original exposure. Finally, using software, the dark frame is subtracted from the image.

Bias frame

Another correction factor for CCD imagers is what is known as a bias frame. This is taken prior to the dark frame (but usually explained after dark frame). In a bias image, the CCD is cleared, and immediately read out without opening the shutter. This is, in effect, a zero-time dark frame. It represents the inherent noise in the electronics and is not zero. Because there is even noise associated with reading the bias frame, a number of bias frames are taken and averaged together to reduce the read noise.

Flat field

Do we have an image yet? Almost. One more correction factor needs to be applied. Since a CCD chip is composed of up to millions of individual pixels, it is not realistic to assume that every pixel is the same. Some may be more sensitive to light (or heat), others a little less so. We must obtain a uniform value for all pixels and this is done by taking a flat field.

A flat field basically averages out all the pixel sensitivities. Flat fields are taken by pointing the telescope at an object with very even illumination. Some CCD imagers use the twilight sky as their target. Others use the interior of their observatories or a large piece of grayish paper board. Whatever is used, the image shows the problems caused by everything – dust, vignetting in the optical system,

NGC 6960 and the star 52 Cygni. (Celestron Fastar 8 at f/1.95, PixCel 237 CCD camera. This is a 300-second image consisting of five 60-second exposures "Track and Accumulate" using the AP900 mount with Periodic Error correction engaged. Unguided at −20.42°C. North is up. (Image by Chris Anderson of Kentucky)

differences in pixel sensitivity, etc. The imaging software then goes to work, comparing each pixel with the flat field and producing a much better image.

The combined process of first subtracting the dark frame and then adjusting by using the flat field is called image calibration. It is almost automatic for CCD imagers and produces higher-quality images.

CCD guiders

I remember the "old days" of astrophotography. Telescope drives were not quite good enough to totally compensate for the rotation of the Earth. The solution was for the observer to sit at the eyepiece of a guide telescope, perhaps for hours, making miniscule corrections in right ascension and/or declination. Nobody thought that was fun or looked forward to it in the least.

Today, an observer can use a guiding CCD, attached to (what used to be) the guide telescope of the optical system. Once the guide star is located and targeted, the controlling software of the CCD will keep the star centered on a certain number of pixels (this depends on the star's apparent size on the chip). A guiding CCD used with a separate telescope can

even allow film astrophotos to be made without the need for observer guiding.

Some manufacturers have eliminated the need for a second CCD camera in one of two ways. (1) Part of the chip is used as a guiding CCD; or (2) a second, much smaller CCD chip, in close proximity to the first, is built into the mounting for the camera and is used as the guiding CCD.

The above description assumes a setup where the telescope's motions are controlled by a computer. It is this computer into which the guiding CCD's information is sent.

M16 and M17. (Chris Woodruff, of Valencia, California used an SBIG ST7e NABG camera with the SBIG CLA Nikon lens adapter on a Vixen GP-DX EQ mount. 28 mm Nikon lens, 10-minute exposure. A set of 2" RGB filters were manually switched between shots)

Stacking images

Since CCDs work by accumulating light and recording the electric current given off by the photoelectric effect, not all the light need be gathered in the same exposure. Multiple exposures can be made and then stacked with the

A stacked image. This is part of the vast Eta Carinae Nebula; the Keyhole Nebula is the dark area above the star Eta Carinae. This image was taken using a hydrogen alpha filter. A total of 28 1-minute images were taken, and the best 20 were stacked to produce the final image. False color was then applied. (Image by Steven Juchnowski, Balliang East in the State of Victoria, Australia)

appropriate software. The software will also align the images – a very important feature!

There are two important reasons amateur astronomers stack images. The first is to overcome the effects of seeing. By taking a large number of short exposures, the best images can be selected and combined into a single image of greater quality than one exposure of equal time.

The second reason to stack images is to overcome the effects of a not-quite-aligned telescope drive. A drive error will be much more apparent in a 50-second exposure than in a 1-second exposure. Taking 50 1-second exposures and stacking them produces, in this example, a drive error only 2% that which would be seen in the longer exposure.

Pixel scale

The formula to determine the pixel scale of your telescope/CCD combination is

$$x = (205 \times s)/f$$

Where s = pixel size in microns, f = focal length in millimetres, and x = pixel scale.

If your telescope's focal length is measured in inches, the formula is

$$x = (8.2 \times s)/f$$

Binning

The term binning refers to the combining of adjacent pixels with the purpose of capturing more light. Pixels are binned 2×2 or 3×3. In the first case, four times the amount of light is captured and in the second case, nine times the amount. This reduces exposure times by those amounts. Unfortunately, binning reduces the scale of the telescope/CCD system. For example, if your unbinned scale is 3 arcseconds per pixel, using 2×2 binning will increase this to 6 arcseconds/pixel.

Binning may prove advantageous to certain systems. If your pixel scale is 0.5 arcseconds, for instance, and your typical image size is 2 arcseconds, you will gain by binning. Binning has been generally used for imaging large celestial objects. It has, however, also been used for asteroid astrometry to allow the right pixel scale to be achieved.

Drift scanning

A relatively new method (as of this writing) of doing asteroid astrometry is by a technique called drift scanning. Sometimes referred to as time-delay-and-integration (TDI) mode, drift scanning utilizes a motionless telescope/CCD setup, pointed at a particular spot on the sky. Attempting to image a celestial object this way would result in a hopeless blurring of the image. But drift scanning counts on the apparent motion of the stars.

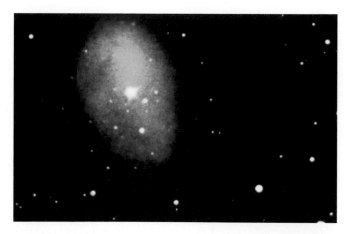

NGC 1360, a planetary nebula in the southern constellation Fornax. (LRGB: 47 m L with an AO-7, 10 m R, 11 m G, 23 m B (R,G,B 2 x 2 binned) with an ST-7 through a Celestron 11. Image by Steven Juchnowski, Balliang East in the State of Victoria, Australia)

The CCD chips of setups in drift-scan mode are aligned so that the Earth's rotation causes stars (and other objects) to move exactly along the columns of pixels comprising the chips. Using imaging software containing options for drift scanning, pixels within columns on the chip are programmed to read out at the same rate at which the stars appear to be moving, called the sidereal rate. Thus, the light from a star (or asteroid) falls on different pixels within the column, but are added together to form a single image. The result is a long, thin image in which point sources appear as points, not streaks. Many new asteroids have been discovered by this method.

The advantage of drift scanning is in the cost of the setup and the types of telescopes which may be used. With drift scanning, only the optical tube is required – no mount! Some amateur astronomers using drift scanning methods have literally bolted telescopes to steel beams. And, for the first time, large Dobsonian-mounted telescopes may be used for asteroid work because the mount plays no part at all in acquiring the image. Be certain to check with your CCD camera manufacturer about the option of drift scanning within their software. As of this writing, some do not provide it.

Integration time

The amount of time that it takes for a star to complete its passage over an entire column is known as the integration time. Of course, the longer the integration time the more electrons can be built up and the fainter the limiting magnitude. The integration time depends on three factors: number of pixels per column, pixel scale, and distance from the celestial equator of the scan.

At the celestial equator the sky drifts at a rate of 15 arcseconds per second. Changing the declination of your scan changes this rate by the cosine of the declination. Thus at a declination of either 30° N or 30° S, the sky is drifting at

The components of a color CCD image of Jupiter. (Images taken and processed by Ed Grafton, Houston, Texas, 22 Mar 2002; Ed used a Celestron 14 and an ST5c CCD camera)

approximately 87% the rate at the equator, or about 13″/second. Let's look at an example.

Imagine a telescope/CCD system with a 2″/pixel scale and in which the CCD chip has 700 pixels per row. The integration time, if the system is pointed at the celestial equator (declination = 0°) will be

$$((2''/\text{pixel}) \times 700 \text{ pixels})/(15''/\text{second})/\cos(0°)$$

or 93 seconds, approximately.

If the same system is pointed at declination 30° (north or south), the equation becomes

$$((2''/\text{pixel}) \times 700 \text{ pixels})/(15''/\text{second})/\cos(30°)$$

or 108 seconds, approximately.

Resources

Hardware

Manufacturer	Internet Address
Apogee Instruments	http://www.apogee-ccd.com
Finger Lakes Instrumentation	http://www.fli-cam.com
Meade Instruments	http://www.meade.com
PixelVision	http://www.pv-inc.com
Santa Barbara Instrument Group	http://www.sbig.com
Starlight Xpress	http://www.starlightxpress-usa.com [a]

[a] Part of the Adirondack Video Astronomy website.

Software

Product	Manufacturer	Internet Address
Maxim DL	Diffraction Limited	http://www.cyanogen.com
MIRA Pro & MIRA AP	Axiom Research	http://www.axres.com
PictorView	Meade Instruments	http://www.meade.com
QMiPS32	Christian Buil	http://www.astrosurf.com/qm32
Stella Image3	AstroArts	http://www.astroarts.co.jp/index.html
RegiStar	Auriga Imaging	http://aurigaimaging.com

Further Reading

The Art & Science of CCD Astronomy by David Ratledge (editor) (Springer-Verlag, New York, 1997)

Handbook of CCD Astronomy by Steve B. Howell (Cambridge University Press, 2000.)

A Practical Guide to CCD Astronomy by Patrick Martinez and Alain Klotz (Cambridge University Press, 1997)

3.5 Photometry

Astronomical photometry is the measure and subsequent study of light from celestial objects. It differs from spectroscopy in that the light is not broken up into its component wavelengths. Sometimes, however, the light is filtered and only the portion of the light passing through the filter is collected.

When I was actively doing photometry (lo, those many years ago!), there were no CCDs, no autoguiders, and no personal computers, although their advent was not far off. The light was collected using a photoelectric photometer and mechanically graphed onto paper. It was a slow process but reasonably accurate.

Today, the story is quite different. CCDs have entered the playing field, making the process simple. Computers abound, allowing reductions of measurements in seconds. But the question remains: are you willing to enter the relatively new area (to amateur astronomers) of astronomical photometry?

The only requirement is that your software must have a photometry option. This means that, after you have taken an image the software is able to derive the magnitude of the star by comparing it to known standards in the same field. The software must also be able to measure and subtract the brightness of the sky near the variable, so that this value is not added in with the star's magnitude. The software you choose may be able to plot the magnitude of the star versus time, or you may have to employ a graphing program for that (or do it by hand, if the number of points is not too great).

Asteroid photometry

For the vast majority of asteroids there is no real photometry available. For these, the only data that exist are sporadic magnitudes reported by astrometrists. If you are an amateur astronomer wishing you could "find a niche" to make a contribution, consider asteroid photometry.

Stellar photometry

I said this in another section and I shall repeat it here: the days of obtaining magnitudes of variable stars through visual observations are nearly at an end. The CCD is faster, more accurate, and tireless. And amateurs who target this as their chosen area of study will find a limitless supply of objects for whom valuable data may be retrieved. For much more about this, see the chapter "Variable stars".

Further reading

Astronomical Photometry by A. A. Henden and R. H. Kaitchuck (Willmann-Bell, Inc., Richmond, VA, 1990)

Introduction to Astronomical Photometry by Edwin Budding (Cambridge University Press, 1993)

Photoelectric Photometry of Variable Stars by Douglas S. Hall and Russell M. Genet (Willmann-Bell, Inc., Richmond, VA, 1988)

3.6 Spectroscopy

Once upon a time, when an astronomer examined the light from a celestial object, he used a triangular glass prism. Isaac Newton, around 1666, was the first to realize that the colors produced when white light is passed through a prism are a property of the light itself, rather than something introduced by the glass. The prism separated the light into its component colors, called a spectrum. The study of the light from celestial objects, therefore, is called spectroscopy. During the twentieth century, the diffraction grating, a flat mirror with a large number of parallel lines scribed onto its surface, replaced the prism. The device used for visual observation is the spectroscope and the spectrograph is used for imaging.

The English chemist William Wollaston (1766–1828) first observed dark lines in the solar spectrum, in 1802. He believed that these were gaps separating the colors of sunlight. In 1817, the German optician Joseph Fraunhofer (1787–1826) independently discovered the lines and in 1823, found dark lines in the spectra of stars.

It was soon learned that each dark line indicated the presence of a particular element in the upper layers of the Sun or a star. Then the process by which the lines were created was determined. As it turns out, the atmosphere of a star absorbs wavelengths of light created in its interior. The dark lines came to be known as absorption lines. The specific wavelengths absorbed depend on the temperature of the star. In our Sun alone, spectroscopy has revealed the presence of over 60 chemical elements.

It was natural, therefore, to classify stars by their temperature, which was determined by the presence of certain lines in their spectra. The system (or spectral sequence) we use today was developed in 1943 by the American astronomers William W. Morgan (1906–1994) and Philip C. Keenan (1908–2000), who was one of my professors when I attended the Ohio State University. The MK spectral sequence uses the letters O, B, A, F, G, K, and M to classify stars, as shown in the following table.

Star classification

Class	Temperature (K)	Absorption lines	Examples
O	>28 000	Relatively few absorption lines except for those of highly ionized atoms; weak hydrogen lines	ζ Pup Mintaka
B	10 000–28 000	Neutral helium and ionized silicon, oxygen and magnesium; hydrogen lines	Rigel Spica
A	7500–10 000	Strong hydrogen lines; singly ionized calcium and magnesium; weak neutral metals	Sirius Vega
F	6000–7500	Weaker hydrogen lines; singly ionized calcium, chromium and iron; neutral metals	Canopus Procyon
G	5000–6000	Ionized calcium strongest; lines of ionized and neutral metals and CH are strong; weak hydrogen lines	Capella the Sun
K	3500–5000	Neutral metal lines strongest	Arcturus Aldebaran
M	<3500	Neutral metals strong and molecular bands such as titanium oxide	Betelgeuse Antares
Special Types			
R	Spectra similar to K stars except for carbon and cyanogen molecular bands; a carbon star		
N	Spectra similar to M stars except that titanium oxide is weak and carbon, cyanogen and CH bands are strong; a carbon star		
S	Similar to M stars with the addition of zirconium oxide and lanthanum oxide bands; titanium oxide may be strong or weak		

To further define the stellar types, each spectral type was divided into tenths, with 0 being the hottest and 9 the coolest. Thus, A0 is next to B9 and only slightly cooler. Morgan and Keenan also combined luminosity classes with the stellar types, as shown in the following table.

Luminosity class of star types

Luminosity class	Star type
Ia	Bright supergiants
Ib	Less luminous supergiants
II	Bright giants
III	Giants
IV	Subgiants
V	Dwarfs (main sequence stars)
VI	Subdwarfs
VII	White dwarfs

Thus, the Sun is classified as MK type G2V, Deneb is spectral class A2I, and Caph (β Cas) is type F2III.

Throughout the nineteenth century, the resolution of the lines improved until astronomers could literally tell what the star was made of. But in addition to chemical composition and temperature, spectroscopy tells us much more about stars.

What we can learn from spectroscopy

Radial velocity

Defined as motion to or away from us, radial velocity is measured by the Doppler shift. Christian Andreas Doppler (1803–1853) discovered that if a source of light is moving towards or away from us, the colors or wavelengths of its spectrum lines are shifted by an amount which depends on the speed. A blue shift means the object is approaching us; a red shift means it is receding.

Rotation

If the spectrum is shifted to the blue and to the red, the star must be rotating. The speed of rotation can easily be worked out.

Magnetic field

If each of the absorption lines of a star are split into two or more lines, the cause is due to the Zeeman effect, which splits the energy levels of atoms.

Shells of gas surrounding stars

Sometimes bright lines are seen in the spectra of hot stars. These lines are called emission lines and are produced by material that has been ejected from the star and has started re-emitting radiation that it has absorbed from the star. A type of star demonstrating this process is the Wolf–Rayet class.

Spectroscopy by amateur astronomers

As a percentage, the number of amateur astronomers doing spectroscopy is quite low. This is partly due to the cost involved and partly due to the few opportunities to make a real contribution, as with asteroid astrometry, for example. Spectroscopy can be rewarding, but there is a lot to learn. Here are a few representative websites.

In England

http://www.astroman.fsnet.co.uk/spectro.htm
http://www.astroman.fsnet.co.uk/

In France

http://www.astrosurf.com/buil
http://valerie.desnoux.free.fr/vspec/

In Germany

http://pollmann.ernst.org/

In the US

http://sunmil1.uml.edu/eyes/veio/index.html
http://members.cts.com/cafe/m/mais/

CHAPTER FOUR

Observing tips

I consider this part to be a key to helping us all become better observers, the main goal of this book. As there are observing tips and techniques included in most of the chapters, this becomes the clearinghouse for all those that were not mentioned. Since there is no "correct" way to present such a group, I have chosen to alphabetize them. Some are short tips while others are longer discussions. A few are tips for beginners and others are quite advanced.

I wish to give a very hearty "thank you" to all the observers who responded to my call for "generalized tips." Each is credited below with their direct quote. To all those anonymous or forgotten amateurs who have provided tips to me in the past – whether by book, video, internet, or in person – and that I have absorbed without remembering the source, please accept my deepest gratitude.

Aperture masks

David Knisely of Lincoln, Nebraska, has a great idea:
I occasionally use an off-axis variable aperture mask I built for my 250 mm f/5.6 Newtonian to judge double star resolution. This mask provides me with 94 mm, 80 mm, 70 mm, 60 mm, and 50 mm clear apertures. This way, I am able to stop down the scope in well-defined steps to see at which point the division between a pair becomes invisible.

Astigmatism

The best corrective device for astigmatism seems to be hard gas permeable contacts because they force the corneal curvature to match the inner surface of the contact. If your eyes are tolerant to hard contacts, they remove the effect of corneal surface aberrations. However some folks can't tolerate hard contacts. Soft contacts take the shape of the cornea they're placed on, so they will not correct for astigmatism unless they are the toric design with weights to keep proper orientation of the correction. A quality pair of glasses is close or equal to hard contacts, superior to plain soft contacts, and equal to or better than the toric design soft contacts.

Averted vision

Paul Alsing of Poway, California, says that averted vision seems to work best when you place the object either 10 o'clock or 2 o'clock in your field of view.
I have relayed this tip to many hundreds of people over the years, mostly people lining up to view faint stuff in my 500mm Obsession, many newbies. As you know, an inexperienced eye has little chance of seeing, for example, the Horsehead or maybe each member of Stephan's Quintet – no matter how good their eyes are – if they do not employ averted vision. This little tip has made these objects, and many others, magically pop into view. I personally have no doubt that this tip really works.

Brian Skiff of Flagstaff, Arizona, adds:
The easy way to remember is to place the object between your nose and the direction you are looking. Folks seem to vary as to whether looking slightly above or below is best – there may be some normal variation, or it may be part of what you get used to. Being left-eyed, I tend to look left-down for averted vision, but I usually also try left-up.

Avoiding eye fatigue

One of the best things you can do for your eyes is to take short breaks. Simple one-minute eye exercises done every 20 minutes will dramatically reduce eye fatigue. Change focus by glancing at somebody's nearby telescope. Then, lightly cup your eyes with your palms, and relax for 60 seconds. Or simply look away from the eyepiece, and roll your eyes up and down, around and side to side for 20 seconds. Then relax, eyes closed, for 30 more.

Batteries

What more need be said? Take the batteries you know you will need and the batteries you think you won't need.

Battery life

Mark Beuttemeier of Sammamish, Washington, offers the following advice:
If you have a choice between plugging your computer or other accessory into an AC inverter or using 12-volt DC, going with straight DC will help prolong battery life. It could be considerable depending on the efficiency of your inverter and AC/DC converter. Also, consider taking the internal battery out of your laptop while running on an external battery. No need to keep that charging. Another tip: use the power-saver features of your laptop. Shut down the display backlight after a few minutes and the hard disk after a few more. Also, some laptops have a lower-power CPU setting which

slows the processor down a few steps – your applications will run a bit slower but it will use less power. Every little bit helps. One additional tip for the SkyCommander used on StarMaster telescopes. Standby mode will save power while still tracking the encoders, and will still communicate coordinate information to your laptop, thus tracking will keep running.

Caffeine and Dramamine

Jeff Medkeff of Sierra Vista, Arizona, states:

One of Dramamine's side effects is to reduce scotopic (low-light) visual acuity. And caffeine is a real faint-limit killer. It has little effect on the visibility of objects well above the threshold, but one can of a caffinated soft drink seems to detract $0.3 - 0.5$ magnitude (telescopic stellar magnitude) for the average observer. Caffeine is the one to watch out for, because more people are going to have caffeine at an observing session than Dramamine.

Camera focusing

This tip is for tripod-mounted or piggyback astrophotography. If you own one of the newer auto-focus lenses which can go past infinity when manually focused, set it at infinity during the daytime. Then tape the lens with one or two wraps around the barrel so it stays in sharp focus. Manual focus lenses do not have this problem but some astrophotographers tape them anyway, making certain of one less problem. Use a tape which will not leave a residue. Stay away from "duct" tape.

Cold weather: camera

My friend Ray Shubinski, who grew up enduring Michigan winters and now resides in Prestonsburg, Kentucky, has a few tips for cold-weather astrophotographers.

If the night is very cold and very dry, as often happens, short periods in and out of a warm vehicle do not create condensation problems for cameras. However, if you really freeze a camera, you should place it in a zip-lock plastic bag before thawing. By doing this, any condensation will form on the outside of the bag and not on the camera itself. Never try for "that last frame" in cold conditions, especially with a motorized camera. Your film might break and by the time you fix the problem your opportunity could be lost.

Comfort

It is also worth talking about comfort at the eyepiece. I see many observers who use various contortions and gyrations to look through the eyepiece. In particular, the one Jeff Medkeff labeled the "monkey squat." It's pretty hard on the back, and requires that all kinds of muscles stay tense to keep the eye at the eyepiece. I find that when I am seated comfortably at the eyepiece, I see a great deal more than I do while standing.

Contrast

Contrast is the difference in brightness between various parts of a telescopic image. When light is scattered in the fov, for whatever reason, it reduces the difference between the dark and bright areas of the image. Contrast is calculated by this formula:

$$c = (b^2 - b^1) / b^2$$

where c = contrast, and b^1 and b^2 = brightness of each of two areas of the object measured in candlepower/m^2, or some equivalent unit.

An example is the difference in brightnesses of Saturn's various rings. Contrast efficiency of telescopes is very important because a planet's surface or atmosphere is composed of various materials that reflect different levels of sunlight.

If we consider two features on Mars, a light area with a brightness of $400\,\text{cd/m}^2$ and a darker one half as bright:

$$c = (400 - 200) / 400 = 0.5$$

The contrast is thus 50%. But what if we scatter just 10% of the light from the bright area into the dark area?

$$c = (360 - 240) / 360 = 0.33$$

The contrast has dropped to only 33%! Thus, a relatively small amount of scatter causes a big change in image contrast.

Dark adaptation

Dark adaptation is the process by which the eyes increase their sensitivity to low levels of illumination. Rhodopsin (sometimes called visual purple) is the substance in the rod cells responsible for light sensitivity. The degree of dark adaptation increases as the amount of visual purple in the rod cells increases through biochemical reactions. Each person adapts to darkness in varying degrees and at different rates. In a darkened theater, the eye adapts quickly to the prevailing level of illumination. Compared to the light level of a moonless night, this level is high.

In the first 30 minutes the eye's sensitivity increases 10 000 fold, with little gain after that. Dark adaptation approaches its maximum level in approximately 30 to 45 minutes under minimal lighting conditions. If the eyes are then exposed to a bright light, their sensitivity is temporarily impaired, the degree of which depends on the intensity and duration of the exposure. Brief flashes from high-intensity strobe lights have been shown to have little effect on night vision. This is because the pulses of energy are of such short duration (milliseconds). Durations of bright light lasting one second or longer, however, can seriously impair night vision.

Rod cells are much more sensitive to blue light and are least affected by the wavelength of a dim red light. Red lights do not significantly impair night vision if proper techniques are used. To minimize the effect of red light on night vision, the intensity should be adjusted to the lowest usable level and the illuminated object should only be viewed for a short time.

Eye miscellany

We have decreased resolution at night, and little color vision as well. Resolution diminishes at night for multiple reasons: reduced numbers of retinal cells firing; the color shift in sensitivity vs. the focusing ability of the eye; chromatic aberration of the eye; variable transparency of the lens and humors of the eye; etc. In dim light, the spectral sensitivity of rods peaks at about 505 nm, and in bright light the peak of the cones is about 560 nm. Cones outnumber rods only in the center of the retina, which is the area of greatest density. It is also the area most heavily used by daytime, direct, vision. But rods do fire during the day or we would have no peripheral vision at all.

Make your "pushing the telescope's limit" observations no earlier than midnight, when your eyes will be at their greatest sensitivity.

Eye patch

A very good tip comes to us from Arild Moland, of Oslo, Norway:

> To ease dark adaptation, wear an eye patch over your observing eye while setting up. Put it on as long before the start of your observing session as possible and you will be awarded with a fully dark-adapted eye right at the beginning of your session. Move the patch to your non-observing eye while at the eyepiece so that you may keep it open while observing. This relieves the strain on your eye muscles and improves observing.

Jeff Medkeff, who does his visual observing under pristine skies in Sierra Vista, Arizona, adds:

> As for experimental data, when I was observing faint stars just after I moved to Arizona, I frequently used an eye patch on the right (observing) eye while checking charts with the left. I rarely recorded more than a 0.1 magnitude difference with the right eye after going back to the eyepiece, and roughly half the time the difference was positive (toward fainter stars).

Faint objects, viewing

Paul Alsing of Poway, California, offers this tip.

> This is pretty widely known and used, but mostly with Dobsonian-mounted telescopes. That is, MOVE THE

One of the best accessories to prevent eye fatigue is a simple eyepatch. (Photo by the author)

TELESCOPE when looking for really faint stuff. With my old Celestron 8, I used to "tap the tube," and I'm sure that still applies, but with the 500 mm Obsession, the first thing I tell folks is, "Grab the front and move the scope a little." More often than not they say something like "Oh, now I see it."

Field fractions

To estimate object size, Brian Skiff recommends using field-fractions. Field fraction may be defined as the ratio of the width of the object being observed to the width of the entire fov. In Brian's words:

> I would, however, not simply estimate by eye the relative fraction (of the fov) covered by an object, but instead actually move the telescope in steps, starting with one edge of the object at the field-stop, then move the telescope successive object-diameters until you get at least halfway across the field. Now that most folks are using wide-apparent-field eyepieces, I found that the fraction estimated by simple inspection was systematically larger than doing the spanning bit. The cluster looks about 1/4 the field diameter, but when you actually do the stepping across, it turns out to be only 1/6 or 1/7 the diameter. You'll see this readily when you first move the object to the field stop: one of those "oh, this is really quite small" reactions.

Finding the sun

You've placed your solar filter on the front of your telescope and want to point it at the Sun, but you're not certain how to get the Sun in the field of view. Robert Kuberek, of Valencia, California, uses the "smallest shadow" method. He inserts the longest focal length eyepiece, then loosens both motion locks on his drive and moves his telescope by hand until the shadow of the tube is smallest on the ground. "If the Sun isn't in the fov at that point, usually a very small movement of the tube will find it."

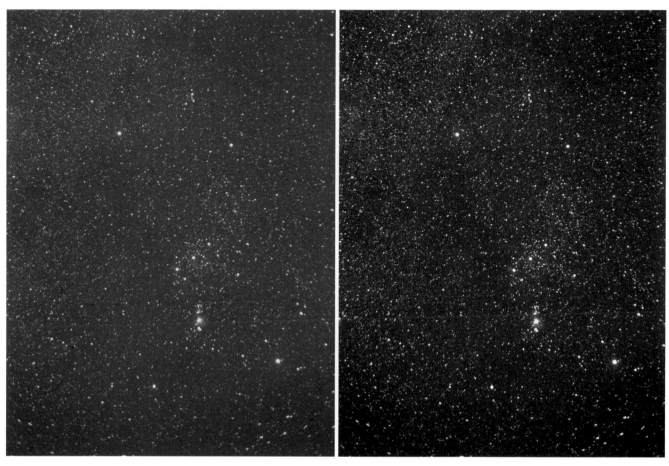

Image processing example. The original image is on the left. The right image shows dramatic improvement. (Photo by the author. Image processing by Holley Y. Bakich)

Focus

FOCUS! Whether it is your scope or someone else's, do not be afraid to focus. Otherwise, you are wasting your time. The great visual observer Brian Skiff, who works and views in Flagstaff, Arizona, cites a little-known fact:

> During a long observing session, always focus the telescope as far out as your vision allows to significantly reduce eye-strain.

Footwear

At your observing site, wear sturdy shoes or hiking boots. A mis-step in the dark could have painful consequences. Beyond actually falling, good footwear will protect your feet from sharp rocks, opportunistic creatures and (at my site) cactus.

Gegenschein

If your observing site has a limiting magnitude of six or better, you have an opportunity to see this elusive sight, which is also known as the counterglow. Simply look toward whatever constellation of the zodiac is opposite the Sun, once that constellation has risen past about 25° altitude. Note that if the Milky Way is also in that direction, you will not see the gegenschein. So if Gemini or the Scorpius–Sagittarius region are opposite the Sun, your search will have to wait a month or so.

High-altitude observing

Jeff Medkeff says that acclimatization is the key:

> A couple studies I've read suggest that visual effects of altitude are measurable at about 600 meters higher than where you live for the average person. But those who live at sea level can go twice that before having any effects, so its not a strictly linear relationship. People in good shape can go higher before encountering any effects than people who aren't.

The lack of oxygen at high-altitude locations results in a degree of hypoxia. More importantly, it has a significant adverse impact on low-light color perception. An excellent article relating to this may be found on the W. M. Keck website:

http://www2.keck.hawaii.edu:3636/realpublic/observing/visitor/hyalt.html

Jeff continues:

> There are strong indications that Viagra is an effective drug against hypoxia and specifically addresses the visual symptoms of hypoxia. The FDA is currently evaluating it as a treatment for this condition.

Image processing

When you display your images electronically, consider examining the levels of each color with software specifically designed for the task. One such software program is Adobe PhotoShop. Even a piggyback photograph of a constellation can be dramatically improved.

Intoxication

When observing in a more "serious" manner (this means different things to different people), do not consume alcohol. It impairs vision, and has other side-effects as well. If, however, the observing is of a more casual nature or if you wish to have a celebratory beer after viewing an elusive object, "Cheers!"

Just in case

One item I recommend you keep with your equipment is a Mylar blanket. Often billed as "survival blankets," that's exactly what they may be to you. Wrapped around you, a Mylar blanket will trap your body heat, allowing you to survive all but the most severe weather conditions. As a cover for your sleeping bag, they will keep you warmer, provide a shield against the wind and protect you from frost. These blankets are also waterproof. Find Mylar blankets at any quality camping store.

Know your telescope

My friend and observing buddy Mark Marcotte, of El Paso, Texas, provided this valuable tip:

For those with new telescopes, or new equipment that you have added to your telescopes, perform a setup at home first. Any problems revealed in the daylight will be ones you won't have later. As a second step, set it up at night out in your yard and observe as if you were at your remote dark-sky site. It's surprising what a test-run like this will teach you about your setup and what it requires.

Laser eye surgery

On July 17, 2000, an article appeared in the Canadian Press: entitled "Laser Eye Surgery May Damage Night Vision Long-Term: Study."

The article reported several studies that show decreased night-vision years after the surgery. One study warns that 58% of patients who had LASIK (Laser-Assisted In Situ Keratomileusis) or PRK (Photorefractive Keratectomy) fail night-vision tests. The study was of 38 persons that had moderate to severe corrections done two to seven years prior. Similar results were found for patients that underwent PRK between 1996 and 1998. 60% had reduced night vision up to two years after surgery. Early results from another study

show 30% of people were likewise affected. An international study of such effects was called for.

The article stated that not all doctors agree that a problem exits. One doctor said night-vision problems are more common in patients with large pupils. A separate study of 1 300 people revealed 50% had night-vision problems the first month, declining to 5% after a year.

I know of only one person who has had LASIK surgery. She had to return for a "tune-up" of one of her eyes but is now doing quite well. Unfortunately, I thought about conducting some "before and after" dark-sky tests on the very day she was scheduled for surgery.

Still, I would be reluctant to undergo this procedure. As I understand it, a key to LASIK is removing a circular flap from the front of your cornea. No matter how well it heals, a cut cornea won't be as clear as the original. The probable outcome, as I understand it, is a decrease in low-contrast detail perception, as well as varying degrees of glare at night.

Lightning

By all means, get to a safe place if lightning is anywhere nearby. Think of yourself and others first, but also remember that electrical discharges from lightning can fry sensitive equipment.

Protecting a laptop

Susan Carroll of Fort Scott, Kansas, explains her technique for protecting and lightproofing her observing field computer:

I have rigged a "dew box" for my laptop. I used one of those plastic milk crates you can get anywhere and cut out one side. I then set it on its side and put it over the laptop. I covered the milk crate with a piece of leftover desert storm shield material. It kept off the dew, and with one sheet of Rubylith® (or dark red Plexiglas) over the screen, and running The Sky in night mode, nobody could see it unless they were right on top of it, and even then it was pretty dim. This system works like a charm for me. I just cut a small hole in the back of the storm shield material for the cord to run through. Easy and cheap.

Limiting magnitude

Many observers estimate the limiting visual magnitude of their observing site by identifying the faintest star they can see, usually near the zenith. Some familiarity with a star chart is necessary for this, and it may take time. Other observers use a counting method. Simply by counting the number of stars in a given area, the limiting magnitude is calculated. A resource toward this, covering sections all over the sky, may be found on the SEDS (Students for the Exploration and Development of Space) website. The specific page is: http://www.seds.org/billa/lm/rjm.html

A scary lightning display at the Kitt Peak National Observatory. (Image by Adam Block/NOAO/AURA/NSF)

Morning caution

On nights when you are staying until morning, when quitting, point your telescope to the west, away from the rising Sun. This is crucial if you are leaving your telescope uncovered. Even with covers, however, the Sun's heat has been known to cause damage.

Mosquitoes

While mosquitoes will attack whoever is handiest, they prefer adults to children, women to men, and pregnant women most. (They find ovulating women more attractive than those menstruating.) Movement attracts mosquitoes, so swatting at them is a good way to get them swarming around you. Contrary to popular belief, colors of clothing are not very important to mosquitoes because the other attractions are so much stronger.

Lotions containing citronella (concentrations of 0.05% percent or 0.1%) are used to repel mosquitoes. More effective is a lotion or spray containing DEET, the acronym for N,N-diethyl-meta-toluamide. Use products with no more than 34% DEET (10% for children aged two or older) as higher concentrations can cause serious toxic reactions. DEET should not be used on children under two, pregnant women or on children's clothing or bedding.

While most people apply repellent only to exposed skin, experts suggest treating clothing as well. Most fabrics are only 1 millimeter thick, if that, but a mosquito's proboscis is 2 millimeters long and can easily bite through clothing, especially clothing that clings to the skin.

DEET won't keep a mosquito from approaching, but it can stop her from biting by jamming the cells in the insect's antennas that are sensitive to lactic acid.

For home control, so-called "bug zappers," which attract insects with UV light, are the most popular and least effective. In a recent study by the University of Delaware, only 31 of 13 789 (0.2%) insects trapped were female mosquitoes or biting gnats. Another study found that zappers could spread bacteria and viruses up to six feet away.

If you have a real problem over a large area, consider the Mosquito Magnet, which emits plumes of carbon dioxide to attract biting insects, then quietly vacuums them into a net and dehydrates them. It operates on a tank of propane. One tank burning 24 hours a day lasts three weeks and is said to control mosquitoes over three-quarters of an acre. Find it on the internet at

http://www.mosquitomagnet.com

These devices are not cheap – each costs about 800 US$.

For control around your backyard observatory, mosquito coils, citronella candles and citronella oil lamps do a reasonable job of keeping mosquitoes away for several hours.

Make certain that you are not in direct view of the light these items emit.

Not-quite-essential?

The following list contains some of the non-astronomy items I take to events where I may be camping or staying for an extended time:

- auxiliary table
- bungee cords
- rope, etc.
- canned air
- claw hammer
- duct tape and electrical tape
- extension cords (minimum of two at 7.5 m each)
- hand sledge
- power outlet strip
- pry bar
- spare parts
- tie-down straps
- tools (especially Allen wrenches)
- trash bags
- wire (assorted)
- wire ties (assorted sizes)
- volt–ohm meter.

Observing alone

If you must observe at a remote site alone, double-check everything. Before going out, be absolutely certain to let someone know exactly where you will be and exactly how long you plan to be out. This is not the time to be picking an unfamiliar location at random. Your life may depend on somebody knowing where you are.

Observing awards

Some amateur astronomers love observing awards. They pursue them with vigor and eagerly await the appearance of the next list of objects. Other amateurs couldn't care less about observing awards. They would rather not be bothered filling out forms or submitting log entries, whatever they may think of the objects on the list itself.

Personally, I like structure. Therefore, I tend to favor observing awards, but that is if – and only if – they are not looked upon as conferring any status upon the observer. Judging by the many e-mails I have seen on the subject, apparently there is an award which is singled out by some as the one all "great" observers acquire: the Herschel 400. See a description of this list in the "Catalogs of the deep-sky objects" chapter. This list has become lodged in the minds of many amateur astronomers as the "next step" once the Messier list has been viewed. Fair enough. But please don't tell me that simply because Observer A has a Herschel 400 certificate s/he is a better observer than Observer B. My

experience does not confirm this.

My friend Jeff Medkeff has thought at length about this subject and makes a number of excellent points.

As a keen observer of the cultural aspects of amateur astronomy, I could be tempted to go on at length about this – I won't, though. I'll just relate an anecdote. I can date the point at which the "immortal observer" gulf first entered my consciousness almost exactly. A friend had gone to Stellafane in 1985 (I did not attend), where, the story goes, John Bortle had acquired Halley's Comet, very early in the apparition as many will recall, in a large instrument. A line developed, and my friend was in it. When he came to the eyepiece, he could see no comet, and said so out loud. According to his report, he was shushed by whoever was operating the scope, who was otherwise encouraging everyone to see Halley's Comet – which had, in fact, been allowed to drift out of the field quite deliberately.

In short, an environment was created in which people were pressured to see what wasn't there. The entrapment required that a person lie, and claim they had seen the comet, or else be abused for revealing the underhandedness of the scope operator. Since then I've noticed the increasing development of an observer culture which is very similar to that of high school athletes (in which some have status, others are of lesser accomplishment, and yet others simply don't belong). Since I see amateur astronomy in many ways as an athletic hobby, this doesn't entirely surprise me.

Most lies in amateur astronomy bug me halfway to death. The above lie is one of them. Another lie that I have a hard time accepting is that there are standards in the hobby. I think there is nothing wrong with checking off lists or pulling off difficult observations, but the notion that these standard achievements confer status is silly.

The single status-conferring standard that I recognize in any hobby is whether the partaker derives enjoyment from it. So if a highly structured observing plan floats your boat, then go for it. But if the boat begins to take on water, don't hesitate to jettison the plan.

Again, let me stress that there is nothing wrong with the awards themselves. Quite the contrary. Many observers have been coaxed back into more regular observing by lists which they found interesting. Such lists have also proven invaluable to many beginning amateurs who have absolutely no idea on which object to begin. And there is nothing wrong with the added incentive of a certificate or a pin. Apparently, I am not alone in thinking this.

John Wagoner, the Observing Chairman at the Texas Star Party, graciously responded to my request for information regarding the number of observing awards (pins, in this case) that have been given out at TSP:

At TSP 2000, we gave out 256 Telescope pins and 88 binocular pins, for a total of 344. At TSP 2001, 218 telescope pins and 70 binocular pins were awarded, for

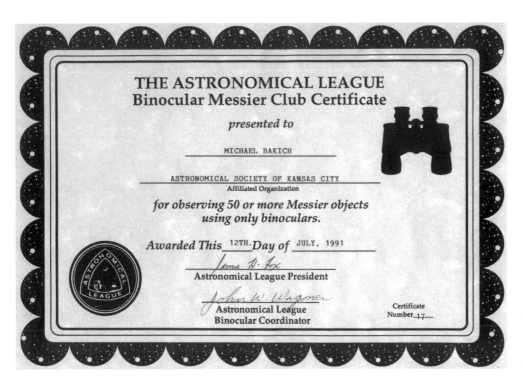

The Astronomical League offers a number of awards. This is an example of an easy one. (Photo by the author)

a total of 288. (TSP 2000 had more pins than TSP 2001 because we had more clear nights.) The total number of pins given out in the six years 1996–2001 is 1593.

Observing lists

Arild Moland of Oslo, Norway, talks about pre-observing session preparation:

Always prepare your observing session by assembling

Coyladene McKean receives her observing pin from John W. Wagoner at the 2001 Texas Star Party. Earning observing awards at this event is very popular. (Photo by the author)

an observing list. Use your star charts, software, the internet, and other sources to find objects. Even if you've got a GoTo system, you will benefit from sorting your objects in a natural order of observing, west to east. It is probable that you will find interesting sights along the way you may want to add to your list. Always give the session a smooth start. No ultra-faint stuff at the beginning, or else you run the risk of getting frustrated. If the skies prove to be absolutely super, either by way of transparency, seeing, or both, keep a challenge list on hand to take advantage of the unexpected conditions. Finally, although it may seem fruitful to concentrate on a small area of sky, beware that clouds may stay in your chosen area all night long. So bring lists for alternative spots!

Steven R. Coe of Phoenix, Arizona, a dedicated observer and author of the book *Deep Sky Observing: The Astronomical Tourist*, adds the following.

I have used a wide variety of methods to create an observing list:

(1) One of the most compelling reasons to belong to an astronomy club is to be in contact with lots of other observers and get a chance to look at someone else's favorite object and add it to your list.

(2) Obviously great books and magazine articles provide good information on objects to point the scope toward.

(3) I have just plain stumbled onto objects while searching near a well-known deep sky goodie. As an example, I "discovered" NGC 1528, a very nice open cluster in Perseus, while scanning with binoculars many years ago.

(4) Some years ago I re-read all my observations so far. Then, as I did that, I made a list of observations that intrigued me and I re-observed those objects to see if I could still see the feature that I wrote about years before. It was a fun project and I would recommend it to anyone. Fun to go visit old friends again.

(5) A few times I have just opened *Uranometria*[†] to a page that was on the meridian and started looking for the objects on that page. That can be lots of fun also.

Author's Note: I especially like Steve's #4 above.

Packing up

A. J. Crayon, of Phoenix, Arizona, warns, "When trying to fold a tarpaulin in windy weather, turn your back to the wind and let it blow the tarp away from you. Be sure not to let it go!"

Author's Note: I can personally attest to the efficacy of this tip.

Prescription painkillers

Jeff Medkeff relates, "Most prescription painkillers will whack your faint magnitude limit by two magnitudes or more."

Question yourself

Constantly ask yourself questions about the objects you are observing. I have heard it suggested that observers, especially those just starting out, take index cards into the field with them. There would be questions written on these cards and one card for each type of object. For example, an index card for planetary nebulae might have these questions on them: "Can the central star be seen?" "Is the nebula symmetrical?" "Is the nebula annular?" And so on. This would jog observers' memories and remind them to look for certain common details. I know friends who have started to do this, and they find the index cards help them out when they are

tired and not necessarily thinking straight.

Recording observations

When doing anything other than just-for-fun observing, I maintain a detailed log of observations. At the telescope, I employ a tape (now digital) recorder. I have done this for over two decades. It has become part of my routine. Other observers, however, have a different view and only write down their observations, either on specific observing logs or right on their star charts.

Satellites, shuttles, etc.

The ruination of many a great image. I tend to refer to all such objects as "manmade crap," however I know many observers who enjoy watching the Space Shuttle or

Recorders at the eyepiece

Advantages	Disadvantages
Recorders are small and light	Batteries can die
No light required	Tape can run out / Memory can become full
Recorded descriptions can be very long	Wrong button pressed
No undecipherable penmanship to decode	Recorder malfunctions
Not susceptible to wind or dew	Recorders can be dropped
You can speak a description as you are viewing the object	Observations are not immediately transcribed, creating a backlog

International Space Station move across the sky. With the new generation of GoTo telescopes, these objects can even be followed through the eyepiece, with medium telescopes even allowing the outline of the vehicles to be discerned. Iridium flares, seen after sunset or before sunrise, are also of interest to some.

A tape recorder (left) and a digital recorder. (Photo by the author)

[†]Uranometria 2000.0: The Northern Hemisphere to Minus 6 Degrees by Wil Tirion, Barry Rappaport, and George Lovi (Willmann-Bell, Inc., Richmond, VA, 1987)

A so-called iridium flare, a reflection from a manmade satellite. This one occurred within the Keystone of Hercules. (Photo by Steve Coe, Phoenix, Arizona)

Sky brightness

The finest skies on Earth are not the darkest. This is simply because the more stars you are able to see, the brighter the overall sky appears. In such locations, I have seen the Milky Way cast a shadow. The zodiacal light, gegenschein, etc., are also contributors to the overall brightness. No, a site rated "excellent" has three characteristics: (1) it is free of light pollution; (2) it contains low amounts of aerosols (dust, air pollution, water vapor); (3) it is at a relatively high altitude, between 1.5–2.5 km.

Sunlight

Exposure to bright sunlight has a cumulative effect on dark adaptation. Reflective surfaces such as sand, snow, or water, intensify this condition. Exposure to intense sunlight for two to five hours decreases visual sensitivity for up to five hours. In addition, the rate of dark adaptation and the degree of night visual acuity decrease. These effects may persist for several days, depending on the person.

Tents

Mark Marcotte of El Paso, Texas, offers the following excellent suggestion.

> When using a tent during an observing run or star party, consider a second tent for your equipment. Such a setup will keep things dry, relatively dust-free and out of direct sunlight. It will also give you a lot more room in your sleeping tent. Also, forget regular tent stakes. Most are flimsy and will bend in hard ground. Instead, use large spikes, which may be found at any good hardware store.
>
> **Author's Note:** The ones Mark recommended to me are 20 cm in length, with a diameter of approximately 9 mm and a head diameter of 19 mm.

Time per object

How much time you spend visually observing each object you see is a decision only you can make. Some observers spend an hour or more on each object, trying to glean every last bit of detail. Sketching may also be involved, which limits the number of objects/hour. Other observers take a leisurely pace of from 5–15 objects/hour. This provides them a chance to compare and contrast many different objects of the same type, or to view different types of objects, if their progression happens to be, say, eastward across the sky. Still other observers choose a single constellation for the night's observing, detailing objects of all types within the chosen star group. On rare instances, observers participate in marathons (Messier or otherwise), to see just how many objects can be viewed in one night.

Tube currents

The degradation of seeing due to tube currents is a widely recognized problem. But how do you know if it's your problem? Check the out-of-focus image of a fairly bright star. This will allow you to actually see air currents. If the image is reasonably steady, you have good seeing. If you see lots of circular motions moving around inside the image, you have very severe tube currents. Jeff Medkeff adds:

> If your out-of-focus star image has "hair" or a "ring of fire" around the outer edges, and these patterns change subtly over a period of 30 seconds to five minutes, then you have image-damaging tube currents – even though no swirling air is visible in the star itself.

The solution is to use a small, low-flow fan to blow the warmer air out of the tube or to get your mirror to the same temperature as the outside air.

Ultraviolet protection

I do a lot of solar observing, both white light and h-alpha. When I am in my observatory, it's easy to forget how much UV I am absorbing (El Paso is about 31° N latitude with the nickname "Sun City"). I like to wear loafers around the house. For a while I wasn't wearing socks. You can guess my surprise when one day I realized that the tops of my feet were painfully sunburned! So now I wear socks and keep some sunscreen lotion near the door.

Universal time

Use it. Memorize your correction to UT with and without Daylight Savings Time. It's really not difficult. Use it in your observing log entries. Use it in your correspondence. It's one area where you can standardize your observations. (I'm certain everyone reading this in the UK is wondering what the big deal is.)

Vitamin A

A diet that is deficient in Vitamin A can cause impairment of night vision. Vitamin A is an essential element in the buildup of rhodopsin in the rod cells. Without this, night vision is severely degraded. An adequate intake of Vitamin A through a balanced diet that includes such foods as eggs, cheese, liver, carrots, and most green vegetables will help ensure proper visual acuity. Keep in mind that excessive quantities of Vitamin A will not improve night vision and may be harmful.

Weather and seeing

A cold air mass (colder than the ground) is likely to display convective clouds such as cumulus. This air is therefore cleaner overall, but it may be unsteady. Probable outcome: good transparency, unsteady seeing. A warm air mass (warmer than the ground) is likely to display stratiform clouds, haze or mist. This air is likely to contain a lot of dust. The air tends to be steadier but less transparent. Probable outcome: lower transparency, steady seeing.

Bad seeing is almost guaranteed at least 24 hours following the passage of a front (the boundary between warm and cool air) or trough (an elongated area of low pressure that can occur at the surface or at higher altitudes).

Seeing is at times very good when thin cirrus clouds are present. This is not true, however, if the cirrus clouds are moving in one direction and there are lower-level cross winds.

Winter observing

Extreme cold weather is hard on equipment, on vehicles and on people. Cameras malfunction, batteries go dead, and there is risk of hypothermia and frostbite. Still, the prospect of a clear winter night is difficult to resist. Note, I specifically say "winter," although I am aware that similar conditions may occur during other seasons if you are observing at high altitude. For such situations, the law is simply stated: Preparation is Everything.

Being prepared does not mean owning the right equipment. It means having what you need at hand for that which you did not expect. If you are going to be observing in winter at a remote site, please, for your own sake, overpack. I am aware that it is highly improbable that you will need to use all of what you bring. But, in that most improbable of situations, having a certain item along could literally be the difference between life and death.

Don't care to think about such things? Fine. Forget harsh-weather observing and move onto the next chapter. If, however, you're keeping this as an option, read on.

It is often stressed that the most important body parts to keep warm are the head and the feet. Most of the heat loss during cold evenings is through the top of the head. And nothing will chill you faster than heat seeping into the ground through thin shoes.

My personal headgear consists of a pullover head cover made of soft fleece. Everything but my eyes can be covered, but in all but the worst conditions I usually have the front pulled down so that my nose and mouth are exposed. Over this I wear a felt "Indiana Jones" type hat. I do not like the down-filled hood of my parka (even though it is very warm) for one reason: when the hood is on, the material rubbing on itself makes so much noise that normal conversation is impossible.

Regarding boots, I decided to go for the best that I could get which would still allow me to drive a vehicle. I usually change into my cold-weather boots at the observing site, but you never can tell what might happen. The boots I chose are Baffin Mountaineers. They are comfortable, not very heavy

Lenticular altocumulus, also known as "Flying Saucer" clouds. (Courtesy NOAA)

Fleece pullover headgear. (Photo by the author)

and have a thick insulated lining in the sole to prevent heat transfer downward. Find them online at the J. Siegel Footwear web site:

http://www.jseigelfootwear.on.ca/baffin.html

Bring hand warmers, and bring extras. They are superb when working but notorious for not lasting the full time specified on the package. Keep an active one in your side pockets. Slip them in and out of your gloves or mittens for a quick warm-up. If you are using gloves, pull your fingers out of the finger slots for more warmth. If your feet get cold, there are also toe warmers which seem to work pretty well, although with my new boots I haven't tried them out yet. Again, camping outlets will have these in quantity.

The experts say to dress in layers. I generally use fleece long underwear and pants. My upper body is covered with a T-shirt, thin, long-sleeved flannel shirt, fleece pullover and down jacket. My wife, who is affected by the cold to a much greater extent than I am, wears a ski rescue suit as her outer layer. When fully zipped with the hood up and boots and gloves in place, there are few places for the wind to chill her delicate frame.

Choose a sleeping bag with the appropriate temperature range for your area. Here in the desert we have different requirements than my friends in Wisconsin and Norway. If possible, do not sleep directly on the ground. We use a lightweight, fold-out platform (Northern Design Queen Size Instant Bed Frame) upon which we place a battery-powered, self-inflating air mattress. This places us about 75 cm above ground and, while it may not sound like much, indeed it makes a big difference.

Jeff Medkeff offers this sound advice:

There is a general consensus among campers that you should add 15 degrees F to a bag's rated temperature to derive the temperature the outside air can fall to before you will feel cold in the bag. For some, that figure can be 20 degrees.

Zodiacal light

Phoenix, Arizona, amateur A. J. Crayon relates this interesting story.

One time while trying to make observations of the Coma–Virgo Galaxy Cluster it appeared that my 200 mm telescope wasn't delivering the images I had come to expect from prior years observing. A quick check of the telescope didn't turn up any problems. Finally, I stopped to examine the western sky and discovered the problem – zodiacal light! So my observing tip is to not observe objects while they are in the zodiacal light, especially with a moderate size telescope. This measurement hasn't been made with my 370 mm, but I would suspect that some degradation of the image would have to be expected.

Jeff Medkeff adds:

Zodiacal light is much easier to see in poor skies at 31° latitude than it is to see under very good skies at 41°, because at the lower latitude, the light makes a significantly steeper angle with the horizon at any given time.

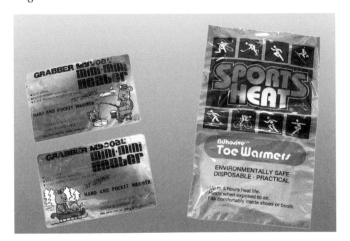

Chemical warmers for the hands and feet. They work very well. (Photo by the author)

CHAPTER FIVE

References, further information, and other points

5.1 Books and atlases

I love astronomy books. I read them, I collect them, I write them. Since this is only the third book I have written, I am far from an expert in writing. But I have a collection of astronomy books that is notable. Specifically, I collect first edition, nineteenth century astronomy books printed in English. As of this writing, I am closing in on 400 distinct first editions. My lifetime goal is 500. (Perhaps I should slow down my acquisitions!) Apart from this "core" collection, I have acquired a reasonable twentieth century reference library. I offer the following titles for your consideration.

Essential books

Astronomical Tables of the Sun, Moon, and Planets (second edition) by Jean Meeus Willmann-Bell, Inc., Richmond, VA, 1995

Astronomical Tables by Meeus.
(Photo by the author)

If there is something going on in the solar system, this book will tell you. *Astronomical Tables* is a detailed and highly accurate collection of lists of events past, present, and future. Oppositions of Mars from the years 0–3010, sunspot numbers from 1749–1981, inferior and superior conjunctions of Venus from 0–2500, solar and lunar eclipses through 2050, even conjunctions of the Sun with twelve bright zodiacal stars from −1000 to +2399 – these lists are but a fraction of the information contained within this book. I have referred to this book as a primary source of information for nearly two decades.

Burnham's Celestial Handbook (three volumes) by Robert Burnham Jr (Dover Publications, New York, 1978)
This is the great reference for amateur astronomers and a copy of it deserves to be on all our shelves. The style of writing is engaging and the layout is great. It's fun to read even when you are not consulting it as a reference. The

The bible of amateur astronomy, *Burnham's Celestial Handbook*. (Photo by the author)

problem is that in the intervening decades since the set was published, things have changed. Lots of things. Many of the facts in the book are woefully outdated. Look at Burnham's as a starting point. The celestial objects he gets excited about are worthy of observation.

High Resolution Astrophotography by Jean Dragesco (Cambridge University Press, 1995)

Dragesco's excellent *High Resolution Astrophotography*. (Photo by the author)

In my opinion, the best book ever written about astrophotography. Contains lots of detailed description on how to get the most out of a telescope/camera setup. Concentrates on solar system objects but once you learn his techniques, the (deep) sky is the limit.

A History of Astronomy by Antonin Pannekoek (George Allen & Unwin, London 1969)
Again, the best of a very large lot. There are many things in

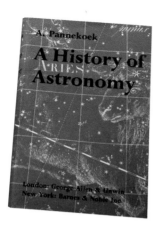

A History of Astronomy by Pannekoek. (Photo by the author)

Star Names by Allen. (Photo by the author)

this book that you won't find in other astronomical histories. Pannekoek gives about equal coverage to ancient astronomy, the Galileo–Copernicus–Tycho era, and the telescopic age. This book is thick and can be a little dry and technical at times, but it is also a fascinating, fact-filled account of the history of the hobby we all love.

Solar Astronomy Handbook by Rainer Beck, Heinz Hilbrecht, Klaus Reinsch, Peter Volker (Willmann-Bell, Inc., Richmond, VA, 1995)

Solar Astronomy Handbook. (Photo by the author)

Rarely does one find such a comprehensive source of information on a single subject. This book pretty much covers the Sun, from choosing a telescope and filters to observing in h-alpha light. There are plenty of images and diagrams and even math, for those so inclined. If you are planning to observe the Sun, get this book.

Star Names: Their Lore and Meaning by Richard Hinckley Allen (Dover Publications, New York, 1963)
Considered quaint by some, this is a fun book to read and the authority on the subject. If you have an interest in the names of the constellations and brighter stars and how they came to be, you will enjoy this book. The descriptions are sometimes a bit tedious but interspersed are enough facts to fill an encyclopedia. Not too often do you find a book with not only a general index, but with Arabic and Greek ones as well.

Star Testing Astronomical Telescopes by Harold Richard Suiter (Willmann-Bell, Inc., Richmond, VA, 1994)
The book that will allow you to find out how good your telescope really is. On the level of advanced amateur astronomer, *Star Testing* is an exact, clearly written, but technical approach to telescope optics and nearly every possible problem they could encounter.

Visual Astronomy of the Deep Sky by Roger N. Clark (Cambridge University Press, 1990)
If you are an advanced amateur astronomer and if you can find a copy of this book, buy it. *Visual Astronomy* is 355 pages in length, but you are buying it – and it is essential – because of the first 63 pages and several of the appendices. Within the first six chapters is a wealth of information that is so valuable and revolutionary that this book can truly be called unique. Chapter 7 (pages 64–244) is "A Visual Atlas of Deep-Sky Objects."

Star atlases

Sky Calendar by Robert C. Victor (Abrams Planetarium)
Not an atlas, but a great reference for beginning amateur astronomers nonetheless. Mailed out quarterly, the *Sky Calendar* informs you of naked-eye occurrences in the sky. On one side of the sheet is a day-by-day calendar of visual events such as conjunctions, close approaches, and appearances of the morning and evening sky. The reverse is a planisphere specifically for that month. Although specific to 40° N latitude, the *Sky Calendar* can be used effectively throughout the US and Europe. Contact: Abrams Planetarium, Michigan State University, East Lansing, MI, 48824, USA. Subscription is 10 US$ per year.

Atlas of the Moon by Antonín Rükl (Kalmbach Books, Brookfield, WI, 1996)
This is a great lunar atlas. As of this writing, it is out of print but I have heard rumblings for a couple years that it may be reprinted. I hope so. The maps are very well done and detailed, with one small exception. There is no overlap between maps. Such a feature would have been very helpful

and here's hoping it will be included in any future edition. There are also 50 photographs of interesting lunar features with excellent descriptions.

Good luck finding a copy of Rükl – it is now out of print. This is one of the best lunar references. (Photo by the author)

Herald–Bobroff Astroatlas, by David Herald and Peter Bobroff (HB2000 Publications, Woden, Australia, 1994)

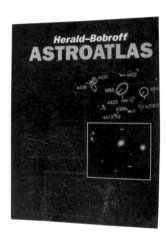

The Herald–Bobroff Astroatlas. (Photo by the author)

If you are more than just a beginning amateur astronomer and if you live in the southern hemisphere, this is the atlas for you. If you live in the northern hemisphere, this may also be the atlas for you. There are 214 charts total with black-on-white stars down to the ninth magnitude. The unusual part of this atlas is the varying map scales. No less than six scales are used within this work. Navigation is fairly straightforward although a list of charts where constellations could be located would have been a nice addition. There are a limited number of dealers in the US who carry this book. Lymax Astronomy is one. Find them on the internet at
http://www.lymax.com

Norton's Star Atlas by Arthur Philip Norton, edited by Ian Ridpath (Prentice-Hall, Englewood Cliffs, N 1998)

First published in 1910, Norton's is the first star atlas I recommend to beginning amateur astronomers. It contains excellent star charts and a great deal of useful information. The 16 black-on-white (with a green milky way area) charts

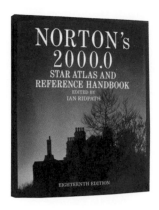

Norton's Star Atlas. (Photo by the author)

cover most stars visible from dark sites, plotting 8700 stars down to a magnitude of 6.5. In addition, 600 of the brighter DSOs are included. Before each two-page star chart are two pages of lists of interesting double stars, variable stars and DSOs. Right ascension and declination lines and constellation boundaries are also plotted. Maps of each of the four quadrants of the Moon are also included. The binding style of Norton's allows the book to lay flat on a table, making it easy to use while observing.

Tip: A large piece of thin, transparent plastic will protect your star atlas against dew.

The Night Sky Observer's Guide (two volumes) by George Robert Kepple and Glen W. Sanner (Willmann-Bell, Inc., Richmond, VA, 1998)

For the serious observer, NSOG (pronounced en' sog) is a terrific reference. All objects are grouped by constellation, with volume 1 covering "Autumn and Winter" and volume 2 "Spring and Summer." Reverse these labels if you live in the southern hemisphere. This set has a decided northern bias, but at least the equatorial constellations are accessible by all. The descriptions are clear and it's great fun at the telescope to compare what you are seeing to what was observed by George and Glen. One object may have four observations through different sized telescopes. Images, finder charts, and sketches abound.

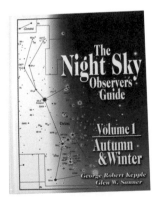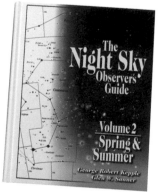

The Night Sky Observer's Guide, commonly called NSOG. An excellent at-the-telescope reference. (Photo by the author)

Sky Atlas 2000 by Wil Tirion (Sky Publishing, Cambridge, MA, 1981)

A very good set of 26 charts for intermediate amateurs and up. 43 000 stars are plotted down to the eighth magnitude. 2 500 DSOs are also plotted. The best part is the size of the charts – each is 18 inches × 13 inches (46 cm × 33 cm). This spreads objects out a little, making them easier to find. *Sky Atlas 2000* comes in five versions. I recommend the Laminated Field Edition. It has white stars on a black background, is spiral bound and is laminated, for those "less-than-dry" nights. Several catalogs for objects plotted in *Sky Atlas 2000* are available, such as the *Sky Atlas 2000.0 Companion*, which lists every DSO plotted in the atlas.

Uranometria 2000.0 (two volumes) by Tirion, Rappaport and Lovi (Willmann-Bell, Inc., Richmond, VA, 2000)

This work is intended for advanced amateurs only. You must know the sky fairly well and you must be able to get reasonably close to your target object before using most of the charts in this set. And charts are here a-plenty – 473 of them, plotting 332 000 stars down to magnitude 9.5. 10 300 DSOs are also plotted. The layout of this atlas, in order of increasing RA, leaves a lot to be desired until you get used to it. A necessary touch is the addition of a set of guide charts to help you find the exact chart containing your desired observing area. As with *Sky Atlas 2000*, there is a companion volume – *The Uranometria 2000.0 Deep Sky Field Guide*.

5.2 Catalogs of the deep sky

The Messier Catalog

Charles Messier (1730–1817) was a French comet hunter. In his searches he occasionally encountered celestial objects which had the appearance of comets in that they appeared fuzzy in his small telescope but which did not move against the background of stars. This last trait completely set them apart from comets. The astronomers of the day called these objects nebulae, a word whose definition is much more specific today. The catalog he eventually compiled contained open and globular star clusters, galaxies, and, well, nebulae, both normal and planetary.

M83, one of Messier's finer galaxies. (Image by Adam Block/NOAO/AURA/NSF, using a 0.4m Meade LX200 telescope)

The Trifid Nebula, M20. (Celestron 14 on a Mountain Instruments MI-250 with a Graflex film holder/Lumicon giant easy guider. Shot on Fuji NPH film. Photo by Rick Thurmond of Mayhill, New Mexico)

During his searches, he discovered a comet-like patch in Taurus on 28 August 1758. This was the first of his non-moving objects and it later became the first entry, M1, in his famous catalog of nebulous objects which might be taken for comets. Messier published three versions of this catalog. The first, in 1769, contained 45 objects. The second, with 68 objects, appeared in 1780. The third, containing 103 objects, appeared the very next year. Later discoveries by Messier and others brought the final tally of objects to 110, the catalog we recognize today.

In all, Messier discovered 13 comets and made seven additional independent co-discoveries. Messier himself discovered 41 of the 110 objects on the current Messier list, the most of any discoverer (Pierre Mechain (1744–1804) is second with 27). Messier has two lunar craters named for him (Messier and Messier A) and asteroid 7359, discovered 16 January 1996. On a celestial globe completed in 1779, Joseph Jerome Francois de Lalande (1732–1807), introduced Custos Messium, a new constellation near the north celestial pole. This was done to honor Messier, who had

become a friend of Lalande. The constellation never received wide acceptance and is now extinct.

The best book I have read on Messier, and a wonderful reference concerning his catalog, is *Messier's Nebulae & Star Clusters* by Kenneth Glyn Jones (Cambridge University Press, 1968, second edition, 1991).

Messier marathons

I remember participating in Messier marathons in the early 1970s. In 1974, from the campus of the Ohio State University in Columbus, Ohio, a friend and I logged 106 Messier objects in one night using his 150mm Newtonian reflector. The first extended mention of the Messier marathon as an event was in an article by Brent Archinal (who I also met at Ohio State) published in the March 1982 issue of *Deep Sky Monthly* magazine.

Possible marathon dates

Tom Polakis, of Arizona City, Arizona, has calculated the

Wide-field view of M101 in Ursa Major. (Photo by Robert Kuberek of Valencia, California)

The region of M24. (8" Celestron Schmidt, f/1.5, 15 minutes on TP 2415 film, 25 Apr 1979, Naco, Arizona. Image by David Healy, Sierra Vista, Arizona)

window of opportunity for a complete Messier marathon. The beginning of the observing window is defined by the date when M30 is high enough to see in a dark sky. Working with the premise that the object has to be at an altitude of 2° to be seen, Tom calculated the Sun's altitude at that time and its distance from M30. On the evening end, the limiting object is M74. Seeing it defines the end of the window of observing dates.

His numbers are for 33° N latitude, and Tom adds that they get more favorable as you work your way south, particularly for M30. So the beginning of the season is about 17 March. Using the same criterion for the evening view of M74, it ends on 3 April.

Can you see them all?

The quick answer is "yes." Many people have viewed all the Messier objects during a single night. The challenge comes from trying to see as many as possible in as small a telescope (or binoculars) as possible. With a good 75 mm telescope, all the Messier objects can be seen from a dark site on a good night. The aforementioned Brent Archinal saw all 110 with 11×70 binoculars on 24/25 March 2000 at the All Arizona Messier Marathon hosted by the Saguaro Astronomy Club. I have done the marathon with my 7×50 Fujinon binoculars. This was on a miserable night, even from our dark-sky site. On 4/5 March 2000, it was partly cloudy, seeing 7/10, transparency 4/10. I saw 75 Messiers and I believe seeing 90 with quality 7×50 binoculars is a strong possibility.

Messiers with the naked eye

Brian Skiff, of Flagstaff, Arizona, maintains a working list of naked-eye deep-sky objects seen by him or reported to him by reputable observers. As of 29 January 1999, the following Messier objects were seen naked-eye: M2, 3, 4, 5, 6, 7, 8, 11, 13, 15, 16, 17, 20*, 21*, 22, 23, 24, 25, 31, 33, 34, 35, 36, 37, 38, 39, 41, 42, 44, 45, 46, 47, 48, 50, 55*, 67, 81, 83*, 92, 93, 101*. The ones marked with an * have not personally been seen by Brian.

Marathon search order

Many resources exist detailing the order you should look for the Messier objects during a marathon (an example is given in Appendix C). It must be stated that most have been done with more northerly latitudes in mind. If you live in Europe or in the northern tier of states in the US, these will be fine. However, for the southern states the sequence is somewhat off. Southerly Messiers rise sooner than indicated by most lists. The best procedure is to use a planetarium charting program such as TheSky to compile a custom Messier sequence based on your latitude.

Catalogs of the Herschels

On 7 September, 1782, William Herschel made his first original discovery of a deep-sky object (DSO), the Saturn Nebula (NGC 7009). How appropriate that the discoverer of the planet Uranus (13 March 1781) would find the Saturn Nebula as his first DSO. Within 20 years, Herschel had discovered (or revisited) 2478 DSOs.

M88. (Image by Adam Block/NOAO/AURA/NSF, using a 0.4 m Meade LX200 telescope)

John Herschel, William's son, was also an astronomer and nebula hunter. He eventually published his *General Catalogue of Nebulae and Clusters* in 1864. It contained 5096 DSOs.

The Herschel 400

In the April 1976 issue of *Sky & Telescope*, James Mullaney, of Pittsburgh, Pennsylvania, (a friend of mine and an excellent observer) suggested that the "best" of William Herschel's list of DSOs, 615 in number, be used as the "next step" for amateur astronomers who wanted a challenging project beyond the Messier list. Mullaney arrived at that number after eliminating the two faintest categories in Herschel's classification.

Several members of the Ancient City Astronomy Club in St. Augustine, Florida, noticed the letter and prepared a Herschel 400 observing list. The Astronomical League made it an official "observing club," complete with award certificate and pin, in 1980. All objects on this list can be seen with a 150 mm telescope, and the entire list has been observed with a 55 mm f/8 refractor.

In August 1997, a second Herschel list, the Herschel II, was developed by the Rose City Astronomers of Portland, Oregon. It has also been adopted by the Astronomical League as an official observing award club.

One of the finest of the Herschel objects is the Hockey Stick, NGC 4656. (Image by Adam Block/NOAO/AURA/NSF, using a 0.4 m Meade LX200 telescope)

NGC 2808 in Carina. (RGB: 600 s R, 600 s G, 900 s B through an SBIG ST-7E CCD on a Celestron 11. Image by Steven Juchnowski, Balliang East in the State of Victoria, Australia)

The NGC

The Danish astronomer John Louis Emil Dreyer (1852–1926), working in Ireland with Lord Rosse and his famous "Leviathan" telescope, published a supplement to Herschel's General Catalogue in 1878. Ten years later, *The Memoirs of the Royal Astronomical Society*, vol. XLIX, Part I, appeared as Dreyer's *A New General Catalog of Nebulae and Clusters of Stars, being the Catalogue of the late Sir John F.W. Herschel, Bart., revised, corrected, and enlarged.* Since its publication, this catalog has been known as the NGC. It lists 7840 objects. In 1895, the *First Index Catalog*, which contained 1520 objects, was published by Dreyer as a supplement. Finally, in 1908, the *Second Index Catalog* appeared,

A far-southern NGC object is the Tarantula Nebula (NGC 2070) in Dorado. Originally catalogued as a star, so it is sometimes known as 30 Doradus. (35-minute exposure on hypered Kodak EGP 400 with Celestron 11, f/6.3. Image by Steven Juchnowski, Balliang East in the State of Victoria, Australia)

containing an additional 3866 DSOs. Within these three catalogs (known hence as the NGC and the IC) were numerous errors.

In 1977, Jack Sulentic and William Tifft published the *Revised New General Catalog* (RNGC). It contained the 7840 objects in the original, but with updated coordinates. Many errors remained.

The most recent revision to the NGC and IC may be found online. See

http://www.ngcic.com

This is a terrific site and my highest commendations go out to the project leaders and observers.

The Caldwell catalogue

I am about to open a can of worms. If you are an amateur astronomer who has been around the hobby a while, you have heard of the famous British popularizer of astronomy, Patrick Moore. In the late 1990s, Moore devised a listing of deep-sky objects which contained 109 of the brightest non-Messier objects. He called the list the "Caldwell catalogue" because his surname is Caldwell-Moore, and he didn't want to use Moore as his list designations might get confused with the Messier list.

As soon as this list appeared, part of the amateur astronomy community lost their mind. "How dare he!" "Such arrogance!" "Many of these objects already have names!" "Look at all the errors!" My reaction was somewhat different: "Why didn't I think of that?" The Bakich Catalog . . . it has a nice ring, doesn't it?

The truth is that it is a great list. I don't know what Dr Moore had in mind, but I try to put personalities and egos aside. The Caldwell objects are worthy of observation by

Caldwell 63, otherwise known as NGC 7293, The Helix Nebula in Aquarius.
(12.5" Ritchey-Chretien, f/7, RRGB = 10:10:10:10 minutes. Taken with a
Finger Lakes Instrumentation IMG1024 CCD camera. Image by Robert
Gendler, Avon, Connecticut)

amateur astronomers and many on the list are within the reach of medium-sized telescopes.

You can quibble with some of the entries, but you could also do that with Messier's list. Messier included a constellation for goodness' sake! Yes, to the ancient Greeks, the Pleiades was a separate constellation.

After Moore originated his list, a follow-up book appeared. *Observing the Caldwell Objects* by David Ratledge (Springer-Verlag, New York, 2000), is an object-by-object guide to all 109 objects, complete with a CCD image of each. I recommend the book. Moreover, I recommend the list. Don't bother trying to memorize the C-numbers, however. Not only could it become confusing but if you have or are planning to acquire a GoTo telescope, both Meade and Celestron have the Caldwell catalogue in their databases.

The entire Caldwell catalogue is given in Appendix D.

5.3 Software

TheSky astronomy software. It does identification, charting, it even controls your telescope. (Photo by the author)

David Bushard, an advanced amateur astronomer from Wisconsin, put it best: "I cannot remember life without the internet." Astronomy-related software has come a long way in the past two decades. There are not a lot of programs out there, all things considered. However, the ones which do exist are of high quality.

Caveat

Upgrade. To software users, this is a helpful word. As an author, it is not. I write this section with the full knowledge that changes in software are inevitable. Some of the software packages that I list here will have changed. Others may no longer be made at all.

"Planetarium" software

Also known as charting software, one of these packages is sure to become indispensable to you. The main features of each are similar. Most can import databases allowing the display of very faint stars. (Using the USNO-SA2.0 database, I have seen a star within Omega Centauri fainter than magnitude 20!)

All of these packages have been upgraded a number of times in the five-plus years that I have been aware of them. The most advanced level of TheSky may be singled out for also providing telescope control. Visit each of their websites for detailed information.

- Carte du Ciel
  freeware
  http://www.stargazing.net/astropc/index.html

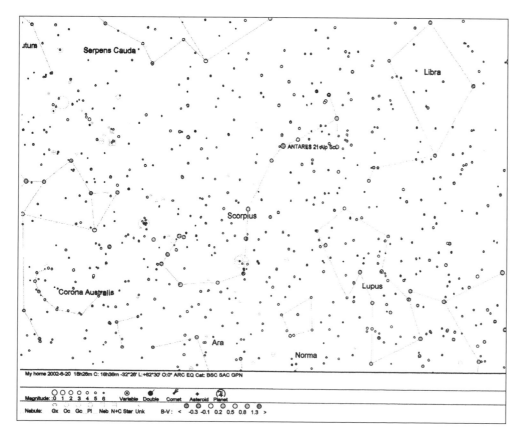

Carte du Ciel star chart, showing the area around Scorpius. (Scan by the author)

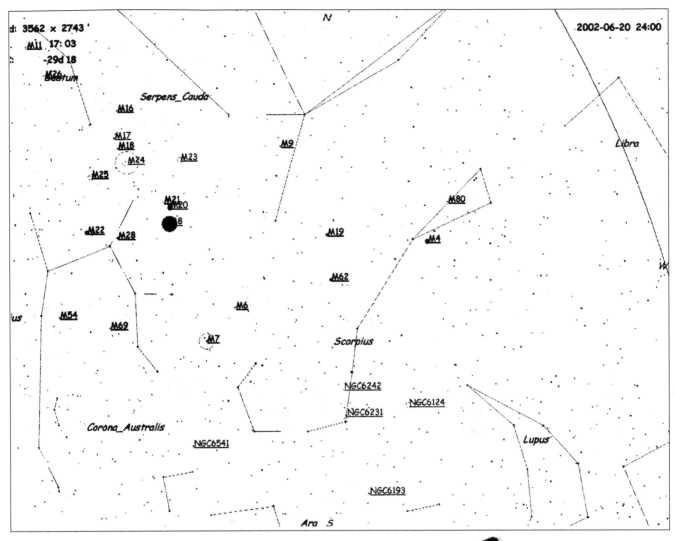

d: 3562 x 2743
M11 17:03
-29d 18
M26

2002-06-20 24:00

Serpens_Cauda

M16
M17
M18
M24
M23
M25
M21
M20
M22
M28
M9
M80
M19
M4
M62
M6
M54
M69
M7
Scorpius
NGC6242
NGC6124
NGC6231
Corona_Australis
NGC6541
Lupus
NGC6193
Ara
Libra

Hallo Northern Sky star chart, showing the area around Scorpius. (Scan by the author)

- Hallo Northern Sky
  freeware
  http://www.hnsky.org/software.htm

- MegaStar
  129.95 US$
  http://www.flash.net/~megastar/

- Star Map Pro
  99 US$
  http://www.skymap.com/index.htm

- TheSky
  129–249 US$
  http://www.bisque.com/

The RealSky software from the Astronomical Society of the Pacific is a great resource. Underlaying actual images of the sky into charting programs brings a whole new era of realism. (Photo by the author)

RealSky

RealSky is a set of CDs distributed by the Astronomical Society of the Pacific. It is based on The Digitized Sky Survey (DSS), a collection of scanned photographic images from the Palomar Schmidt and the UK Schmidt photographic plates. The uncompressed version takes up no less than 102 CDs.

RealSky is a version of the same data compressed by a factor of 100. Eight CDs contain images from the Palomar

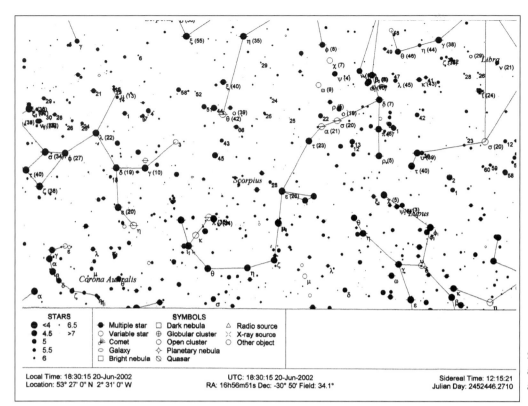

Star MapPro star chart, showing the area around Scorpius. (Scan by the author)

Sky Survey, down to declination 12° S. An additional eight CDs contain images from the UK Schmidt Telescope up to about 3° N declination.

When used with planetarium software, the images from RealSky underlay the display. The effect is utterly mesmerizing. Prepare to lose some time just checking different areas of the sky. With the approximately 1.5 GHz processor in my computer, I can underlay a 3° × 3° field in 10 seconds.

Using deep-sky objects as test subjects, the accuracy of the charting software can be measured. RealSky is a set of WCS (World Coordinate System) equipped images, with astrometry good to approximately 1.6 arcseconds. The most accurate planetarium software (at this writing) seems to be Sky Map Pro.

Other software

Besides planetarium programs, other astronomical software has been written. Here are a few choices that I have found useful.

- Aberrator
  Generates images that show the effects of telescope aberrations
  freeware
  http://aberrator.astronomy.net

- JupSat95
  Graphically shows the positions of the Galilean satellites
  freeware
  http://indigo.ie/~gnugent/JupSat95

- LunarPhase
  Provides information about the Moon in real time
  30 US$
  http://indigo.ie/~gnugent/LunarPhase/

- Mars Previewer
  An excellent program showing what features are visible on Mars, currently or at your upcoming observing time
  freeware
  http://www.astronomysight.com/as/start/books.html

- Satellites of Saturn
  Titan is easy to view, but what were those other moons? This DOS-based program will tell you
  freeware
  http://www.physics.sfasu.edu/astro/dansoftware.html

- Tracker6
  An excellent program to show the position of features in the atmosphere of Jupiter
  freeware
  http://www.physics.sfasu.edu/astro/dansoftware.html

5.4 Observatories

Circumpolar star trails at the Grasslands Observatory. The bright spot just to the left of top center is Polaris. Note that it is hardly trailed at all. The observatory is a cooperative effort between Tim Hunter (the owner) and James McGaha (the Observatory Director), both of Tucson, Arizona (see http://www.3towers.com). (Photo by Tim Hunter)

The best thing I have ever done in amateur astronomy, without exception, is to construct a small observatory in the backyard. If I am seated at my computer and decide I want to observe, I can be at the eyepiece in three minutes. Since the telescope stays outdoors, it's already at the right temperature. It's also polar aligned. And the best thing is that I'm not lugging a tube, mount and counterweights anywhere. In fact, from a seated start, I'm ready to observe before my eyes are even dark-adapted. Shall I continue? Well, there's also the fact that all the "stuff" which would normally be in the house is safely stored in the observatory.

A friend once told me that his setup on average would take up to three hours from the start of setup to the point of imaging. "The three hours were spent lugging out the gear, setting up, doing a rough align followed by 30 to 60 minutes of drift aligning. At that point I would begin setting up the computer and CCD during the final stages of drift aligning. By this point the scope had reached or was very close to thermal equilibrium. Now a check of the collimation and any adjustments and then I am finally working."

In this chapter, I will list some of the considerations involved in constructing an observatory.

Choice of observing site

The first task is to select the best location for the observatory. This can be 10 m out the back door or 50 km away at a dark-sky site. The important point is that it works for you, that it allows you the freedom to observe whenever it is clear.

Generally a closer location works best. This will almost certainly involve some compromises: limiting visual magnitude, buildings and trees blocking some sections of sky, even direct light from security and street lights.

Size of observatory

The Everstar Observatory. Nestled away in the backyard of a house in Olathe, Kansas, this roll-off roof observatory is very functional, even if it can't see half the sky! (Photo by Mark Abraham)

A choice now must be made as to whether to build an observatory which simply houses the telescope or to make it larger and able to accommodate people. If you are interested in the astrometry or photometry of asteroids and know that a CCD will be connected to the telescope 100% of the time, there is no need to consider a larger structure. If, however, you are more of a social observer who likes to have friends over, a larger observatory is probably best.

There are three considerations to take into account when considering the actual size of your future observatory. The first involves the available space. How big *can* you build? This was the size-limiting factor for my own observatory. With our fenced-in backyard, I simply could not make the observatory any larger in either dimension.

The number of telescopes is the second consideration. If you are planning to have more than one telescope set up permanently, or if you have in mind one permanent pier with space for at least one other telescope on a portable mount, you will have to account for that when planning the size of the observatory.

The third consideration related to size is money. It isn't the first consideration because without dealing with the others, you will have no idea how much, even in general terms, the

Tim Hunter's 3towers Observatory (see http://www.3towers.com). Of the surroundings, Tim says, "The towers insure that nobody will be developing the land near here anytime soon." (Photo by Tim Hunter)

observatory will cost. I will discuss individual cost items in a bit. And while it is true that larger observatories cost more, the cost-to-size ratio is not linear.

Roofs

Probably the most personal consideration for your observatory will be the choice between placing a dome on top or using a roll-off roof design. One of the primary factors to consider is aesthetics. Non-astronomy people usually associate a traditional observatory building with a dome. This can be good or bad. The appearance of the dome itself can clash with its surroundings. Also, a dome can draw unwanted attention from people (primarily youth) with less-than-admirable intentions.

Zoning or planned neighborhood ordinances may forbid the construction of a dome in some areas. In one instance, I know of selection of a dome because of existing tax laws. This was done to minimize the area coverage of the building, upon which taxes were based. In this particular location, the area covered by a roll-off roof when it was open was also subject to additional property taxes!

Domes are better at reducing stray light, having only the slit open to the sky. Roll-off roofs are better at radiating trapped heat quickly. Domes will protect you much better on windy nights. Roll-offs provide a more impressive vista of the sky from within. Either can be made to be essentially rainproof.

Dome

Most amateur astronomers who select a domed roof purchase a ready-made dome from a manufacturer. The option remains as to whether to have the manufacturer install the dome or to do it yourself. The choice depends upon your construction skills and level of confidence with such a project.

Of course, you can't just order a dome and be done with it. The building beneath the dome must be constructed first, with the "mating ring" made to exact specifications. The three major dome manufacturers (as of this writing) for personal observatories are as follows.

- Ash Domes
  Plainfield, Illinois
  http://www.ashdome.com

- Technical Innovations
  makers of Home Dome and Pro Dome models Gaithersburg, Maryland
  http://www.homedome.com

- Astro Domes
  Yandina, Queensland, Australia
  http://www.astrodomes.com

In rare cases, amateurs have built their own domes. Certainly if you have the skill you can build a form for a single "orange slice" panel, construct the required number, then bolt the sections together to yield a dome. However, getting the curvature right so that it assembles into a smooth dome is no easy task.

Others have started with a form made of metal rings over thin, curved girders (usually aluminum). Over this you could stretch either plastic or a water-resistant fabric. If fabric, the cloth can be sealed and then resin can be applied in layers to add strength.

Roll-off

If you choose to build (or have built) a roll-off roof, there are a few considerations to discuss. First, plan to build a gabled (triangular) roof. Such a roof will not allow rain or (much) snow to accumulate on it. Pre-made triangular roof trusses may be purchased if you can find them in the right size.

Next, consider how you want the roof to roll, mechanically speaking. A number of styles of casters are available in plastic or metal. Generally these are made to roll in tracks or troughs of metal. Be certain to obtain casters which have a large enough weight rating for your roof load, plus snow, if any. For my observatory, I chose solid steel V-groove casters which ride on an inverted steel L-bracket. There are four casters on each side of the roof. This is more load capacity than I will ever need but I found that the roof rolled more smoothly on eight casters than on six. I only need to remember to lubricate them three or four times a year. The roof opens and closes quite smoothly.

Be certain to design your roof so that entrance from wind, rain, etc., is minimized. The overhang of the roof should be at least 30 cm in each direction. Some amateur astronomers also drop a panel straight down from near the outside edge of the roof to cut down on the entry of dust into the observatory.

Securing the roof when closed is an important consideration. I use two solid steel turnbuckles. Here in the southwestern desert, spring winds are often strong. The roof

has had no problems surviving wind speeds of 30 m/s. Other observatory builders have employed steel hasps, usually one in each corner.

Some amateurs, purely for aesthetic reasons, design their roofs so that they overhang the ends of the track by a small amount (approximately 1/3 m). This shortens the length of the outriggers, the wooden rails that the angle iron in which the casters ride are attached to. Others make the outriggers longer than they really need to be to minimize the amount of sky that the roof blocks in the direction of its opening.

Flooring

You have two basic options for flooring, wood or cement. A poured cement floor is virtually maintenance-free. Make certain a quality grade of cement is used. I wish that I had. The only complaint I have about my observatory is that the people who poured the cement floor used far too much sand.

If you are using wood for your flooring material, choose pressure-treated lumber. Elevate the floor to allow for sufficient air circulation to prevent mold and mildew. Having the structure up off the ground is also beneficial in that rodents will not nest in an unconfined, open space. Allow approximately 6–8 mm spacing between top floor boards so that cold air does not pool in the observatory.

Wood flooring is more forgiving on your feet than cement, especially during those long sessions when you are observing Mars the week before opposition. Consider standing on a reasonably thick (temporary) carpet. The extra padding afforded your feet will make you feel much better.

Walls

The main consideration for the walls is their height. If people are going to be using the observatory, you will want the walls high enough to block out any direct light while still allowing access to a large area of the sky. Two meters (about 7 feet) in height is a standard size. For large observatories, slightly higher walls may be necessary to attain the angle needed to eliminate direct lights.

Observatory walls, especially those for personal use, range from thick to relatively thin. Observatories exist with walls made of stucco, brick, concrete block, tile, shingles, siding, and plastic. I chose plastic. Thicker walls retain heat. There are paints and protective overcoatings which reflect most light and heat but here in the desert coatings abrade rather quickly.

For the outside walls of my observatory, I used an ultraviolet-resistant PVC paneling called PalRuf. I could not recommend this material more highly. It is guaranteed for 20 years. The panels are corrugated, approximately 3 mm thick, and come in 0.66 m × 2.4 m (26 inches × 96 inches) sections. They can be easily cut with a pair of utility scissors. The edges of the walls are covered with a galvanized ridge cap. The panels are held in place with roofing screws which

Grasslands Observatory, owned by Tim Hunter and managed by James McGaha, both of Tucson, Arizona. A great combination of elevation, latitude and good weather. (Photo by Tim Hunter)

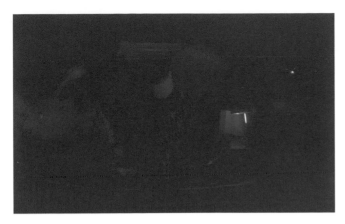

A nighttime panorama made by swinging the dome slit completely through the camera's field of view during a time exposure. This is the 3towers Observatory (see http://www.3towers.com). (Photo by Tim Hunter)

Jeff Medkeff at the computer console of Junk Bond Observatory. From this observatory Jeff, and the observatory owner David Healy, have discovered over 60 asteroids. (Photo by the author)

self-seal via attached Neoprene washers. The panels are very effective at reducing solar heating, the major consideration here in the desert.

Tip: Some PVC is only protected against ultraviolet light on one side – install that side facing outward.

The PalRuf used for the walls is white. I painted the interior of the observatory flat black. The paint I used is Krylon Ultra Flat Black Enamel. When I stay inside the observatory for an extended period of time, I gain between 0.5–0.75 on the visual limiting magnitude due to the black walls.

The PalRuf is attached to a wooden framing material (pine 2×4s) which measures 3.8 cm × 8.9 cm. The choice is yours whether or not you apply a facing material to the interior walls. If you do not (my personal recommendation), you can use the bracing for shelving, etc. The only reason to finish the interior walls is for appearance.

Door

The door of an observatory will get plenty of use. I suggest a quality solid core door with good hinges, a tough lock, and a good seal, top, sides, and bottom. This will also help with security. Add a dead-bolt if you are so inclined.

Attached rooms

Some amateur astronomers who live in colder climates have attached a "warm room" to their observatories. This is a good idea when the observatory is not very close to home. For a backyard observatory with a computerized telescope, this can be a better option than running remote control lines from the house.

If you are planning a warm room, be sure to insulate the wall between the warm room and the observatory. Heat currents are the bane of seeing and even the roof of the warm room will give off enough heat, when the room is in use, to seriously degrade any images in the part of the sky above it.

Interior considerations

The most important interior consideration is power. Alternating current must be run to the observatory. Whether you have a sub-panel within the observatory (the best option) or run a line from your house, allow for at least two circuits, 15 A each. If you are adding a warm room, add a minimum of two additional circuits, and one of them should provide enough amperage to operate whatever model heater you are planning to install.

Lighting is the second interior consideration. Dimmable red lights are mandatory. Dimmable white lights are optional. If you are a dedicated Moon observer, you will probably want the white lights.

Next is ventilation. If your observatory is sealed too tightly, you will have heat retention problems no matter how well your roof and outer walls are insulated. Some amateurs use passive venting, others use quiet, low-flow fans.

If you have the space, do yourself a big favor and build in or purchase some storage. This can be shelves or cabinets. If space is limited, consider a rolling cabinet such as those used for storing tools. If you choose a roll-off roof style of observatory, make use of the corner spaces of the room. Be sure to leave space around the telescope(s) for observing chairs. A great way to acquire some extra storage without taking up floor space is to use pegboard panels. It is amazing how many items can hang on the hooks which snap into the holes of the pegboard.

Finally, be sure there is a clock. I suggest one which has large numerals, illuminated by red light emitting diodes (LEDs).

Pier

Most amateurs install a pier upon which to place at least their main telescope. If you are doing this, a footing is essential. This is a block made of reinforced concrete set into the ground. Any concrete column (or, in certain cases, steel tube) above the ground is called the pier. The mount for the telescope attaches to the pier using bolts which are pre-set

The 600 mm f/5 telescope at the Grasslands Observatory, during an evening's work. The observatory is a cooperative effort between Tim Hunter (the owner) and James McGaha (the Observatory Director), both of Tucson, Arizona (see http://www.3towers.com). (Photo by Tim Hunter)

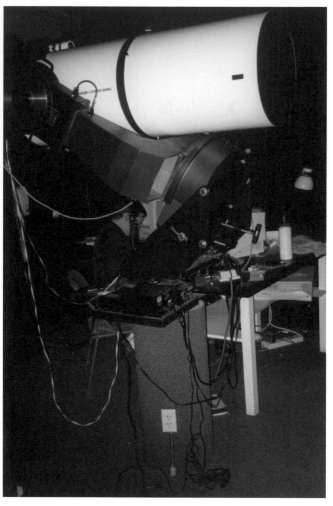

The center of Junk Bond Observatory (JBO), circa 2000. The 400 mm Meade SCT is the heart of an automated asteroid astrometry program. As a side-note, over 60 asteroids have been discovered at JBO. A tour of the observatory was being given as this photo was taken. Because of the automation, the presence of Jeff Medkeff (at the computer console in the background) is not required for operation.

into the concrete. Alternately, a thick steel plate may be attached to the bolts in the concrete and the mount attached to the plate. The bolts should not be too close to the pier edge.

The standard rule is that a footing should be twice the width of the pier and as thick as the pier in depth. Many builders go beyond this rule. Make the footing square. The stiffness of a square section is 1.70 times more rigid than a circular section of the same length. Check the local code for the proper depth of the footer.

When a round concrete pier is planned, many builders use a form into which the concrete will be poured. In the US, this is a thick-walled tube sold under the name Sonotube. The tube comes in widths of 8–56 inches (20–140 cm) and lengths up to 18 feet (5.5 m). The tubing can then be removed if desired, or it can be left in place. If installing a Sonotube pier above a footing, "pin" the pier to the footing with #4 or #5 rebar or similar material which is left protruding 30 cm or so out of the footing.

Whether the observatory has a floor made of cement or wood, it should be totally isolated from the pier. This is especially important for an observer who is planning any type of imaging. A 2 cm gap is enough space all around the pier, but this is not a hard rule. If you do this, you will be able to dance in your observatory without affecting the images.

A lot of builders hand-mix their piers for economic reasons. Unfortunately you can never hand-mix bagged concrete to the strength of quality-controlled ready-mixed concrete which comes from a batch plant.

Some piers are not centered within the observatory. This is intentional, and done to maximize the view in a certain direction. In the northern hemisphere, the direction is almost always south.

It has happened that, instead of a completely immobile mount, observers have noticed an image that seemed to chatter continuously when the drive motor ran. In one case, the mount was found to be *too* rigid, failing to dampen the natural resonance vibrations of the fork arms of the telescope. Some rubberized material between the mount and the pier solved the problem.

In my observatory, I made a shelf which I slipped over the pier prior to installing the plate for the mount. It is square with a 4 cm lip all round and only protrudes about 8 cm at most from the pier.

Cost

Finally, after considering all of the above, the ultimate question is whether you do it yourself or have the observatory built for you. This may or may not be a matter of finance. You could be wealthy beyond imagination and still choose to build it yourself. Whichever path you choose, become familiar with all of the above points. Use quality materials. Don't rush. Have an overall plan. Have a blueprint – or at least a diagram. Good luck!

One man's quest for dark skies

Jim Sheets, McPherson, Kansas

Aperture, location and convenience: I had it all. It was a 16 inch classic Newtonian on an equatorial pier, adjacent to my home in rural Kansas. Then came the move to town, lots of street lights, tall trees, and neighbors. A seven hundred pound telescope is not very portable.

The solution to my dilemma came when I spotted a used ambulance for sale. I began the conversion in December of 2000 and had "first light" for my new mobile observatory, StarTracker on 17 August 2001.

First I had to clean the vehicle of all equipment and logos. Then I had a local weld shop cut a 6 foot hole in the roof with a plasma torch. The weld shop helped fabricate a base ring flange that would support the dome assembly I had purchased. I mounted the pier to the floor of the chassis and built a suspended viewing floor above the original. I had box-beams welded to the chassis to mount four screw jacks to the corners of the frame to bring the vehicle up off the sprung weight when on location.

All of the electronics for operating the observatory are located in a panel in the coach. I also moved the red exterior lights into the coach for night vision. Once on location, it takes about 8 to 10 minutes to stabilize and level the observatory and be ready for operation.

Because of space confinement under a 6 foot dome I bought a new Celestron 11 GPS and mounted it on the pier. This GPS system lends itself perfectly to this application.

The StarTracker was once an ambulance. Jim Sheets, of McPherson, Kansas, has turned it into a fantastic mobile observatory.

The mobility factor alone has made the project worthwhile. Since "first light" I have made numerous excursions to dark sky environs and have enjoyed many hours of viewing. In October 2001 I attended the Great Plains Star Party and the Okie-Tex Star Party; in May 2002 I attended the Texas Star Party and plan to attend many more in the future.

I did give up some aperture. However, I gained a great deal of convenience and location is no longer a problem. The best part was unexpected – a chance to meet new people and make friendships with others who have the same passion for astronomy.

5.5 The social astronomer

Astronomy on the amateur level is a hobby, which is defined as "a pursuit outside one's regular occupation engaged in especially for relaxation." So relax. Relax alone or relax in the company of others. It is the latter on which this chapter is focused. And rather than me telling you about all the many options open to you, I thought it best to provide internet addresses and encourage you to read about them in their own words.

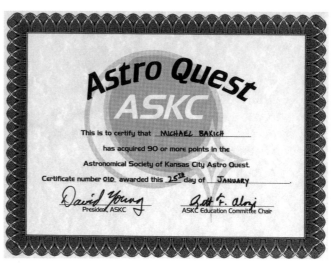

Your local astronomy club can set up an observing award of its own. The Astronomical Society of Kansas City called theirs "Astro Quest." It was very popular with members. (Photo by the author)

Saguaro Astronomy Club members observing the partial phase of the total solar eclipse of 26 Feb 1998, from the deck of the cruise ship Dawn Princess, near the island of Aruba. (Photo by Steve Coe, Phoenix, Arizona)

Large organizations

- American Association of Variable Star Observers (AAVSO)
  http://www.aavso.org/

- Association of Lunar and Planetary Observers (ALPO)
  http://www.lpl.arizona.edu/alpo/

- The Astronomical League (AL)
  http://www.astroleague.org/

- The British Astronomical Association (BAA)
  http://www.britastro.org/index.html

Astronomy clubs

Without question, the best step you can take is to join a local astronomy club and attend its meetings and star parties/observing sessions. This will place you into a group of like-minded individuals who can either answer your questions or help you figure out where to get them answered. Most astronomy clubs have a number of members who look for opportunities to share information about the hobby we all love. But not all clubs are paradise.

I remember the Tucson Amateur Astronomy Association meetings of the mid-1980s. Members would literally stand on their chairs to scream obscenities at one another, trying to make their point. In a way, it was very funny...in a sad way.

Your club will not be like that. It simply wouldn't be tolerated in this day and age. Oh, once you become involved you'll see the politics, the cliques, the differences of opinion, etc. But you'll also experience the camaraderie that comes in sharing a starry sky with newly-found friends. You'll hear talks about a variety of subjects. And soon, you will be given the ultimate privilege – to share the wonders of the heavens with those who were once just like you.

e-mailing lists / e-groups

A number of moderated lists (or groups) exist on the internet which allow the posting of messages to the entire group through the submission of e-mails. Personally, this has been a tremendous resource when I've had a question. Such groups have members with years of experience in all areas of amateur astronomy.

The most popular way to join a group is to go to http://groups.yahoo.com

From this page, you can search out subjects which interest you. A word of caution, if I may. Start by subscribing to one or two groups, and build up from there. In 2001, I received an average of 136 e-mails per day (that's right, nearly 50,000 for the year!). Here are some of the more popular e-groups (all of which I subscribe to) . . .

- amastro
  369 members through Yahoo, 706 through subscribed lists
  Advanced
  A retromoderated mailing list dedicated to the discussion of amateur astronomy, with an emphasis on observing. The first alphabetically and the best. Created 18 May 1999. For the first three years, the total number of postings was 7679.

- ap-ug
  971 members
  Intermediate–advanced
  Short for "Astro-Physics Users Group." General discussions concerning Astro-Physics astronomical products – scopes and accessories.

- historiacoelestis
  74 members
  Advanced
  historiacoelestis is a discussion group and resource for all those with an interest in the History of Astronomy, Astronomers, Telescopes, Observatories, Books and Star Lore, whether amateur or professional.

- meteorobs
  73 members
  Advanced
  The e-mail mailing list meteorobs is dedicated to email discussions relating to amateur meteor astronomy.

- Planetary_Nebulae
  174 members
  Advanced
  This group is for anyone who loves planetary nebulae. Observations can be logged here as well as general discussion concerning any planetary. This group will offer a database of sorts for those to exploit when planning observing lists.

- Refractors
  900 members
  Beginning–intermediate
  This site is for all refractor users to enjoy. Discussions of equipment and observing are encouraged. Owners of all types and apertures are welcome.

- SoftBisqUser
  989 members
  Beginning–advanced
  This group is not the official page of Software Bisque, but has been set up for users to help each other. Requests for technical support from Software Bisque should be sent to them. However, other users may be able to assist you on this list with the use of the software.

Robert Kuberek ("Flash" on IRC) with his 130 mm AstroPhysics refractor set up for solar viewing next to the author's observatory, May, 2002. (Photo by the author)

- Starmaster_scopes
  710 members
  A place for owners of Starmaster telescopes to talk about tips and tricks they may have learned.

Internet news groups

Acting something like e-groups, news groups allow you to post comments. Rather than sending them out as e-mails, they are posted to a special area where you need software called a newsgroup reader (I use Forte's Free Agent) to download and/or respond to them. One of the most popular groups is sci.astro.amateur. Admittedly, there is a lot of fluff to wade through, but there is some valuable information to be had. One advantage is the sheer number of subscribers. If you have a question, you are almost certain to get an answer (or 50!). My recommendation regarding e-groups is to lay low for a while. Wait at least a month before posting a question or a reply. By then, you will have a feel for the subject matter and level of the group.

Internet chat groups

Starting with the proliferation of personal computers, chat rooms have been the rage. Such internet meeting places have a stigma attached to them because of the proliferation of illegal, pornographic, or sex-related activities. However, there are a lot of quality chat rooms out there. One is a specific astronomy-related chat room that I frequent (especially late at night). It is the #sciastro channel on the Star Link server.

To access any chat channel, you need what is known as an Internet Relay Chat (IRC) client. The two most often used in the #sciastro chat room are mIRC and Pirch. Both are shareware – you download them, try them, and then send a nominal fee to their developers. See these (and more) IRC clients on the Stroud's CWSApps site, on the internet at http://cws.internet.com/irc.html

After installation, add the Star Link server. You will need the following information:

Description = Star Link Houston Server

IRC Server = Houston.TX.US.Star Link–IRC.org

Port = 6667

Alternately, the #sciastro web site has a separate interface. It may be found on the sciastro.net web site, maintained by Steve Carroll. Go to

http://sciastro.net/

Simply click on #sciastro Chat and follow the instructions.

Astronomy vacations

- Advanced Observing Program
  Kitt Peak National Observatory
  http://www.noao.edu/outreach/aop

- Arizona Sky Village
  Portal, Arizona
  http://arizonaskyvillage.com/

Rick Singmaster at the Texas Star Party 2001. Here, Rick is trying to find a moment's rest. It's not easy when you make the most popular large amateur telescope ever. (Photo by the author)

- New Mexico Skies Guest Observatory
  50 km east of Alamogordo, New Mexico
  http://www.nmskies.com

- Newton Observatory Bed & breakfast
  Osoyoos, British Columbia, Canada
  http://www.jacknewton.com

- Vega-Bray Observatory Bed & breakfast
  Benson, Arizona
  http://www.communiverse.com/skywatcher

Large star parties

- Astrofest
  Just northwest of Kankakee, Illinois
  Elevation: 175 m
  http://www.chicagoastro.org/aindex.html

- Chiefland Star Party
  10 km south of Chiefland, Florida
  Elevation: 16 m (yes, 16!)
  http://www.c-av.com/

- Enchanted Skies Star Party
  Socorro, New Mexico
  Elevation: 1400 m
  http://www.socorro-nm.com/starparty/

- Grand Canyon Star Party
  Grand Canyon National Park, Arizona
  Elevation: 2135 m (South Rim); 2440 m (North Rim)
  http://www.tucsonastronomy.org/gcsp.html

- Great Plains Star Party
  Scopeville, near the town of Parker, Kansas
  Elevation: 305 m
  http://members.tripod.com/ciorg1/

- Nebraska Star Party
  25 miles southwest of Valentine, Nebraska
  Elevation: 945 m
  http://www.nebraskastarparty.org/

- Okie-Tex Star Party
  Black Mesa, Oklahoma
  Elevation: 1515 m
  http://www.okie-tex.com/

- Oregon Star Party
  Indian Trail Spring, Oregon
  Elevation: 1525 m
  http://www.oregonstarparty.org/

- Riverside Telescope Makers Conference
  8 km south of Big Bear City, California
  Elevation: 2315 m
  http://www.rtmc-inc.org/

- Rocky Mountain Star Stare
  15 km northwest of Lake George, Colorado
  Elevation: 2690 m
  http://www.rmss.org/rmss2001.htm

- Table Mountain Star Party
  32 km northwest of Ellensburg, Washington
  Elevation: 1938 m
  http://www.tmspa.com/

- Texas Star Party
  Fort Davis, Texas
  Elevation: 1540 m
  http://feenix.metronet.com/~tsp/index1.html

One of the most social amateur astronomers you could ever hope to meet. Kathy Machin, of Kansas City, Missouri, next to her superb homemade 300 mm Dobsonian-mounted reflector. The author took this photo at the Texas Star Party 2001, just before Kathy left for the Southern Skies Star Party in Bolivia.

Star party do's and dont's

- Follow all posted instructions. Many star parties provide a sheet of general nighttime rules. If you're new to the star party circuit, read and memorize the list.
- Avoid using light. Note that I did not say "avoid using white light." While it is true that white light is far more detrimental to dark-adapted eyes, it is also true that most red lights are also too bright. And, in most cases, it is the brightness of the light, not its color, which un-adapts your eye. So if your star party light is a big red spotlight, save it for Full Moon. Finally, in advance of any star party or observing session, deal with any vehicle lights, especially those activated by open doors or trunk lids. The easiest method? Simply pull the bulbs.
- This next point is so important, I chose not to combine it

with the one above. It involves light from the now ever-present laptop computers. Most (but not all) amateurs who use these devices are imaging and may not be as attuned to the needs of visual observers. Yes, all have red filters over the screens, however, nearly every one I've seen has still been too bright. When I walk across an observing field and can see someone sitting in a chair enveloped in a faint glow, I know the screen's light is excessive. The kinder, gentler laptop setups have a box (painted black) over them, with only the front side cut out.

- Check with the organizers about restroom facilities before going. You can save yourself a lot of misery, and probably a lot of unnecessary steps, if you know what to expect.
- Clean up after yourself. Do not leave litter of any kind lying around where it can be blown about by the wind or encountered by someone else. If there is no designated trash receptacle, bring one of your own. This is especially important when the group is a guest on public or private land. At the end of each observing session at our dark-sky site, we do a "white-light survey" to check for trash and other items which could be inadvertently left behind.
- Take your time when looking through someone else's telescope. A quick glance does you no good and, quite frankly, it doesn't leave the best impression on the telescope owner either. Questions are not only permitted, they are encouraged. A beginning amateur astronomer may ask, "Am I seeing two lines on Jupiter?" An amateur with some experience might query, "What is the true field of view of this eyepiece?" And an advanced amateur's question may be something like, "Do you know what time the snack bar closes?" The advanced observers are always one step ahead.
- With regard to viewing through someone else's telescope – FOCUS! I have suggested this to thousands of people, young and old, beginning and advanced, and I repeat it here. Our eyes are not all the same. Even a minute amount of focusing can reveal details within Saturn's rings that were invisible before. It is that critical. If you are unfamiliar with the telescope through which you're observing, simply ask, "Excuse me, how do I focus your telescope?"
- Bring your children. Control your children. Star parties are family affairs and children are welcome. However, make certain they stay with you at all times. Many children become cranky if they stay up too late (some adults, too!), but are happy to sleep in the car. Always remember that their comfort and safety is your first concern.
- If you must have music, you may be in the wrong place. Realize that individual taste in music is so widely variant that it can almost be guaranteed that somebody nearby doesn't like what you're listening to. If you must have baseball, well now, that's a totally different story! Use headphones.
- Regarding driving (or even walking), try to minimize the amount of dust you are sending into the air. Many star

parties, especially in drier climates, are held at heavily used sites with little nearby vegetation.

- Unless you have a motor home, leave your pets at home. Amateur astronomers seem to have a strong distaste for stepping in – or setting a case or piece of equipment in – a pile of crap.
- Have your favorite binoculars nearby. Even experienced observers do this. They can be useful to pin down a star field or to check to see if the object you're observing through the telescope is visible in binoculars. They are also handy to have in case of a "wide-angle" event, such as the appearance of a bright meteor which leaves a smoke trail. If your binoculars are higher power, placing them on a tripod or binocular mount can provide some stunning views of the larger DSOs (the Andromeda Galaxy, the Orion Nebula, the Pleiades, etc.) throughout the night.
- Bring a chair to sit in. Others may have an extra chair, but don't count on it. If you're attending an observing session or evening gaze, bring liquid (water is best, but I cannot abide the stuff) and a light snack. If you're attending a several-night star party, you may want to bring a lot more. In warmer seasons, insect repellant may be necessary. Bring your own.
- Never move somebody else's telescope without permission. If the object seems to be drifting out of the field of view, briefly mention this to the telescope's owner and s/he will more than likely show you how to adjust for that, either manually or with slow-motion controls.
- Check the rules before you smoke. There may be a fire hazard. Note the direction the breeze is blowing and be mindful of others who do not appreciate second-hand smoke.
- Beware the words, "imaging" and "astrophotography."

People involved in such endeavors are likely (as a general rule) to be less sociable, at least while performing those tasks. Use no lights in the area, don't walk in front of their telescope or even too closely, and only engage in the amount of conversation they are comfortable with. Some amateurs who live in harsher climates look at the generally clearer, darker, drier sites that star parties provide as a way to acquire new images, rather than as a nighttime social event.

- Arrive before dark if at all possible. There are two reasons for this. (1) After telescopes have been set up, lights from vehicles are frowned upon – strongly! (2) You may literally not be able to find the location in the dark. Star parties are generally held at remote sites, making them nearly impossible to find after sunset (some even with directions!). If you have any doubt about your ability to get there, a good idea is to visit the location during the daytime.
- Let others know that you're planning to leave. Somebody may be in the middle of imaging. For the cost of a few minutes time you can avoid making an enemy. Use double the care with vehicle lights when leaving as you did when arriving.
- Finally, and most importantly, be courteous! The people that you meet are not being paid, and they're especially not being paid to serve you. Most are either members of an astronomy club or their spouses. The rest of the people you meet will be visitors just like you.

The death of a star party

The following is an unsolicited letter which appeared in our astronomy club newsletter following the 2001 Texas Star Party. TSP 2001 to be kind, was inadequate. If you are

NGC 1318. (Kodak Elite Chrome 200, pushed 1 stop. Nikon F2 with a 105 mm Nikkor lens. Photo by Ulrich Beinert of Kronberg, Germany)

I anticipated a great week of observing as Michael Bakich and I prepared for our trip to the 23rd annual Texas Star Party. This was my second journey to this legendary gathering of astronomers from all over the world, and if my first time there was any indication, this really promised to be a great gig. Michael and I had planned since last year to spend the whole week there.

As we set up our camp on the upper observing field the weather was cloudy and somewhat cool, but my spirits were still high. We could at least get some sleep in the cooler weather if the tents were not sitting in the blazing Sun and hopefully the skies would clear as darkness approached. Several of Michael's friends from all over the country had also made the journey to Ft. Davis and had brought some awesome hardware. Unfortunately, the skies did not clear and this was pretty much the pattern for the week. We had 2¹/₂ nights of clear skies out of a total of 7 nights and those skies did not have what one would call good seeing. Bad weather is one of the unavoidable hazards of this hobby. Bad administration of the event is not.

Dirty bathrooms, overflowing trash cans, backed up septic tanks, and non-responsive or unavailable staff quickly became the order of the day. This all became even more irritating on Friday afternoon when I attended a meeting of the Astronomical League's Southwest Region, in which our Club is a member (more on that later). The meeting seemed to be orchestrated as an ego feeding for the organizers of the TSP. The Chairman of the event didn't even show. The only excuse given was that she was indisposed, yet the Vice-Chairman had a walkie-talkie and said he could reach her if somebody really needed to talk to her…go figure that one out. I knew we had trouble as soon as the Vice-Chairman asked if everything was okay with the bathrooms and services. Here it was, the second-to-last day, and one of the highest ranking staff members did not have a clue as to what was happening on the grounds. And even as several questions and complaints were put forth about the lack of services, he quickly brushed them aside with a quick "Sorry, but better luck next year" answer, so he could proceed with the REAL business of bragging about the amount of money they were making and how great they were for putting on this wonderful event! I left the meeting somewhat disturbed, and with more questions than answers. Questions like, "If they have so much money, why don't they have anybody cleaning the bathrooms?" "Where is the staff, and why are they not walking around getting the Prude Ranch to fix the things that are not working?" "If our Club is a member of the organization that is putting this event on, why don't we have anybody on the staff?"

As the President of this Club, I am planning on addressing these problems with the organizers of the TSP. Hopefully, we can find some answers and try to assure that this event is run in a manner that will attract people instead of repelling them. I will keep you informed.

As for a recommendation, I would have to give it two thumbs down. I am going to address some of the problems that we encountered with the organizers, but I hold little hope that we will get much action from them. Most people I have discussed this with seem to agree that the organizers have placed themselves in a select, detached group and are not very willing to open themselves to criticism or help from "outsiders." Something to do with "Command and Control." Well, I guess we will see.

Mark Marcotte
President, EPAC

Steve Carroll of Fort Scott, Kansas, is often seen at star parties throughout the midwest US. Steve, a master woodworker, is seen here finishing a piece for a 500mm StarMaster telescope. (Photo by the author)

organizing an event for amateur astronomers, you may wish to read this letter several times. Let me stress that I am not saying that this was the only star party I have been to which was a bad experience. Far from it. It is simply the one which is closest in time.

5.6 Light pollution

I would be remiss as an amateur astronomer not to say something about excess nighttime lighting, now generally called light pollution. Over the past several decades, nighttime outdoor lighting has increased at a high rate. Today, amateur astronomers view with dismay satellite images which no longer show the night side of our planet as dark.

With all the really bad lighting, it is tough to pick out the one light which conforms to the full-cutoff standard. (Photo by the author)

Forget about astronomy for a moment. These El Paso streetlights produce so much glare that safety is the issue. (Photo by the author)

For much of the populations of North America and Europe the night sky is no longer black, or even dark. It is instead, a bright yellow–orange, aglow from poorly designed light fixtures and almost bereft of stars. For some people born in the 1980s and later, it is possible that the only object they have seen in the night sky is the Moon. In today's world, the most immediately endangered natural resource may be the dark night sky.

Few, if any, of us are for the utter abolition of outdoor lighting. Such lighting is necessary for the safety and security of people during work or recreational activities. Other lighting which highlights historical settings, parks, shopping areas, or that used for advertising may not be necessary, but it is desirable by many. Such is the state of our world today.

But often lighting exists that seems to have no purpose whatever. And even the lighting that has purpose – important purpose – is poorly designed or in need of repair. Take security lighting for example. A study by the US Department of Justice concluded that there is no statistically significant evidence that street lighting impacts the level of crime, but that there is a strong indication that increased lighting decreases the fear of crime. That is an amazing statement. In essence, poor lighting can only make you feel safer even when you are not, actually safer.

Loss of the night

During the time I was writing this book, the world news agency CNN reported that a study, conducted by scientists at the University of Padua, Italy, and the US National Oceanic and Atmospheric Administration, said that a truly dark, starry sky is unavailable to two-thirds of the world's population, including 99% of people in the continental United States and Western Europe.

The survey measured for the first time how light degrades the view of the stars in specific places around the globe. The report described regions of the world where true night never occurs because it is blocked by lights from neighboring cities and towns. In addition, the study concluded that 40% of the US population and one-tenth of the world's people live in places where it is never dark enough for their eyes to become adapted to night vision.

What the eye needs

Lower levels of lighting are perfectly acceptable once you become dark adapted. Problems arise when the eye must move from dark to light, then light to dark – back and forth until it really isn't functioning well in either situation.

A major problem is encountered by people who are older. Changes within the eye, such as diminished dark adaptation speed, can prove dangerous. Entering your car under brilliant service station lights and then driving out onto a highway with much dimmer or no lights places the elderly at a terrible disadvantage. Older eyes are also more sensitive to glare, particularly from bluer light sources.

When will everyone learn that bluer light sources for nighttime use are inefficient? The eye, especially at night, does

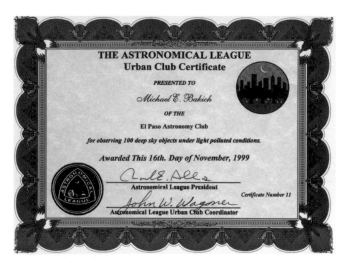

Think you can't see anything from your light-polluted backyard? The Astronomical League says differently. A great observing award for beginning amateur astronomers to try for. (Photo by the author)

Methods of saving energy (and money)

Decrease excessive Illumination	Brighter lights than recommended by engineers waste money because our eyes quickly adapt to either level of illumination
Choose the best lamps	Selecting high-pressure sodium lights saves 40% over metal halide and 150% over mercury vapor lights
Shield all lights	The installation of full "cutoffs" will direct the light to where it is needed, reducing dangerous glare with the side benefit of preserving the sky

not see blue light well. So for blue light sources like mercury vapor lights to look as bright as yellowish sources such as sodium vapor lighting, they must consume more wattage.

Emphasize the pocketbook

What can we as amateur astronomers do? How can we encourage our elected representatives and the people in our communities to change? My best advice is to leave the astronomy-related arguments at home. Emphasize what everyone understands: MONEY. The International Dark Sky Association (IDA), an organization that has been leading the anti-light pollution movement, estimates that over a billion dollars is wasted each year in unnecessary energy use. Find the IDA on the internet at
http://www.darksky.org

A strategy

From personal experience of trying to have a lighting ordinance enacted in El Paso, Texas (as of this writing the City Council has yet to vote on it . . . but it's only been in their hands for 11 months!) I can suggest a four-part strategy. First, focus on economics. You can mention other benefits such as security and night-sky issues, but the clearest note will be sounded by the Almighty Dollar (or Euro).

Second, make it as easy as possible for the political administration. Have copies of lighting ordinances from other cities, copies of statistics and economic benefit reports, copies of engineering and safety reports, etc., and compose a first draft of the ordinance that your group is proposing.

Third, get as many people involved as possible and keep them interested. Draft letters which can be mailed or e-mailed to all members of the governing body. Make them

Good thing there's a mountain in the background or the El Paso lights would go on forever! (Photo by the author)

This light, installed when a local Chinese restaurant opened, is one of the worst examples of light pollution I have ever seen. (Photo by the author)

short, non-confrontational and be sure to identify yourself as a voter. Write letters to the editor of your local newspaper. Host astronomy events in town, and pass out light-pollution information to everyone who attends.

Fourth, choose a "point-person" who is tactful, charming, sweet, is not easily irritated, and who will be able to attend the meetings of the governing authority. Do not choose a vocal activist, however helpful s/he has been.

Conclusion

Finally, realize that light pollution is not just bad for the astronomical community. The IDA's *Lighting Code Handbook* says it best.

> Bad lighting hurts everyone. It starts a cascade of negative consequences – beginning with the loss of our views of the heavens, continuing through falling levels of safety and utility, irritation of neighbors and wildlife, disturbance of the rhythms of day and night that are vital to many natural systems, damage to the aesthetic appearance of our communities, wasted monetary and natural resources used to produce wasted light, and increased air pollution and carbon dioxide levels from wasted fossil fuels. There is nothing good that comes from bad lighting.

CHAPTER SIX

What to observe: solar system

Nightfall

by Susan S. Carroll

Fort Scott, Kansas

I stand, in late twilight, squinting above and waiting for darkness to fall on the field below. All the preparations have been made, and the 18-inch stands next to me, waiting, expectant. The eyepiece is already in the focuser. Now all I have to do is wait for the deep turquoise twilight to turn to black night.

Around me are scores of others, waiting for the same thing. Voices raised in jolly repartee fill the air, punctuated by hearty laughter. But as the Sun dips below the horizon, the voices are quieter. When the first stars appear, the voices drop and here and there the light from a red flashlight can be seen, piercing through the oncoming dark.

I flex my arms and legs and stretch; it will be a long night. I search the sky above me for the two stars that mark the path to the first object I have planned. As soon as I see them, I grab the dowel on the bottom of the secondary cage and swing the 18 into position, sighting the object in my finder. I press my eye against the eyepiece and spot the wondrous nebula that is, even now, giving birth to new stars. I squint and stare at the object to see how many stars I can find within it.

As the nebula rises in the sky, so do I; and curse my petite stature. Up, up, up – until I stand, en pointe, at the eyepiece, straining to see. Finally I must concede defeat, and grab the ladder/chair I use for putting my eye level with the eyepiece.

The dark deepens. The voices around me have become murmurs, and I hop down off my perch to sight another object. Now everyone is faceless; identification of any one of the people around me will be based on voice or shape familiarity.

While I peer into the scope, the inevitable "what are you looking at?" comes from somewhere below me and to my left. "Nothing," I say truthfully. "I'm still looking for it." I don't recognize either the voice or the shape, so my answer comes out more easily, even automatically. There are times when I love nothing more than to share my views with others; but this isn't one of those times.

As midnight approaches, I sit and take a small break. My husband walks by, keys and change jingling in his pockets. "I'm going to hit the sack," he announces. "Okay," I say, and stand back up and stretch again. He knows that it is fruitless to ask me when I will be

ready; somnolence is far from me. And the object I have waited months to see does not appear above the southern horizon for at least another two hours.

Finally, the first stars in my deep southern object peek above the horizon; it won't be long now. I stare at them, hoping that I can raise them higher with my own eagerness. This doesn't happen, and I content myself with scanning the horizon yet again for the enemy. If clouds move in and obscure my view, the southern deep sky will be lost to me. Fortunately, there are no enemy troops in sight.

Now the object is standing, just on the horizon, and I swing the 18 down. This one will be a knee-biter; no chair or ladder is required. I sight the object through the finder, and eagerly drop to my knees. The ground is grass over sharp coral, and I congratulate myself for the forethought to wrap heavy socks around each knee. I cannot, however, see the object in my eyepiece, so I consult the finder again. I nudge the 18 down just a millimeter, put my eye back to the eyepiece – and gasp. There it is. It fills the field of view of the medium-high magnification eyepiece I am using – but it is all there. I stare at it so intensely that it feels as though my eyeball will jump out of its socket. Show me your secrets, I plead. Just a hint will do. As though it heard my silent plea, the object shines brighter and I gasp again.

Now a blanket of adrenaline and sheer wonder covers me. That one split second has given me the thing I treasure most – enlightenment. One more tiny piece of the puzzle of the universe has been implanted in my brain, to be retrieved and lived over and over again. As the object sinks below the horizon in farewell, I pat the 18's secondary cage and again feel the gratitude and humility I have for the exquisite primary mirror she has.

Like a dog sniffing the breeze, the 18 rises up a little of her own accord, as if to say "What's next? Let's go!" I reluctantly turn back to reality, and sight my next object. Now the voices are much fewer, and only occasionally does the red beam of an astronomer's flashlight shine. As I peer once more through the eyepiece, I shake my head in wonder at the many people who have already gone to bed. How could they have missed this, I ask myself. But I shake off the thought. After all, what others do when observing is not my concern.

As night finally begins to fade, and a faint pink glow is visible in the east, I carefully put the 18 to bed, and slide her long silver cover over her. Silent now, she doesn't protest. I close my eyepiece case and look again to the east. A beautiful sunrise is just beginning, so I make my way to the beach, find a suitable rock to sit on, and watch. The glorious sunlight becomes brighter and suddenly I feel the signs of exhaustion; the stiff knees, the protesting feet, the sore back. I am the only one still up, so I watch the sunrise in silence, hugging my stiff knees in front of me. Then I reluctantly make my way back to our trailer for some much needed sleep. After all, behind this sunrise is another sunset, and another opportunity to look heavenward at the wonders contained within. I fall into bed; to sleep, and to dream, of what I will find after that sunset.

Robert Kuberek of Valencia, California, captured the essence of nightfall with this photo taken 24 Jul 1999 at the Kennedy Ranch in Leona Valley, California.

6.2 The Sun

The Sun: statistics

Central temperature	15 600 000 K
Surface temperature	5800 K
Distance from Earth	149 597 870 km
Size	1 390 000 km diameter
Mass	1.989 x 10^{30} kilograms (333 000 × the Earth's mass)
Average density	1400 kg/m³ (1.4 × water)
Solar irradiance (energy reaching Earth)	1380 Watts/m²

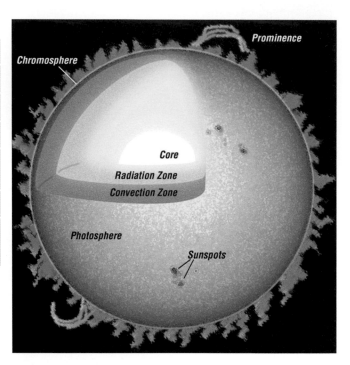

The interior of the Sun. (Illustration by Holley Y. Bakich)

The Sun's interior

The core of the Sun has a radius of 200 000 km. Its temperature is on the order of 15 600 000 K. The core is the only area where energy is being produced. Just outside the core is a region called the radiation zone, extending from 200 000 to 500 000 km and having a temperature of 7 000 000 K. This is a region too cool for nuclear fusion. Energy moves from the core outward by being absorbed by the atoms in this area and then re-radiated. Because the direction of the re-radiated energy is random, it takes on the order of 170 000 years for energy to move through this region.

Between the radiation zone and the surface lies the convection zone. This region extends to 696 000 km. The temperature near the radiation zone is 2 000 000 K but falls progressively as the surface is approached. The visible surface of the Sun, the photosphere, is only about 500 km thick with a temperature of 5800 K. Above the photosphere is a 1500 km thick layer known as the chromosphere. Its temperature varies from about 6000 K to 20 000 K.

The next 8000 km is called the transition zone. Near the chromosphere, the temperature is 8000 K, but it rises rapidly. The outer atmosphere of the Sun, called the corona, extends out from the transition zone. The temperature of the corona is an incredible 2 000 000 K.

Energy production

The Sun shines by nuclear fusion. Every second in the core of the Sun approximately 640 000 000 000 kg of hydrogen is being converted into 635 000 000 000 kg of helium. The "extra" 5 000 000 000 kg has not been lost, but rather has been converted into energy in accordance with Einstein's equation $E = mc^2$.

Mechanically, what's really happening is that, trillions of times per second, two hydrogen nuclei collide. A hydrogen atom is the simplest of all, consisting of a positively-charged proton in the nucleus and a negatively-charged electron outside the nucleus moving in quantum motion (don't think of the Moon orbiting the Earth – it's much more complex). At the pressures and temperatures within the Sun's core, all of the electrons are stripped from all of the hydrogen atoms.

Two protons, moving very fast, collide and, if the conditions are right (and they often are) they will stick together, or fuse, from which the word fusion comes. In the process, two energy particles are created, a positively-charged anti-electron and a neutrino. The neutrino immediately heads for space, essentially interacting with none of the Sun's mass. The anti-electron will eventually collide with an electron, causing the annihilation of both. This produces two photons of energy.

The twin proton originally created eventually collides with a single proton and forms a light helium nucleus. Again, energy is released in the form of a photon. Also, one of the protons has been converted into an uncharged neutron.

The final step involves the collision of two light helium nuclei. A normal helium nucleus is created (two protons, two neutrons) with the extra two protons being released. Because of the way this energy-creating process begins, it is called the proton-proton reaction.

Don't look at the Sun through your telescope without a proper solar filter! You hear it over and over. Well, this is the reason. Ron Lambert of El Paso, Texas, had his TeleVue eyepiece capped when he accidently moved the Sun into the field of view of his telescope for less than a second. The heat caused the eyecap to melt. (Photo by the author)

Composition of the Sun

Using spectroscopy, 67 elements have been identified in the Sun. By mass, hydrogen comprises approximately 75% and helium 25%. This doesn't leave much mass for the other 65 elements. In terms of number of atoms, the most common elements are listed in the table below.

Most common elements in the Sun

Element	Percent of Atoms
Hydrogen	92.1
Helium	7.8
Oxygen	0.078
Carbon	0.043
Nitrogen	0.0088
Silicon	0.0045
Magnesium	0.0038
Neon	0.0035
Iron	0.0030
Sulfur	0.0015

Rotation

The Sun exhibits differential rotation, meaning that its speed of rotation is different at different latitudes. At the equator the period is 25.4 days (the mean synodic equatorial rotation period is 27.2753 days) while near the poles it is 36 days.

In early June and December we see sunspots travel in straight lines across the solar disk. This is because, at those times, the Earth lies in the plane of the Sun's equator. From January to May sunspots follow a more curved path to the north, indicating that the Sun's south pole is tipped towards us. From July to November the reverse happens and the spots curve to the south.

Solar cycle

Every 11 years, solar activity peaks, resulting in greater numbers of sunspots and flares as well as changes in the solar magnetic field. In fact, this "11-year cycle" varies from as few as 9.5 years to as many as 12.5 years. Since the cycle is variable in length, its start and extent are not really known until after they occur. The start of any given solar cycle is defined as the minimum of solar activity. The sunspot cycle was discovered in 1843, by the German astronomer Heinrich Schwabe, who noticed that a maximum number of spots was visible every decade. 15 years later, in 1858, the astronomer Wolf published his formula for determining the daily sunspot number:

$$R = k(10g + f)$$

where g is the number of sunspot groups, f is the number of individual sunspots, and k is a calibration factor, different for each observer.

Photosphere

The surface of the Sun, as we see it, is called the photosphere. Of course, this is not really the Sun's "surface," but rather a region far from the core where the gas becomes opaque. The word photosphere means "sphere of light." The photosphere is the lowest layer of solar atmosphere observable. It is about 500 km thick. Observing the photosphere is easy, either naked-eye or telescopically with visual solar filters.

While observing the photosphere telescopically, if your seeing is good (on the order of 1–2 arcseconds) you may be able to note the presence of granulation. Due to less-than-great seeing, granulation is most often observed as a

The Sun. (Unitron 100 mm, Pentax ME Super 35mm camera. Kodak Ektachrome 100 film. Photo by the author)

Observed solar cycles

Number	Approximate Start
1	February 1755
2	June 1766
3	June 1775
4	August 1784
5	March 1798
6	July 1810
7	March 1823
8	December 1833
9	June 1843
10	January 1856
11	February 1867
12	December 1878
13	July 1889
14	August 1901
15	July 1913
16	July 1923
17	October 1933
18	February 1944
19	March 1954
20	December 1964
21	June 1976
22	September 1986
23	May 1996

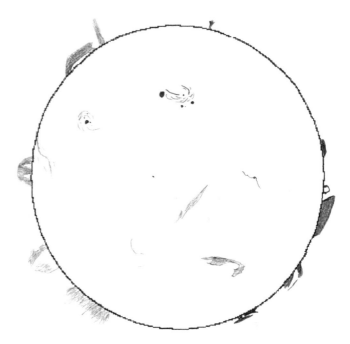

The Sun through a hydrogen alpha filter. (5 Sep 2000. 100 mm f/15 Unitron refractor. Sketch by the author)

mottling effect due to vast bubbles of gas whose centers are rising and whose edges are sinking.

Faculae are bright areas visible on the photosphere. "Facula" is Latin for "little torch." A superb solar filter will reveal their presence all over the solar disk, but they are most often observed near the solar limb. There, the contrast between the faculae and the darkened limb is most noticeable. Limb darkening is caused by the Sun being a sphere. Near what we see as the edge of the solar disk, the light reaching us here on Earth must travel further through the solar atmosphere, causing that area to be less bright. Faculae are roughly ten percent brighter than other areas of the photosphere.

Chromosphere

The "sphere of color" as it is called is found in a region slightly above the photosphere. In this region the temperature varies from about 6000 K (near the photosphere) to around 20 000 K. Because of the higher temperature, hydrogen emits a type of light found in the red end of the spectrum known as hydrogen alpha light. Spectrally, the h-alpha line is at 656.28 nm, and a filter which passes only this light is called a hydrogen alpha filter. (See the section on "Solar filters" in the chapter "Filters".)

With a quality h-alpha filter, expect to see a number of features associated with the solar chromosphere. In fact, expect to be amazed. I had observed the Sun through visual filters for years before I had my first look through an h-α filter. What a difference! There is so much more to be seen. At times, features change in real time, as you are looking at them. I still observe the Sun using an h-alpha on every clear day. And while I'm on this subject, I will mention that h-α filters are much more sensitive to thin, high clouds than are visual solar filters.

Observing Tip: After all precautions have been taken at the telescope, for lengthy sessions, remember to apply sunscreen to yourself.

Many of the features that you will see through an h-alpha filter are prominences, bright clouds of gas ejected from the Sun and shaped by its magnetic field. Prominences come in many different shapes and sizes. You may see spikes, loops, trees, detached prominences, and many more. It is also easy to see prominences silhouetted against the solar disk, but in this case they look like dark lines and are called filaments. The chromospheric network is a web-like pattern that is quite easy to see. Very bright areas around sunspots are called plages, or sometimes plage flares. If you're not in a hurry you can see plages change shape in real time.

Solar flares

A solar flare occurs when magnetic energy that has built up in the solar atmosphere is suddenly released. A large amount

The author using his classic 100 mm Unitron refractor with a 0.7 Å DayStar hydrogen alpha filter at the Texas Star Party 2001. (Photo by Susan Carroll)

A brightness qualifier F, N, or B is generally appended to the importance character to indicate faint, normal, or brilliant respectively (for example, 4N).

Astronomers also classify solar flares by their X-ray brightness. To do this, special X-ray detectors (some as part of the instrument packages of weather satellites) measure the flares' wavelengths in the range from 0.1 to 0.88 nm. There are three categories. X-class flares are the most powerful. Such flares are major events that can affect us here on Earth by triggering radio blackouts and long-lasting radiation storms. M-class flares are less powerful than X-class flares. They can cause brief radio blackouts that affect higher latitudes. C-class flares sometimes produce minor radiation storms but these flares are small with few noticeable consequences here on Earth.

Corona

This region has a very high temperature, at around 2 000 000° (but it is so tenuous that there is very little "heat"). Within the corona, intrusive features called solar prominences can be seen. These are large glowing masses of hydrogen gas coming from the chromosphere (a lower region of the solar atmosphere, but above that of the photosphere). Prominences can be small or extremely large affairs, some fairly quiet and some extremely violent, reaching very high speeds and travelling great distances. These can be seen using special equipment and this will also be explained later.

Sunspots

Sunspots, which are features of the photosphere, come in many shapes and sizes, according to the whim of the Sun's magnetic field. The field traps gases within the sunspot areas, slowing the gases' motion and making the sunspot cooler than the surrounding area.

Usually, sunspots consist of a dark central region called the umbra, surrounded by a lighter region known as the penumbra. Sunspots appear dark because they are at a lower temperature than the surrounding photosphere. This is only a contrast effect, however, as even the darkest umbra of a sunspot is 10–12% as bright as the photosphere. (Penumbras range from 50–85% of the brightness of the photosphere.) The temperature of the penumbra is typically 1000 degrees below that of the photosphere and the umbra between 1500–2000 degrees cooler than the photosphere.

It has been observed that sunspots begin their existence as very small features. These are called pores. Specifically, pores are sunspots without a penumbra and which undergo rapid changes. Sunspots grow for a widely variable time and then begin to shrink. Some spots last a day. Others have been known to last over a month. Sunspots generally are found in complexes of several or dozens. Usually these groups have one or two major spots.

While observing sunspots, you may notice that when a

of radiation is emitted across the whole electromagnetic spectrum. Solar flares are the largest explosions in the solar system.

A solar flare usually has three stages. First is the precursor stage, where the release of magnetic energy begins. Astronomers can detect low-energy X-ray emission during this stage. The second stage is known as the impulsive stage. Here, protons and electrons are accelerated to high energies. During the impulsive stage, radio waves, high-energy X-rays, and gamma rays can be detected. The third stage of a solar flare is the decay stage. In this stage, the gradual buildup and decay of low-energy X-rays can be detected. Each of these stages can last from a few seconds to as much as an hour.

Flares are classified on the basis of area at the time of maximum brightness, measured in hydrogen alpha light:

Importance 0 (Subflare): $\leq$ 2.0 square degrees
Importance 1: 2.1–5.1 square degrees
Importance 2: 5.2–12.4 square degrees
Importance 3: 12.5–24.7 square degrees
Importance 4: $\geq$ 24.8 square degrees
(One square degree is equal to $(1.214 \times 10^4 \text{ km})^2 = 48.5$ millionths of the visible solar hemisphere.)

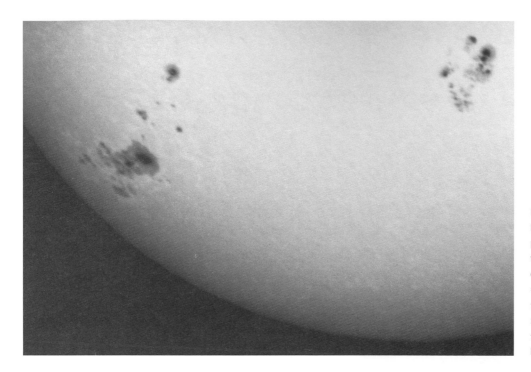

Huge sunspots. Ulrich Beinert, of Kronberg, Germany, took this photograph on 19 Sep 2000, through a Vixen R-114S with eyepiece projection (f = 18mm), F_{eff} = 3100 mm at f/27. Exposure was 1/20 second on Kodak Technical Pan 2415. Developed 8 minutes in HC-110, Dil. B, 23°C. Shot through a Baader AstroSolar filter.

spot is near the limb of the Sun, it looks concave. In fact, the first to notice this was Alexander Wilson (1714–1786), a Scottish astronomer after whom the phenomenon is named. The Wilson Effect describes sunspots which are symmetrical when at the center of the solar disc but which become asymmetrical as they approach the Sun's limb. Also, the part of the penumbra closest to the center of the Sun's disk becomes narrower before it disappears.

Observing the Sun in visible light

You must be very careful when observing the Sun. Never look directly at the Sun with the naked eye or with any optical instrument and always cover your finder when observing the Sun. That said, there are a number of safe ways to observe the Sun. (See the "Solar filters" section in the "Filters" chapter for suggestions.)

If even a portion of your days are free, I highly recommend the Astronomical League's Sunspotter's Observing Club. (Photo by the author)

Solar projection

Solar projection can be as simple as a pinhole or as complex as an attachment through which your telescope can project an image of the Sun. Some observers use an adjustable arm which holds a metal plate to which a white sheet of paper is attached. Others use a box-like assembly, which surrounds the paper. A box is a better choice because it will darken the surrounding area and increase contrast, allowing smaller detail to be seen. Whichever setup you choose, attach it to the telescope so it will remain behind the eyepiece as the telescope is moved. In either case, on the sheet of paper is a circle. The size of the circle (and thus the projected image) should be 150 mm, which is the standard size of image used by observers all over the world.

Mark the four cardinal points (N, S, E, W). Focus the Sun and ensure it fits your circle. If it does not, either the distance between the eyepiece and your sheet of paper will need adjusting or you will have to use an eyepiece of a different focal length. At this point you can observe the Sun, possibly obtaining a sunspot count. Other observers trace the sunspots that are visible. A few others actually photograph the image produced. It's up to you. Remember to note the date and time of your observation, as well as sky conditions and instrument/eyepiece used. If you choose to use your telescope in this way, do not use cemented eyepieces as the heat of the Sun will damage them.

The principle of pinhole projection. (Photo by the author)

Sun pillar, Tucson, Arizona, 7 Aug 1986. (Photo by the author)

Uncoated mirror solar telescope

Some solar observers have employed Newtonian reflectors with uncoated mirrors in conjunction with welder's glass to observe the Sun. Please note: BOTH the primary and secondary are left uncoated. Each uncoated mirror reflects approximately 4% of the Sun's light. As this is still too much infrared and ultraviolet radiation, a #8 welder's glass is used at the eyepiece to eliminate the danger. Some observers have gone as far as replacing a threaded filter's glass with the welder's glass. To reduce scattered light, the mirror cell for the primary and the mirror holder for the secondary are both painted black.

Direct viewing

As indicated above, the "Filters" chapter covers the types of filters you can obtain for direct viewing of the Sun. A filter which fits over the objective of the telescope is called a pre-filter. If you use a pre-filter, be absolutely certain that it cannot be dislodged accidentally by the wind or by being bumped. If you are in doubt, securely tape the side flange of the filter to the telescope tube. When I construct a mounting cell for a solar pre-filter, I always include a large thumb screw which acts to secure the cell to the tube.

In the past, I used soft brass as the bolt material for this thumb screw. Lately, however, I have switched to a Neoprene bolt. I find that such a plastic material tends not to leave marks around the edge of the dew cap.

White-light flares

With a visual solar filter (projection does not show these), observers have infrequently reported the appearance of white-light flares. These last for very short periods of time, on the order of 5–15 minutes. A blue filter or a solar filter which slightly tints the Sun's disk blue will provide your best chance for seeing white-light flares.

Observing sunspots

Apart from being fun, sunspot counts provide a gauge of solar activity. Numbers of sunspots have been recorded on a daily basis since the middle of the eighteenth century. The standard for counting sunspots is to obtain a Wolf number, as detailed earlier in this chapter. Serious solar observers, however, increase the accuracy of the Wolf numbers they obtain by calculating their personal k number. The k number is a correction factor (or reduction coefficient) used to equalize observations by different observers. Such observations can be quite different due to different telescopes being used (with different quality), atmospheric conditions, and factors specifically related to each observer.

Your k number is given by the simple formula

$$k = r/n$$

where r is the Wolf number you calculate, and n is the published value for the same date.

For the "official" number of spots for any date, check the Sunspot Index Data Center (SIDC) which is on the internet at http://sidc.oma.be/index.php3

To get a reasonable estimate for your k number, calculate it for ten straight observing sessions and average the results. The actual number you get is not important. What is important is that it is relatively consistent. If your first observation gives you a k number of 0.8, your second 0.4, your third 1.1, and your fourth 0.5, you may want to wait a few months to calculate your k number. Don't stop observing the Sun. In fact, you may wish to start sketching what you see. Looking for detail in sunspots, memorizing it, and then transferring what you see to paper is an excellent way to improve your observing consistency. The calculation of a k number for a specific telescope and specific observing location need only be done once.

Sunspot classification

There have been several classification systems for sunspots over the years. To state this simply, I suggest that you use the

A multiple exposure of the Sun over the course of an entire day. Ulrich Beinert, of Kronberg, Germany, says about his photo: "First, I metered the correct exposure for a single image: 1/60 sec. @ f/22. Then I thought about how often I wanted to have the Sun in the picture. The picture was to begin at 11:10 and end at 20:40, and I wanted to take an image every 30 minutes, so 20 images in total, all on one piece of film. So, 20 images had to add up to 1/60 second at f/22. 1/60 divided by 20 resulted in an exposure of 1/1200 second at f/22 per image. Since I don't have a 1/1200 setting, I exposed 1/1500 second. That meant I was 1/300 sec. off at the end. 1/300 sec. is 1/5 of the required 1/60 sec. Since the exposure on slide film needs to be very exact, I exposed (after metering the new correct exposure and dividing by 5) 1/750 sec. at f/5.6 in the early evening twilight."(Lens: Canon EF 15 mm fisheye. F-stop: 22. Exposure: 17 x 1/1500 sec. @ f/22, 1x 1/750 sec. @ f/5.6. Film: Kodak Select Series Elite Chrome 100. Filter: none)

McIntosh classification system. It is a three-letter system which will classify any type of sunspot, as follows.

First letter (indicates the overall group structure)

A — An individual spot or group of spots without a penumbra or bipolar structure.

B — Group of sunspots without a penumbra in a bipolar arrangement.

C — Bipolar sunspot group, the principal spot of which is surrounded by a penumbra.

D — Bipolar group, the principal spots of which have a penumbra. At least one of the two principal spots has a simple structure. The length of the group is in general less than 10°.

E — Large bipolar group in which the two principal spots, which are surrounded by penumbrae, generally exhibit a

THE SUN | 163

complex structure. Numerous smaller spots between principal spots. Length of group at least 10°.

F – Very large bipolar or complex sunspot group. Length at least 15°.

H – Unipolar spot with penumbra.

Second letter (indicates the appearance of the penumbra of the largest spot in the group)

x – No penumbra.

r – Rudimentary (incomplete) penumbra, irregular boundaries, width only around 2000 km (0.2° on the Sun or 39 in the sky); brighter than normal penumbrae, granular fine structure (r-penumbrae are a transition between photospheric granulation and normal penumbra filaments).

s – Symmetrical, almost circular penumbrae with a typical filament structure which is directed outwards, diameter less than 2.5° on the Sun (30 000 km). (The umbrae of the sunspot form a compact group in the vicinity of the center of the penumbra. Included in this class are elliptical penumbrae around a single umbra. Spots with s-penumbrae change only slowly.)

h – Symmetrical penumbrae like type s, but with a diameter of more than 2.5°.

k – Asymmetrical penumbrae like type r but with a diameter of more than 2.5° measured in N–S direction to avoid elongated leaders which are decaying and inactive. (If the diameter exceeds 5°, it can be assumed with certainty that both magnetic polarities occur within a penumbra (bipolar group) so that the group can be classified as Dkc, Ekc, or Fkc.)

Third letter (indicates the distribution of the spots within the group)

x – Individual spot.

o – Open distribution. The area between the preceding and following principal spots is free of sunspots so that the group clearly comprises two parts with a differing magnetic polarity. (The gradient of the magnetic field strength along the line of connection between the two principal spots is correspondingly low.)

c – Compact distribution. The area between the main spots is populated with many large spots, of which at least one has a penumbra. In extreme cases the entire sunspot area can be surrounded by one huge penumbra. (In groups with a c-distribution, there are steep local gradients in the magnetic field strength.

i – Intermediate type between o and c. (Some sunspots without a penumbra can be observed between the principal spots.)

Two mock Suns, also known as sundogs. (Courtesy NOAA)

6.3   The aurora

The aurora borealis and aurora australis are beautiful, dynamic, luminous displays seen in the night sky near the poles. Although called the northern lights, "aurora borealis" actually means "northern dawn." The aurorae cover vast areas known as auroral ovals, centered on the magnetic north and south poles.

A beautiful aurora northwest of Fairbanks, Alaska, 6 Mar 2000. (Gene Dolphin of San Diego, California, used an Olympus OM-10, 50 mm, f/1.8, 5-second exposure on 800 ASA Kodak print film)

Cause of aurorae

Auroras occur because the Sun is emitting a stream of charged particles called the solar wind. As these particles pass the Earth, our magnetic field channels them in the directions of the two magnetic poles. In our atmosphere, the particles from the solar wind interact with different gases, causing colorful displays. We see another effect of the solar wind on the tails of comets. Because of the particles flowing outward from the Sun, comet tails always point in the direction opposite the Sun. At our distance from the Sun, the solar wind has an average density of eight particles per cubic centimeter and an average speed of 400 km/s.

Auroral colors and shapes

As the particles from the solar wind interact with our atmosphere each type of gas glows with a particular color. This is actually due to the atom or molecule of gas absorbing the energy from the charged particle and then re-emitting it a short time later as a distinct wavelength.

Because of this absorption/re-emission, high-altitude oxygen, at about 350 km altitude, glows with a deep red color. Oxygen at lower altitudes, about 100 km up, emits a

green light, the brightest and most common auroral color. Ionized nitrogen molecules produce blue light; neutral nitrogen glows red (another common auroral color). Together, the nitrogens create the purplish-red lower borders and rippled edges of the aurora.

Auroral shapes are varied and ever-changing. Most appear as curtains or very thick ribbons. Streamers and "light beam" shapes are also common. The extent of the aurorae, and the intricacy of their shapes, depend totally on the degree of solar activity.

Seeing aurorae

The aurora is a common sight from northern Europe, Canada and Alaska. In many locations, there are more nights with aurorae than without! In the southern hemisphere, the auroral oval falls mostly over Antarctica, so the aurora australis is not often seen. Auroral displays reach as low as 65 km above the ground with their upper extent being higher than 750 km, although that is rare.

Aurorae may appear anywhere from the pole to latitude 30°. In fact, during the past four years, I have personally seen nice auroral displays twice from El Paso, Texas, latitude 31°45′ N. The displays were "nice" here, but judging by some of the photographs I've seen, they were phenomenal further north!

Those who live near the northern auroral oval often see aurorae along the northern horizon early in the evening. The aurora gradually moves south, caused by that location on Earth rotating under the thicker part of the auroral oval. If you live

Aurora east of Fairbanks, Alaska, 6 Mar 2000. Olympus OM-10, 50mm, f/1.8, 5-second exposure on 800 ASA Kodak print film. (Image by Gene Dolphin of San Diego, California)

Mark Cunningham, just outside the city of Craig, Colorado, recorded this aurora on 30 Mar 2001.

Photographing aurorae

I don't know of anyone using CCD astro-cameras to photograph aurorae, so we'll concentrate on film cameras. For digital still cameras, many of the same rules apply.

For tripod-mounted shots, which are the most common, choose a wide lens, one with a focal length shorter than 50 mm. You will capture more of the aurora and any stars registered will not be as badly trailed. Bracket your exposures with the lens at its widest setting (smallest focal ratio). If you have a fast lens (faster than f/2), start with exposures of 10 seconds and move up to 30 seconds. Tend to longer exposures on moonless nights. If the Moon is near Full, don't shoot, just watch.

Lens filters are not recommended when photographing aurorae. The green atomic oxygen emission line causes interference fringes due to the parallel faces of the filter. This will reveal itself as a series of concentric rings on the final photograph.

The brighter the aurora, in general, the faster it is moving. Use faster shutter speeds and faster film. ISO 800–1600 will be sufficient in most cases, or your could have your developer push-process your film. For more normal displays, use a fine-grained film with ISO 200–400.

south of the auroral oval, look for aurorae from about 10 p.m. to 3 a.m., during periods of known high solar activity.

For predictions of auroral activity via the internet, see http://www.sec.noaa.gov/Aurora or http://www.sel.noaa.gov/today.html especially the 3-Day Solar Geophysical Forecast link. For specific views of the up-to-the-minute auroral ovals, see http://www.sec.noaa.gov/pmap

Wide-angle view of the northern lights east of Fairbanks, Alaska, 6 Mar 2000. (Olympus OM-10, 28 mm, f/2.2, 7-second exposure on 400 ASA Kodak Gold print film. Image by Gene Dolphin of San Diego, California)

6.4 The Moon

I admit it. I am a "deep-sky snob." I don't like the Moon. Oh, I appreciate what it's done for us over the years, slowing our rotation rate and helping mix things up via tides. And, yes, I know that the Moon offers something for any observer, that its face is ever-changing, that following the terminator through a lunar month can be a fascinating endeavor, etc., etc. Still, over the years when the Moon has been bright, I've been more interested in having observing friends over for supper rather than meeting them in the field.

The Moon: statistics

Size	3476km
Mass	7.15×10^{22} kg
Density	3.340 g/cm^3
Distance from Earth	3.844×10.5km
Eccentricity	0.0549
Inclination of orbit	5.145°
Orbital period	27d07h43.7m
Synodic period	29d12h44.1m

Unfortunately, the time when the Moon is brightest and most intrusive to the rest of the sky – that is to say at Full Moon – is also the worst time to view the Moon. At that phase the Sun is behind the Earth (as we face the Moon) shining directly down on the Moon and any shadows are minimized, allowing little detail to be seen.

Conversely, the best times for viewing the Moon are from shortly after New Moon until about two days after First Quarter (evening sky) and from about two days before Last Quarter to almost New Moon (morning sky). During these times, shadows are longer and features stand out in sharp relief.

The area of the Moon where this is especially true is along the line dividing the light and dark portions. This line is called the terminator. Prior to Full Moon the terminator is where sunrise is occurring and after Full Moon it is where sunset is occurring.

Along the terminator you can catch the tops of mountains protruding high enough to catch the sunlight while being surrounded by dark lower terrain. On the floors of large craters you can follow the "wall shadows" cast by sides of craters hundreds of meters high. All these features seem to change in real time, and over even a partial night's worth of observing the differences are striking.

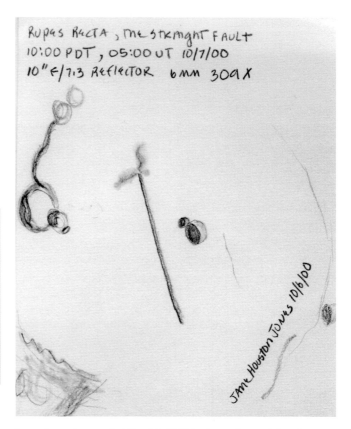

Rupes Recta, the so-called "Straight Wall" is a lunar observer's favorite. (Sketch by Jane Houston Jones of San Francisco, California)

During the crescent phase, you'll also observe "earthshine" on the unlit portion. This is sunlight reflecting off the Earth and on to the Moon. It is easily visible with the naked eye.

The formation of the Moon

In the early part of the nineteenth century, it was suggested that the origin of the Moon was a simple one: when the Earth formed, the Moon formed nearby. This is called the "common condensation" theory or sometimes the "double planet" theory. The main problem with this theory is the very different densities of the Earth and Moon. The Earth's density is 5.515 g/cm^3, while that of the Moon is only 3.34 g/cm^3. If the Moon formed next to the Earth, and at the same time, its density should match almost exactly.

In 1878, British astronomer Sir George Howard Darwin (1845–1912) proposed that the Moon was once part of the Earth. According to Darwin, this was a time near the formation of the Earth, when the planet was still molten.

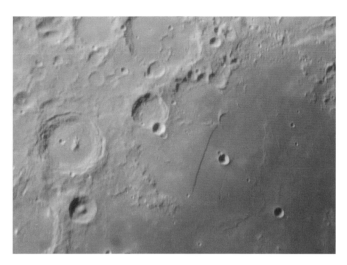

Rupes Recta, the Straight Wall. (Imaged by Arpad Kovacsy with a Nikon CoolPix 950 through an AP 155 EDT refractor)

Because of the rapid rotation of the Earth, a large piece broke off, eventually becoming the Moon. This is known as the "fission" theory. Darwin proposed the Pacific Ocean basin as the place from which the Moon was ejected. Unfortunately, no satisfactory mechanism has ever been suggested to account for such an event. Also, if the Moon were thrown off of a spinning Earth, it would orbit in the plane of the Earth's equator, and not be inclined (5.145°) as it is.

The Moon. (Unitron 100 mm, Pentax ME Super 35 mm camera. Kodak Ektachrome 100 film. Photo by the author)

A third possibility, proposed at the start of the twentieth century is known as the "capture" theory. As the name implies, this theory states that a separate astronomical object encountered our planet at a very close distance and was captured by the gravitational field of the Earth. This sounds plausible, but the mechanics of the situation almost demand the presence of a third body. In such a chance encounter, the interaction of three objects results in one of them (the Moon) being slowed to an orbital speed.

New light was shed on this problem when the American Apollo astronauts returned samples of rock and soil gathered from the surface of the Moon during six missions from 1969–1972. Suddenly, not only the density of the Moon was known but also its chemical composition. It was learned that the Moon has a similar composition to the Earth's crust. Both have approximately the same proportions of silicon, magnesium, manganese, and iron. The Moon has far smaller proportions of volatiles, but higher proportions of non-volatiles such as aluminum and titanium. In the minds of most astronomers, the differences in chemical composition, along with the Moon's lack of an iron core, ruled out both the common condensation theory and the fission theory.

In the mid-1970s, American astronomers William K. Hartmann and Donald R. Davis proposed an alternative theory of the Moon's formation. According to this new hypothesis, the Moon formed from debris blasted out of the Earth by the impact of a Mars-sized body. The great age of lunar rocks and the absence of any impact feature on Earth indicate that this event must have occurred during the Earth's own formation, some 4.5 billion years ago. The "Big Splat" theory, as it is called, answers many questions that have been raised.

Such an impact would vaporize elements with low melting points and disperse them. Since only the crust and outer layer of the Earth's mantle would be blasted out, Earth's iron core would remain intact. This explains the low density and low iron content of lunar material. The differences in composition between Earth and Moon can be

Interesting facts about the Moon

The Moon has 1940 named features, of which 1545 are craters.

The largest crater on the Moon, Hertzsprung, has a diameter of 591 km. It is the second largest crater in the solar system (next to Beethoven on Mercury) and the largest crater on any satellite.

The Full Moon, though bright is only 1/400 000 as bright as the Sun. If the entire sky were covered with Full Moons, we would receive only about one fifth the illumination of the Sun on a bright day.

According to the Belgian mathematician Jean Meeus, the extreme distances between the centers of the Earth and the Moon are minimum = 356371 km and maximum = 406720 km. Meeus evaluated ten centuries worth of data from 1500–2500 AD.

The First and Last Quarter Moons are only about 10% as bright as the Full Moon.

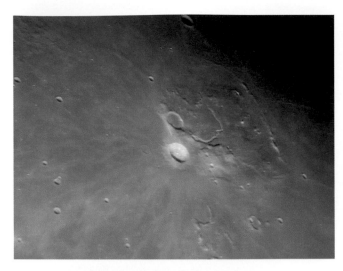

Schroter Valley showing the lightly colored Wood's Spot. The feature known as the "Cobra Head" is easily seen. Very few images have captured the subtle color of this area. 24 Feb 2002. (Imaged by Arpad Kovacsy with an AP 155 EDT refractor using a Sony digital video camera)

accounted for by the additional material of the impacting body. Even the Earth's 23.5° tilt can be explained: it was knocked over by the blast!

In 1997, Sigeru Ida, Robin M. Canup, and Glen R. Stewart presented data regarding the impact. They found that the cloud of vaporized rock ejected by the blast flattens into a disk after about a few months. According to their calculations, about two-thirds of this material falls back to Earth. To produce a satellite the size of the Moon, they propose a minimum impactor mass of two to three times as massive as Mars. They point out, however, that such an impact would leave the Earth–Moon system with twice as much angular momentum as it has today. They offered no mechanism to reduce this initial angular momentum.

Phases of the Moon

The Moon orbits the Earth every 27.3 days, approximately. Because the Earth is also in orbit about the Sun, the Moon

and Sun line up roughly every 29.5 days. The changing position of the Moon with respect to the Sun cause the Moon, as seen here on Earth, to cycle through a series of phases. One complete set of phases is sometimes known as a lunar month.

Traditionally, the lunar month begins at New Moon. It was called New Moon long ago because people thought that it was at these times that the Moon was being reborn. We can't actually see New Moon from Earth, as the lit side is in the same direction as the Sun and, in the sky, the Moon's position is also quite near to that of the Sun's. From New Moon to Full Moon, when the entire side of the Moon facing us is illuminated, is a continuous growth of the sunlit portion of the Moon, but only as we see it. The Moon is half in light and half in darkness all the time. It is simply its position which determines how much of the bright, sunlit part we see.

So, from New, the Moon progresses through Crescent, First Quarter, Waxing Gibbous, Full, Waning Gibbous, Last or Third Quarter, Crescent and back to New again to start another lunar month. Waxing is used to denote that the illuminated portion is getting larger while waning is used when the illuminated portion is getting smaller.

Observing the Moon

The Moon is a brilliant object when viewed through a telescope. Many observers employ either neutral density filters or variable polarizing filters to reduce the Moon's light. Between these two methods, I prefer the latter because of its variability. I have recently been introduced to a better way, however. Throughout history, observers have used a simple method to help them observe the Moon in comfort: turn on white lights when observing the Moon between First Quarter and Full. The addition of white lights suppresses the eyes' tendency to dark adapt at night and, in fact, causes the eye to use normal scotopic vision which is of much higher quality than dark-adapted photopic vision.

Color filters are almost never used by lunar observers. Some have noted a darkening of the lunar basalt when

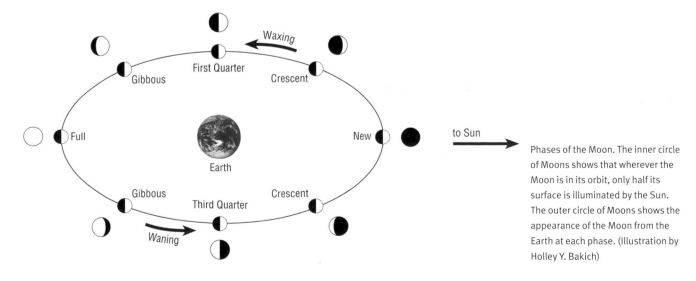

Phases of the Moon. The inner circle of Moons shows that wherever the Moon is in its orbit, only half its surface is illuminated by the Sun. The outer circle of Moons shows the appearance of the Moon from the Earth at each phase. (Illustration by Holley Y. Bakich)

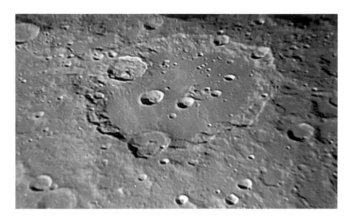

Clavius. (Imaged by Arpad Kovacsy with a Nikon CoolPix 950 through an AP 155 EDT refractor)

reddish filters are used. A red filter can also help moderate seeing problems, with the added benefit of cutting down the brightness of the image.

Two other methods for reducing the brightness of the Moon are the use of high magnification and an aperture mask. The first restricts the field of view to a very small area of the lunar surface and reduces light throughput. The second causes your telescope to act like one of much smaller aperture, but at the same focal length.

One of the best ways to familiarize yourself with the Moon is to undertake an observing project. One such project, known as the Lunar Observing Club, is offered through the Astronomical League. To receive the certificate, you must be a member of the League, either individually or through an astronomy club. For details, see the specific page on the internet at

http://www.astroleague.org/al/obsclubs/lunar/lunar1.html

In the UK, lunar observing is coordinated by the British Astronomical Association. See the specific page on the internet at http://mysite.freeserve.com/lunar/index.html

The near side of the Moon is divided into lighter areas called highlands and darker areas called maria (the Latin word for seas). The maria are lower in altitude than the highlands. The dark material filling the maria is solidified basaltic lava from earlier periods of lunar volcanism. Essentially anything not covered by lunar basalt is a highland. The highlands consist of ancient lunar surface rock, anorthosite, and materials thrown out during the creation of the impact basins. For observing, the highlands are a treasure trove of mountains and valleys, bright areas and shadows.

Of the 1940 named features on the Moon, 1545 (nearly 80%) are craters. There are many more craters in the highlands than in the maria. The size range of craters is large and some observers make it a personal challenge to see either (1) how small a crater they can see or (2) how many small craters in a given area they can observe with a particular telescope. A detailed map of the Moon is, of course, a necessity for a project of this type. For #2, generally, a large mare (the Latin word for sea) or a crater with a large, flat bottom is chosen. The observer then searches for craterlets, the name given to small craters. For example, if you search the large crater Plato, on its floor are four craterlets, each about 2 km across. This is considered a test for an upper-range medium-sized telescope.

When observing larger craters, note whether you can see rays emanating from them. These features were formed when crushed rock sprayed out from an impact of a meteor. They form streaks in a radial pattern and which can be a great distance from the crater itself. A good example is the crater Copernicus.

Imaging the Moon

Certainly the easiest of all celestial objects to image, the Moon is nevertheless difficult to image well. The Moon is large and bright and essentially any camera (film, digital, CCD, video) may be used. That makes it easy. But the Moon is

Gassendi. (Imaged by Arpad Kovacsy with a Nikon CoolPix 950 through an AP 155 EDT refractor)

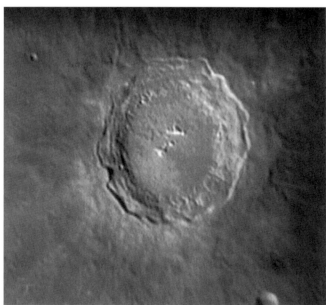

Copernicus. (Imaged by Arpad Kovacsy with a Nikon CoolPix 950 through an AP 155 EDT refractor)

This series was taken 30 Dec 2001, by professional photographer Shay Stephens of Seattle, Washington, through a Sony F707 Digital camera. In this time-lapse sequence, the Moon was re-imaged every 2 minutes 35 seconds, with the last exposure of longer duration to bring up a magnificent panorama of the city of Seattle. The exposures were then combined using Photoshop Elements. Note the Moon appears nearly the same size no matter its location. Jupiter is seen in the upper right.

also composed of vast areas of low contrast and very little color differential. That makes it difficult.

Another way to capture the Moon (almost exclusively on film) is to shoot with a regular or telephoto lens with the Moon low in the sky and with mountains, trees, city skyline, etc., in the foreground. I have found that such a photograph can be a good opening image for a talk about the Moon.

CCD users who desire to image the Moon almost always have to employ some type of filter or aperture mask along with a fast exposure to decrease the brightness of the lunar surface. The keys to good CCD images are great seeing and the

A lunar halo. The Moon is behind one of the towers near the 3towers Observatory, operated by Tim Hunter of Tucson, Arizona (see http://www.3towers.com)

correct image manipulation to highlight low contrast details. Alternatively, a large number of short-duration images may be examined for the few which show the best seeing.

6.5 Eclipses and transits

Eclipses and transits are dramatic events loved by amateur astronomers around the world. There was a time – even as recently as a century ago – when much could be learned from eclipses and transits. Studies of the Sun's outer atmosphere were conducted during eclipses and measurements were taken during transits to determine the distance scale of the solar system. Today, these events yield no great science, and yet they are still anticipated with great fervor. Eclipse trips and cruises are big business; public observing sessions of eclipses help astronomy clubs promote their existence and programs; and images and videos of eclipses are among the most numerous produced by amateur astronomers.

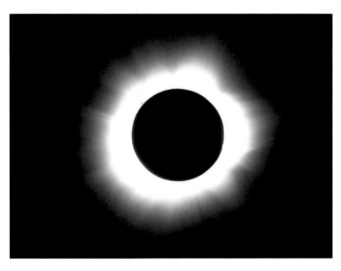

This image shows the full extent of the corona of the 21 Jun 2001 total solar eclipse. (Imaged from Chisamba, Zambia, with a Canon D30 by Charles Manske of Watsonville, California)

If I had one astronomy-related wish for each of you reading this book, it would be that, at some time in your life, you could stand underneath a totally eclipsed Sun. As of this writing I have witnessed six total eclipses, each one an amazing testimony to sublime celestial geometry. The remainder of this chapter will give you the details, but for the feeling of the moment, I quote from *Corona and Coronet* by Mabel Loomis Todd (Cambridge University Press, 1898):

> With an indescribable out-flashing at the same instant the corona burst forth in mysterious radiance. But dimly seen through thin cloud, it was nevertheless beautiful beyond description, a celestial flame from some unimaginable heaven. Simultaneously the whole northwestern sky, nearly to the zenith, was flooded with lurid and startlingly brilliant orange, across

which drifted clouds slightly darker, like flecks of liquid flame, or huge ejecta from some vast volcanic Hades. The west and southwest gleamed in shining lemon yellow.

> Least like a sunset, it was too sombre and terrible. The pale, broken circle of coronal light still glowed on with thrilling peacefulness, which nature held her breath for another stage in this majestic spectacle. Well might it have been a prelude to the shriveling and disappearance of the whole world – weird to horror, and beautiful to heartbreak, heaven and hell in the same sky.

> Absolute silence reigned. No human being spoke. No bird twittered. Even sighing of the surf breathed into utter repose, and not a ripple stirred the leaden sea. One human being seemed so small, so helpless, so slight a part of all this strangeness and mystery! It was as if the hand of Deity had been visibly laid upon space and worlds, to allow one momentary glimpse of the awfulness of creation.

> Hours might have passed – time was annihilated; and yet when the tiniest globule of sunlight, a drop, a needle-shaft, a pinhole, reappeared, even before it had become the slenderest possible crescent, the fair corona and all color in sky and cloud withdrew, and a natural aspect of stormy twilight returned. Then the two minutes and a half in memory seemed but a few seconds – a breath, the briefest tale ever told.

Shadows

Shadows of the Earth and Moon have two parts. The penumbra is the fainter outer shadow. The umbra is the dark inner shadow. The reason there are two shadow densities is that the Sun, which is casting the shadows, is not a point source. Thus, some light from the top of the solar disk enters the shadow cast by the sunlight originating from the bottom of the solar disk.

Lunar eclipses

Eclipses of the Moon occur when the Moon passes into the shadow cast by Earth. They always take place at Full Moon. Lunar eclipses can be observed from any part of the Earth where the Moon is in the sky at the time (essentially, at night). This means that, over time, far more lunar eclipses than solar eclipses can be seen from any particular location.

Circumstances of a solar eclipse. Where the Moon's umbra touches the Earth, the eclipse will be total. Within the penumbra, the eclipse is partial. If the umbra does not quite reach the Earth, the eclipse there will be annular. (Illustration by Holley Y. Bakich)

New Moon occurs every 29.5 days, approximately. However, we don't experience a lunar eclipse (or a solar eclipse at every New Moon) because the Moon's orbit is tilted to the Earth's orbit around the Sun by about five degrees. The two intersections of the Moon's orbit around the Earth and the Earth's orbit around the Sun are called nodes. Only when the Sun is at one of the lunar nodes and the Moon is at its Full (or New, in the case of a solar eclipse) phase can an eclipse occur.

The Moon does not disappear during a total eclipse

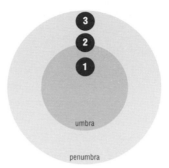

The circumstances of lunar eclipses are shown here. (1) A total eclipse. The entire Moon passes through the umbra. (2) A partial eclipse. A portion of the Moon passes through the umbra. (3) A penumbral eclipse. A portion (or all) of the Moon passes through the penumbra but does not pass through the umbra. (Illustration by Holley Y. Bakich)

Total lunar eclipse of 21 Jan 2000, here just getting started. (Image (from video) by Charles Manske of Watsonville, California)

because some sunlight is scattered onto the Moon's surface by the atmosphere of the Earth. Total lunar eclipses have a wide range of colors, from a dim yellowish-white to orange, copper, reddish-brown and nearly black. The color depends on factors such as the amount of dust and clouds in our atmosphere at the time.

Three types of lunar eclipses are possible: total, partial, and penumbral. A total lunar eclipse occurs when the entire visible surface of the Moon is covered by the umbra of the Earth's shadow. The contrast between the brilliant lunar face and the dark portion of the Earth's shadow causes the Full Moon's circular surface to gradually change as if it is going through phases. Only during and near totality does the covered portion of the Moon become visible, and then in color.

By the way, the Sun, Earth, and Moon do not have to be

exactly aligned for a total eclipse to occur. The Earth's umbral shadow measures approximately 9000 km at the distance of the Moon, so the Moon can pass through its center, or quite a bit above or below that line. The closer to the centerline of the umbra the Moon passes, the longer the eclipse. Another variable contributing to the length of a lunar eclipse is the Moon–Earth distance (and, to a much lesser extent, the Sun–Earth distance) at the time of the eclipse. If all conditions are maximized, the total phase of a lunar eclipse can last up to 1 hour 45 minutes.

A partial lunar eclipse occurs when only a fraction of the Moon's surface is obscured by the umbra. This is still a rather spectacular view. The color can still vary dramatically, although less dark than in the case of a total umbral eclipse. At the Moon's distance, the diameter of the Earth's penumbral shadow is approximately 16 000 km.

A penumbral eclipse occurs when the Moon enters only the penumbra of the Earth's shadow. The brightness of the Moon's surface gradually decreases and also some extremely subtle color changes can occur. The smaller the percentage of the Moon's surface entering the penumbra, the more difficult the observation. Most penumbral eclipses are noticeable (and thus worth observing) if the percentage of the Moon's surface covered by the penumbra is greater than about 40%. Atmospheric conditions and elevation of the Moon above the horizon should also be taken into account.

Observing a lunar eclipse

Lunar eclipses are easy and safe to observe. No equipment at all is necessary, although some amateurs prefer the view through binoculars or a telescope with a wide field of view, wide enough to encompass the entire Moon.

During total lunar eclipses amateur astronomers are often asked to estimate the brightness of the lunar disk at mid-

Danjon's rating system for lunar eclipses

Danjon number	Overall coloration	Details of eclipse
L = 0	Very dark eclipse	Moon almost invisible, especially at mid-totality
L = 1	Dark eclipse, gray or brownish in coloration	Details distinguishable only with difficulty
L = 2	Deep red or rust-colored eclipse	Very dark central shadow, while outer edge of umbra is relatively bright
L = 3	Brick-red eclipse	Umbral shadow usually has a bright or yellow rim
L = 4	Very bright copper-red or orange eclipse	Umbral shadow has a bluish, very bright rim

totality. The French astronomer Andre Danjon (1890–1967) developed a rating system for the color and brightness of lunar eclipses. His scale assigns a value for the luminosity (L) of the fully eclipsed Moon.

During lunar eclipses, amateurs can also time the entrance of craters into the shadow. This helps determine the extent of

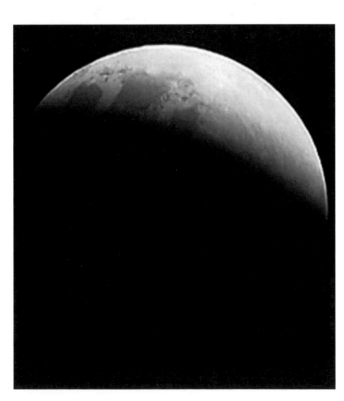

Total lunar eclipse of 21 Jan 2000, much closer to totality. (Image (from video) by Charles Manske of Watsonville, California)

the Earth's atmosphere by providing a rough measurement of the amount of suspended dust and volcanic ash.

Imaging

Lunar eclipse photography is among the easiest an amateur can do. Film works fine for this and it doesn't have to be the fastest emulsion around. You have plenty of time to shoot and can image through the telescope or with a telephoto lens on a camera, tripod-mounted or guided with your telescope drive. My recommendation would be for you to shoot through the telescope. If you are shooting with a 35 mm camera, then I suggest at least a 300 mm telephoto lens. Such a lens will provide a reasonable scale for your images. A 300 mm lens has a field width of 8° so the Full Moon will be approximately 1/16, or 8.25% of the width of the field of view. Of course, a lens with a longer focal length would be even better. Many amateurs use their telescopes as such lenses. In all cases, bracket your exposures.

Upcoming total lunar eclipses

Date	Time (UT)	Duration of totality
16 May 2003	03h41m	00h52m
9 Nov 2003	01h20m	00h22m
4 May 2004	20h32m	01h16m
28 Oct 2004	03h05m	01h20m
3 Mar 2007	23h22m	01h14m
28 Aug 2007	10h38m	01h30m
21 Feb 2008	03h27m	00h50m
21 Dec 2010	08h18m	01h12m
15 Jun 2011	20h13m	01h40m
10 Dec 2011	14h33m	00h50m
15 Apr 2014	07h48m	01h18m
8 Oct 2014	10h55m	00h58m
28 Sep 2015	02h48m	01h12m
31 Jan 2018	13h31m	01h16m
27 Jul 2018	20h23m	01h42m
21 Jan 2019	05h13m	01h02m

Solar eclipses

Solar eclipses occur at New Moon, but not every New Moon, as described above. Solar eclipses may be total, partial, or annular. Total solar eclipses occur when the umbra of the Moon's shadow passes over the Earth. At best, the linear diameter of the Moon's shadow is 273 km, so it is a very small (and very fortunate) swath of the Earth's surface which experiences each total solar eclipse, even though the path of the eclipse is typically 15 000 km long. The track of the Moon's shadow across Earth's surface is called the path of totality.

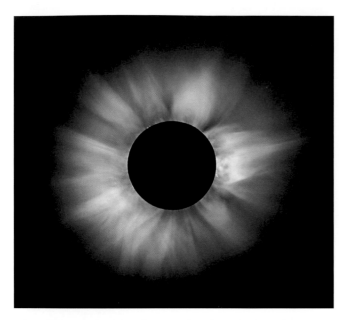

A beautiful composite of the solar corona from the total eclipse of 21 Jun 2001. (Imaged from Chisamba, Zambia, with a Canon D30 by Charles Manske of Watsonville, California)

The Saros Cycle

A similar eclipse, be it lunar or solar, to one that has just been experienced will occur over the same region every 54 years and 34 days. That length of time is exactly three Saros Cycles. One Saros Cycle is equal to approximately 18 years 11 days 8 hours. The reason that eclipses recur is due to a relationship between three different orbital periods of the Moon, as detailed in the following table.

Orbital periods of the Moon

Name of period	How determined?	Length (days)	Number in Saros Cycle
Anomalistic month	Perigee to Perigee	27.5545 5	239
Draconic month	Node to (same) Node	27.2122 2	242
Synodic month	New Moon to New Moon	29.5305 9	223

Thus, to a remarkable degree of coincidence, there are 239 anomalistic months in one Saros Cycle, 242 draconic months and 223 synodic months.

Two eclipses separated by one Saros Cycle are very similar. However, since the Saros Cycle is not equal to a whole number of days, there is a "remainder" of 8 hours. So after one Saros Cycle, a similar eclipse will occur, but 8 hours (or roughly 120° in longitude) later. After three Saros Cycles, however, the Sun and Moon are back to where they started and an eclipse occurs which looks remarkably similar to one

that was experienced in that location 54 years and 34 days prior.

Partial and annular eclipses

Partial solar eclipses occur over much larger areas of the Earth's surface. A partial eclipse can be partial everywhere if the Moon's umbra misses the Earth or it can be partial due to an observer not being on the centerline of either a total or an annular eclipse.

Annular eclipses result from the fact that the Earth is not always the same distance from the Sun and the fact that the Moon is not always the same distance from the Earth. The Earth–Sun distance varies by 3% and the Moon–Earth distance by 12%. The result of this is that the apparent diameter of the Moon can range from 7% larger to 10% smaller than that of the Sun.

Set up to observe the annular eclipse of 30 May 1984. Left to right are the author (using a 60 mm Unitron refractor), Bradley Unruh (visually observing the Sun through optical Mylar), and Raymond Shubinski (using a Celestron 8). Note the author's dog to the far left, ready for a nap as "evening" approaches. (Timer photo by the author)

When the Moon's apparent diameter is smaller than the Sun's, there is a ring of light around the Moon at mid-eclipse. The Latin word for ring is annulus, thus this is known as an annular eclipse. In this case, the umbral shadow of the Moon falls short of the Earth's surface. Because the sunlight reaching us is still very bright, no trace of the corona can be observed during annular eclipses. The maximum diameter of what is called the anti-umbra (the light cone of an annular eclipse) is 313 km. The maximum length of an annular eclipse is approximately 12 minutes.

There is a type of eclipse known as an annular–total. This occurs when the Moon's shadow only contacts the Earth's surface at the center of the eclipse. The Earth's spherical shape is the cause of this type of eclipse.

Solar eclipses are categorized in terms of their magnitude and obscuration. The magnitude of a solar eclipse is the percentage of the Sun's diameter that the Moon covers during maximum eclipse. The obscuration is the percent of the Sun's total surface area covered at maximum.

Observing a total solar eclipse

The same precautions you would use to observe the Sun at any time are in effect during eclipses of the Sun. The only exception is that, during totality, no safety measures are necessary. The following is a chronological listing of observations you can make during a total eclipse of the Sun.

Partial phases
First contact occurs at the moment the eclipse begins. The black outline of the Moon's disk slowly covers the Sun. Standard or digital photography can be done during this time at a leisurely pace. Many observers with still cameras image at regular intervals, combining them later as an animation.

Pinhole projection of the partially eclipsed Sun is fun to see. The effect is better if hundreds of images are being simultaneously projected onto the ground. The shafts of light shining through the leaves of a tree, for example, show up on the ground as images of the Sun during different stages of the eclipse.

Sun crescents. During the partial eclipse of 10 Jun 2002, a nearby tree (out of view to the right) produced a huge number of "pinhole cameras." The result was the images of the 60% eclipsed Sun on our outdoor workshop. (Photo by the author)

You'll probably notice that as totality approaches, shadows sharpen and darken. This is due to the fact that the Sun, normally a large disk in our sky, is approaching a point source. Shadows darken, quite simply, because there is less scattered light reflected into the dark areas.

Planets and bright stars
During the early partial stages of the eclipse, the sky is still very bright. But as totality approaches, the light fades very quickly and brighter objects may be seen. Always know the

positions of the planets and brightest stars in the sky. I have observed zero magnitude and brighter objects prior to totality and third magnitude stars naked-eye once totality begins. Telescopically, brighter deep-sky objects such as the Orion Nebula may be seen during totality if you choose to look for them. (OK, what I actually mean to say is "if you are able to tear your eyes away from the Sun.")

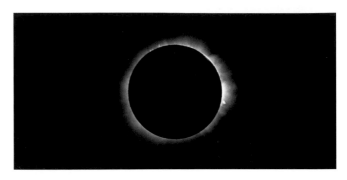

Baily's Beads. Total solar eclipse of 21 Jun 2001. (Imaged from Chisamba, Zambia, with a Canon D30 by Charles Manske of Watsonville, California)

Shadow bands
While a sliver of the bright Sun remains in the sky, many observers have reported long, straight bands of shadows moving across the ground. These wavy lines of alternating light and dark are called shadow bands. They are the result of refraction, that is, sunlight being bent by irregularities in the Earth's atmosphere. Shadow bands are very low in contrast. Many observers have spread a large, white sheet in front of them to help them view this phenomenon. They have been photographed, but only rarely.

Approaching shadow
Occasionally, just moments before totality begins, observers have reported seeing the shadow approach. This is most easily seen from a flat, arid landscape or when the observer is next to a large body of water which lies in the direction of the shadow's approach. For the 3 November 1994 total eclipse, I was in Peru. There were very thin, high cirrus clouds in the sky. Just prior to totality, I observed the shadow approaching – from overhead! I can tell you with certainty that that is a sight I will never forget.

Baily's Beads
In 1836, the English astronomer Francis Baily (1774–1844) noticed a string of irregularly-spaced, brightly-lit points at the edge of the Moon just prior to totality. Now known as Baily's Beads, this phenomenon may be easily seen several seconds before and after totality. The beads are actually the last few rays of sunlight shining through valleys on the edge of the Moon.

Diamond ring
Immediately before totality, and just prior to the last vestige of sunlight disappearing, the corona begins to appear. This

combination of one brilliant point of light and the round corona has been aptly dubbed the diamond ring. Once viewed, it is unforgettable.

At this point, (quickly!) remove any solar filters from optical equipment. The Sun is safe to look at until the next diamond ring.

The diamond ring. Total solar eclipse of 21 Jun 2001. (Imaged from Chisamba, Zambia, with a Canon D30 by Charles Manske of Watsonville, California)

Chromosphere

Immediately after the diamond ring vanishes, you may be able to glimpse the reddish chromosphere. This is a difficult observation, as it is only visible for about a second. With the disappearance of the chromosphere behind the Moon's disk, second contact (the beginning of totality) begins.

At this time, and throughout totality, you may be able to see any number of large solar prominences. Smaller ones will be visible through binoculars or a telescope. The number and size of visible prominences depends upon the solar activity at that time. Observers who have seen a number of total solar eclipses tend to remember each by the size of the largest prominences and by the overall shape of the corona. Take advantage of this view. The only other way to see such features is through a hydrogen alpha filter.

Corona

The wispy outer atmosphere of the Sun is now in full view. Even experienced eclipse-goers have no idea how the corona will appear prior to totality. It is always a great surprise. Note the shape and extent of the corona. Is it longer in one dimension than the other?

Nature at rest

I know – believe me, I know – what you are feeling at this point. The awesome eclipsed Sun is overhead and you're trying to catch your breath. If you have been imaging (more on this later), the pressure is on. You may or may not be aware of the frantic activities and conversations of observers nearby. Stop! Take a moment and make yourself aware of the bigger picture.

The spectacle is not only in the sky, it is also right here on Earth. Look. You will see that there is a resemblance to the onset of night, though not exactly. Around the horizon you will see areas much lighter than the sky near the Sun. Shadows look different. Listen. Usually, any breeze will dissipate and birds (many of whom will come into roost) will stop chirping. It is quiet. Feel. Notice a difference in temperature? A 5–10°C drop in temperature is not unusual.

The end of totality signals third contact. The above chronology now works itself in reverse. Time to catch your breath! Finally, the last bit of the lunar disk passes away from the Sun at fourth contact.

Upcoming total solar eclipses

Date	Time (UT)	Duration of totality
23 Nov 2003	23^h	01^{m}57^s
8 Apr 2005	21^h	00^{m}42^s
29 Mar 2006	10^h	04^{m}07^s
1 Aug 2008	10^h	02^{m}28^s
22 Jul 2009	03^h	06^{m}40^s
11 Jul 2010	20^h	05^{m}20^s
13 Nov 2012	22^h	04^{m}02^s
3 Nov 2013	13^h	01^{m}40^s
20 Mar 2015	10^h	02^{m}47^s
9 Mar 2016	02^h	04^{m}10^s
21 Aug 2017	18^h	02^{m}40^s
2 Jul 2019	19^h	04^{m}32^s
14 Dec 2020	16^h	02^{m}10^s

Eclipse photography

The urge of many amateur astronomers is to record such a fantastic event either digitally, on film or by video. Personally, I will never image during a total solar eclipse. The spectacle is simply too overwhelming and I want to absorb all that I can without worrying about shutter speeds and focal ratios and switches and . . . well, you get the idea. If you are absolutely dead-set on imaging a total solar eclipse, at least wait until your second one. Enjoy the first one for what it is. Trust me, if you want images there will be more than enough available in all formats taken by professional photographers.

Transits

Transits occur when either Mercury or Venus passes across the face of the Sun as seen from the Earth. This can only occur when either is at inferior conjunction. However, for the same reason we don't see eclipses every two weeks, transits do not occur at each inferior conjunction. Because the orbits of both Mercury and Venus are tilted with respect

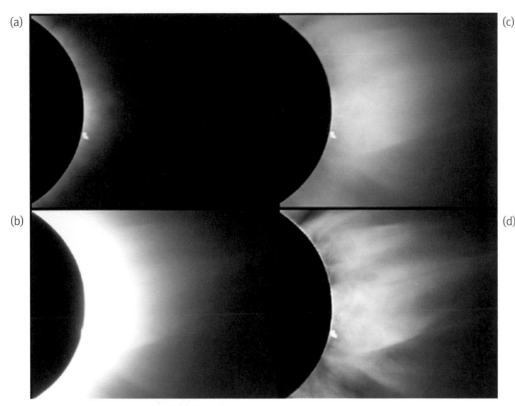

(a)

(b)

(c)

(d)

Four images, including two composites, of the solar corona from the total solar eclipse of 21 Jun 2001. (a) 1/125-sec exposure showing inner corona. (b) 1/2 sec showing outer corona (inner is totally saturated). (c) A composite of 9 images that show the correct exposure for the corona but not much detail. (d) Same as (c), but adding a "radial unsharp mask" technique to bring out much more detail in the corona. (Imaged with a Canon D30 by Charles Manske of Watsonville, California)

to the ecliptic, most conjunctions occur with the planets passing either above or below the Sun. If inferior conjunction is within a day or so of the planet crossing the ecliptic, a transit will result. Transits of Mercury are uncommon. Transits of Venus are rare.

All transits of Mercury fall within several days of the calendar dates of 8 May and 10 November. This is because the orbit of Mercury crosses the ecliptic at two points, known as nodes. (Their longitudes, measured from the Sun, are approximately 48° and 228°.) If Mercury passes through inferior conjunction around that time, a transit will occur. During November transits, Mercury is also near perihelion. The apparent diameter of its disk is only 10 arcseconds. In May, the planet is near aphelion and appears 12 arcseconds across. At aphelion, however, Mercury is moving more slowly in its orbit, so the chance of a transit during May is only one-half that of one occurring in November. November transits recur at intervals of 7, 13, or 33 years. May transits recur over intervals of 13 or 33 years.

For the same reasons, transits of Venus are only possible around 9 December and 8 June (the longitudes of the nodes of Venus being 77° and 257°). Transits of Venus always occur in pairs eight years apart. There is also a recurring pattern of either 105.5 or 121.5 years between the pairs. Only five transits of Venus have been observed: in 1639, 1761, 1769, 1874, and 1882.

> **Note:** On 11 May 1984, the Earth transited the disk of the Sun for (theoretical) observers on Mars. The Earth took about 8 hours to traverse the face of the Sun. The Moon was about 6 hours behind, so for

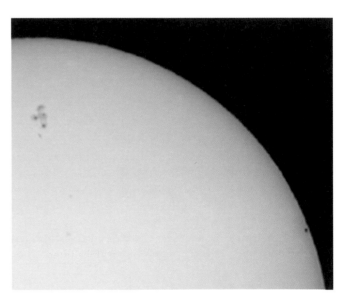

Transit of Mercury, 15 Nov 1999. Mercury is on the far-right edge. (Image (from video) by Charles Manske of Watsonville, California)

two hours both Earth and Moon would have been seen as black dots crossing the Sun.

Transits of Mercury

A small telescope equipped with a solar filter is required to view a transit of Mercury. During the transit, Mercury will be quite small, a black spot less than $1/2$ of 1% the diameter of the Sun. A lot of amateurs perform imaging and video during transits of Mercury.

In the following table, the duration is measured from first contact to fourth contact, that is, the entire duration of the transit. PA 1st Contact and PA 4th Contact are the position angles at the points of first and fourth contacts, measured in degrees counterclockwise from the north point on the Sun's disk.

Upcoming transits of Mercury

Date	Time	Duration	PA 1st Contact	PA 4th Contact
8 Nov 2006	$21^h 42^m$	$04^h 58^m$	141	269
9 May 2016	$15^h 00^m$	$07^h 30^m$	83	224
11 Nov 2019	$15^h 22^m$	$05^h 31^m$	110	299
13 Nov 2032	$08^h 58^m$	$04^h 28^m$	77	330
7 Nov 2039	$08^h 48^m$	$02^h 57^m$	174	237
7 May 2049	$14^h 31^m$	$06^h 42^m$	31	276
9 Nov 2052	$02^h 31^m$	$05^h 12^m$	134	275
10 May 2062	$21^h 41^m$	$06^h 41^m$	97	211
11 Nov 2065	$20^h 10^m$	$05^h 24^m$	103	305
14 Nov 2078	$13^h 45^m$	$02^h 57^m$	69	337

Transits of Venus

The first recorded transit of Venus occurred in the year 1639. Only two observers saw it: the English astronomer Jeremiah Horrocks and his friend, William Crabtree. The transit which had occurred eight years earlier – in 1631 – had gone unobserved, as far as is known.

Transits of Venus have been regarded as having great scientific importance. This is because transits enabled astronomers to determine the scale on which our solar system was constructed.

It was comparatively easy to learn the shape of the solar system, to measure the relative distances of the planets from the Sun, and even the relative sizes of the planets themselves. From this information, astronomers constructed a map of the solar system. This included the orbits of the planets, their satellites, asteroids, and comets. The easy part was to correctly lay out the relative scale of all these objects. One only had to perform the most basic observations for this. But it was not at all easy to accurately assign the correct scale to this map of the solar system.

In 1716, the English astronomer Edmund Halley proposed a method by which this scale could be measured. Halley explained his method of finding the distance to the Sun by using the transit of Venus which would occur in 1761 (or the following transit, set to take place in 1769). The results of the first of this pair of transits were not very successful, in spite of the arduous labors of those who undertook the observations.

On 3 June 1769, however, the transit of Venus was carefully observed and measured by different observers including the party of Captain Cook in Tahiti. Other measurements were being obtained in Europe and North America, and from the combination of these the first (reasonably) accurate knowledge of the Sun's distance was determined, but not for quite a while.

It wasn't until 1824 that the German astronomer Johann Franz Encke computed the distance of the Sun from the result of the 1769 transit. He gave 95 000 000 miles as the definite (albeit incorrect) result. It would take another transit of Venus before the Sun's distance was revised.

The two transits of Venus which occurred in the nineteenth century attained an importance unsurpassed by any observed occurrence in the solar system, and, in fact, received a degree of attention never before accorded to any astronomical phenomenon. Observers all over the world had an army of telescopes aimed at the Sun for these two events. Unfortunately, the results obtained were less than satisfying.

The main problem that arose was the reconciliation of timings of the entrance of Venus onto the disk of the Sun (and, at the end of the transit, its exit from the disk) from one observer to the next. It might be assumed that because Venus is a black circle and the solar disk is bright, the moment when the entire sphere of Venus crosses onto the Sun's face would be easy to determine. In real circumstances, however, the disk of Venus seems almost to attach itself to the limb of the Sun similar to a water droplet emerging from a faucet. Then, all at once, the contact is broken and Venus stands some distance in from the solar limb. Nineteenth century observers christened this phenomenon the "Black Drop" or the "Black Ligament." Due to this situation, the observations of skilled observers with decades of experience showed significant differences in timing this important event.

Still, averaging out these differences allowed nineteenth century astronomers to refine Encke's calculation of the Sun–Earth distance. Following the two transits of Venus, a figure of 149 182 110 km was published and generally accepted. As can be seen, this number is much closer to the

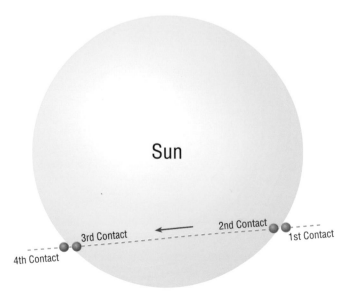

The four "contacts" of a transit. (Illustration by Holley Y. Bakich)

presently accepted value (149 597 892 km) and its calculation was a significant step forward in the refinement of the scale of our solar system.

Transits of Venus have also allowed astronomers to learn two important facts about our "twin" world. First, they learned with a great deal of assurance that Venus had no moon. It had been conjectured by some that if Venus were attended by a small body in close proximity, the brilliancy of the planet would overwhelm the light of any satellite, and thus a moon would remain undiscovered. It was, therefore, a matter of some importance to carefully examine the vicinity of the planet during a transit. If a satellite of any appreciable dimensions had existed, it would have been detected against the brilliant background of the Sun.

Another fact first gleaned about Venus as a result of a transit was the existence of an atmosphere. This was first discovered by M. V. Lomonosov during the transit of 1761. It was thought that if Venus had no atmosphere it would be totally invisible just before beginning its crossing of the solar disk, and would relapse into total invisibility immediately after the transit. The observations proved otherwise, however. As Venus gradually moved off the Sun, the circular edge of the planet extending out into the darkness was seen to be bounded by an arc of light. Some observers, under extremely favorable conditions, have been able to follow the planet until it passed entirely away from the brilliant solar background. At this point the globe of Venus, though itself invisible, was distinctly marked by the circle of light surrounding it. The only explanation possible was that Venus was surrounded by an atmosphere.

Unless lifespans get a tremendous boost in the next few decades, those of you reading this will have exactly two opportunities to view a transit of Venus. Those will occur on 8 June 2004 and 6 June 2012.

The diameter of Venus' disk will be approximately 1/32 (3.125%) that of the Sun, so astute observers will be able to detect the transit naked-eye (with a proper filter, of course).

To observe the upcoming transits of Venus, all that is required is a comfortable chair and a solar filter. A telescopic view, however, will be preferable if you want to try your hand at timing the four contacts. Watch out for the Black Drop!

Upcoming transits of Venus

Date	Time (UT)	Duration	PA 1st Contact	PA 4th Contact
8 Jun 2004	08^h 24^m	06^h 12^m	118	215
6 Jun 2012	01^h 36^m	06^h 40^m	40	291
11 Dec 2117	02^h 51^m	05^h 41^m	56	332
8 Dec 2125	16^h 01^m	05^h 31^m	154	235
11 Jun 2247	11^h 43^m	05^h 44^m	124	210
9 Jun 2255	04^h 50^m	07^h 00^m	45	288
13 Dec 2360	01^h 52^m	06^h 26^m	63	323
10 Dec 2368	14^h 58^m	04^h 30^m	164	224
12 Jun 2498	14^h 46^m	05^h 14^m	130	206
10 Jun 2498	07^h 52^m	07^h 14^m	49	285

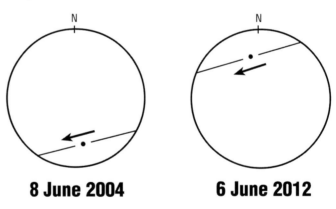

8 June 2004   **6 June 2012**

The upcoming transits of Venus as they will be seen. The Sun and Venus are to scale. (Illustration by Holley Y. Bakich)

6.6 Mercury and Venus

Mercury and Venus – physical data

	Mercury	Venus
Size	4879.4 km	12 104 km
Mass	3.303×10^{23} kg	4.869×10^{24} kg
Oblateness	0	0
Volume	6.084×10^{10} km^3 (5.8% of Earth)	9.284×10^{11} km^3 (85.4% of Earth)
Albedo	0.10	0.65
Density	5.429 g/cm^3	5.25 g/cm^3
Solar Irradiance	3566 W/m^2	2660 W/m^2

Mercury and Venus – orbital data

	Mercury	Venus
Period of rotation	$58^d15^h30.5^m$	$243^d0^h36.5^m$
Period of revolution	$0^y87^d23.3^h$	$0^y224^d16.8^h$
Synodic period	$115^d21^h07.2^m$	$583^d22^h05^m$
Velocity of revolution	47.88 km/s	35.02 km/s
Distance from the Sun	0.3871 AU 0.7233 AU	57 910 000 km 108 200 000 km
Minimum distance from Earth	0.517 AU 0.26 AU	77 269 200 km 38 150 900 km
Maximum distance from Earth	1.483 AU 1.75 AU	221 920 200 km 261 039 880 km
Inclination of equator to orbit	0°	177.36°

Mercury and Venus – observational data

	Mercury	Venus
Max. angular distance from Sun	28°	47°19′
Greatest brilliancy	−1.3	−4.4
Maximum angular size	10″	64″
Minimum angular size	4.9″	10″

Mercury

The Sumerians named the planet Mercury Ubu-idim-gud-ud. The Babylonians called it gu-ad or gu-utu, and it was this early group of people who recorded the first detailed observations of the planet. There are existing tablets which show that the Babylonians were very careful observers of Mercury. They recorded six dates of importance in each of Mercury's cycles: the first visible heliacal rising of the planet; the start of its retrogression; the beginning of the period in which the planet is too near the Sun to be seen; the end of the "invisible" period; the end of retrogression; and, the last visible heliacal setting of the planet.

For centuries, Mercury remained an enigma. It was a small planet, and visual observations were difficult due to the fact that it never ventured far beyond the region of the Sun. Observations of detail on Mercury were first announced by the German amateur astronomer Johann H. Schroeter (1745–1816) from his private observatory at Lilienthal near Bremen. Schroeter claimed to have seen one of the horns of a crescent Mercury blunted. (He made the same observational claim for Venus.) He attributed this to the presence of a mountain 20 km in height. Sir William Herschel (1738–1822) attempted, but was unable, to verify this sighting. From Schroeter's observations, F. W. Bessel (1784–1846) obtained a rotational period of $24^h00^m53^s$. Bessel also calculated the tilt of Mercury's axis to be 70°.

During the nineteenth century, a number of observers claimed to see surface detail on Mercury. The British astronomer William Frederick Denning (1848–1931) made a series of sketches in 1881, and obtained a rotational period of 25 hours. In 1892, the French astronomer Leopold Trouvelot (1827–1895) seemed to obtain the same features as Schroeter had, but his observations were never confirmed.

It was not until the Italian astronomer Giovanni Virginio

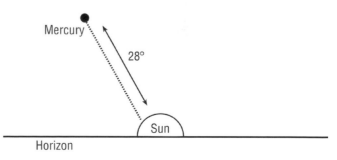

The maximum observable angular distance between Mercury and the Sun is 28°. As this illustration demonstrates, however, Mercury's distance above the horizon is seldom the full 28°. (Illustration by Holley Y. Bakich)

Schiaparelli (1835–1910) made a detailed map of Mercury in 1889, that all doubt as to the reality of surface features on Mercury was erased. At the Brera Observatory in Milan, Schiaparelli used 22 cm and 49 cm refractors and made his observations during daylight hours, when Mercury was high in the sky. He concluded that Mercury was tidally locked to the Sun, always keeping the same face toward it, as the Moon does to the Earth. He published a rotational period for Mercury of 88 days, thus concluding that Mercury was tidally locked to the Sun. This meant that the planet rotated once for each of its revolutions around the Sun.

In 1896, at Lowell Observatory, Percival Lowell (1855–1916) started a series of observations of Mercury with a 61 cm refractor. His results agreed with Schiaparelli's with regard to the synchronous orbital period of Mercury. The map which he created was covered with dark lines and patches. Lowell explained the markings as the result of planet-wide cooling. Unfortunately, as with his observations of Mars, no other observers saw what Lowell imagined.

Interesting facts about Mercury

Using predictions made by Johannes Kepler (1571–1630), the French astronomer Pierre Gassendi (1592–1655) became the first to observe a transit of Mercury on 7 Nov 1631.

Mercury has a total of 299 named features, of which 239 are craters.

From Mercury, the Sun is 63 times brighter than from Earth.

The largest crater on Mercury (in fact, in the entire solar system) is Beethoven, with a diameter of 643 km.

In 1878, American astronomer Samuel Pierpont Langley was the first to see Mercury against the solar corona. Langley used a 33cm refractor at the Allegheny Observatory in Pittsburgh, Pennsylvania, and noted an angular diameter of 15″ for Mercury just prior to its transit across the solar disk.

The light from the Sun takes, on average, 3 minutes 13 seconds to reach Mercury. At Mercury's perihelion, this time is reduced to 2 minutes 33 seconds, and at aphelion it is 3 minutes 53 seconds.

T. J. J. See, working with a 66 cm refractor at the US Naval Observatory, claimed to have observed a large number of craters on Mercury. In June 1901, See made a drawing of Mercury showing, among others, a very large crater. Some modern-day astronomers have suggested that See actually observed the crater Beethoven, the largest on Mercury. Other astronomers point out that at the time Mercury's angular diameter was only 6.6 seconds of arc, and that such an observation would be impossible.

Perhaps the greatest of the twentieth century observers of Mercury was the Greek-born French astronomer Eugène Marie Antoniadi (1870–1944). His numerous observations of Mercury were always made during daylight hours, just like Schiaparelli. Antoniadi's *La Planète Mercure*, an in-depth look at the planet, was published in 1934. The book contained maps of various features that Antoniadi claimed to have seen. However, as with those of See, the observations of Antoniadi have been questioned by modern astronomers.

In 1962, radio astronomers examined radio emissions from Mercury and determined that the dark side was too warm for the planet to be tidally locked. That part of Mercury would be much colder if it always faced away from the Sun. In 1965, Gordon Pettengill and Rolf B. Dyce broke with the traditional synchronous rotation theory. They determined Mercury's period of rotation to be 59 ± 5 days based upon radar observations. This agrees well with the current value, obtained by the Mariner 10 spacecraft, of 58.646 ± 0.005 days.

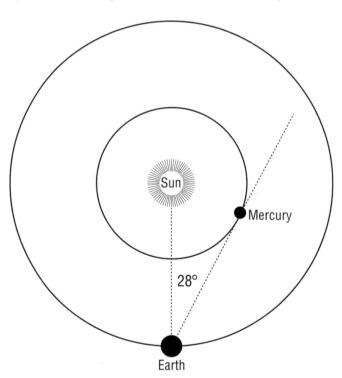

Greatest elongation of Mercury. (Illustration by Holley Y. Bakich)

Observing Mercury

As may be inferred from the above description, telescopically, there's not much to see on Mercury. Phases may be seen, due to the fact that Mercury lies between the Sun and the Earth. Most people that have an interest in astronomy are usually glad just to see Mercury naked-eye. The best times for this are when Mercury is at one of its elongations. Greatest western elongation occurs when Mercury is at a maximum distance west of the Sun. However, in the sky this means that Mercury will be seen in the eastern sky in the morning. Likewise, greatest eastern elongation has the planet east of the Sun in the evening sky.

Elongations of Mercury

East (evening sky)	West (morning sky)
	3 Jun 2003
14 Aug 2003	27 Sep 2003
9 Dec 2003	17 Jan 2004
29 Mar 2004	14 May 2004
27 Jul 2004	9 Sep 2004
21 Nov 2004	29 Dec 2004
12 Mar 2005	26 Apr 2005
9 Jul 2005	23 Aug 2005
3 Nov 2005	12 Dec 2005
24 Feb 2006	9 Apr 2006
20 Jun 2006	7 Aug 2006
17 Oct 2006	26 Nov 2006
7 Feb 2007	21 Mar 2007
2 Jun 2007	20 Jul 2007
30 Sep 2007	9 Nov 2007
22 Jan 2008	4 Mar 2008
13 May 2008	1 Jul 2008
10 Sep 2008	22 Oct 2008
4 Jan 2009	14 Feb 2009
26 Apr 2009	13 Jun 2009
25 Aug 2009	6 Oct 2009
18 Dec 2009	27 Jan 2010
9 Apr 2010	26 May 2010
6 Aug 2010	20 Sep 2010
2 Dec 2010	

Conjunctions of Mercury

Inferior	Superior
11 Jan 2003	5 Jul 2003
(Transit) 7 May 2003	25 Oct 2003
11 Sep 2003	4 Mar 2004
27 Dec 2003	18 Jun 2004
17 Apr 2004	5 Oct 2004
23 Aug 2004	14 Feb 2005
10 Dec 2004	3 Jun 2005
29 Mar 2005	18 Sep 2005
5 Aug 2005	27 Jan 2006
24 Nov 2005	19 May 2006
12 Mar 2006	1 Sep 2006
18 Jul 2006	7 Jan 2007
9 Nov 2006	3 May 2007
23 Feb 2007	16 Aug 2007
28 Jun 2007	17 Dec 2007
24 Oct 2007	16 Apr 2008
6 Feb 2008	29 Jul 2008
7 Jun 2008	25 Nov 2008
7 Oct 2008	31 Mar 2009
20 Jan 2009	14 Jul 2009
18 May 2009	5 Nov 2009
20 Sep 2009	14 Mar 2010
4 Jan 2010	28 Jun 2010
29 Apr 2010	17 Oct 2010
3 Sep 2010	25 Feb 2010
20 Dec 2010	23 May 2011

Since Mercury is an inferior planet (closer to the Sun than the Earth), there are two distinct times during its orbit when it is in line with the Sun. These are called inferior conjunction and superior conjunction. Inferior conjunction occurs when Mercury is between the Earth and Sun. Superior conjunction happens when the Sun is in the middle. Occasionally at inferior conjunction Mercury can cross in front of the solar disk. This is known as a transit. For more on transits of Mercury (and Venus) see the earlier chapter "Eclipses and transits".

The best telescopic views of Mercury by serious amateur astronomers are made at midday when the planet is high in the sky. Be very careful about the position of your telescope with respect to the Sun. Use a yellow, orange, or even red (if your telescope is large enough to gather enough light) filter to cut down most of the sky's blue light. Observers describe the easiest "marking" observed on Mercury as a slight blunting of the planet's southern cusp. This may be due to Mercury's surface being darker at high southern latitudes or it may be caused by a different phenomenon.

Venus

One of the oldest surviving astronomical documents regarding Venus comes to us from the first Babylonian dynasty, being at least as old as 1600 BC. It is from the library of Ashurbanipal and is a series of observational time intervals relating to Venus, which the early Babylonians called Nindar-anna. This tablet is a 21-year record which refers to the appearances and disappearances of Venus in the morning and evening skies, giving the correct time intervals.

In the fourth century BC, the Turkish-born Greek astronomer Heraclides Ponticus (388–315 BC), a contemporary of Aristotle, put forth an idea related to the placement of Venus (along with Mercury) in the solar system. Up to that time, most thought was geared to an immobile Earth with the Sun, Moon, and planets revolving around it. There had always been some thought that perhaps Mercury and Venus did not follow this norm. Heraclides became the first to suggest that Venus (along with Mercury) traveled in circles around the Sun, and not around the Earth.

Intersting facts about Venus

Nearly 90% of the surface of Venus is covered by volcanic landforms.

The atmospheric pressure on Venus (9 321 900 Pa) is equivalent to being 914 m under the surface of Earth's oceans.

Venus has more dry land than any other planet in our solar system. It has three times the dry land of Earth.

The orbit of Venus is the most circular of any planet in the solar system. The eccentricity of its orbit is only 0.0068.

For an observer on Venus, the Earth at opposition would have an apparent visual magnitude of −6.7. The Moon would have an apparent visual magnitude of −2.5. The maximum separation of the Earth–Moon pair would be 31.9 arc minutes.

Maxwell Montes is the only feature on Venus not named after a female.

Due to the thickness of the atmosphere of Venus, causing meteors to decelerate as they fall toward the surface, no impact crater smaller than about 3.2 km across can form.

On 3 January 1818, at 21:51 UT, Venus occulted Jupiter. This was the last mutual occultation by planets until 22 November 2065, when, at 12:47 UT, Venus will again pass in front of Jupiter. There are no reports of this event being previously observed.

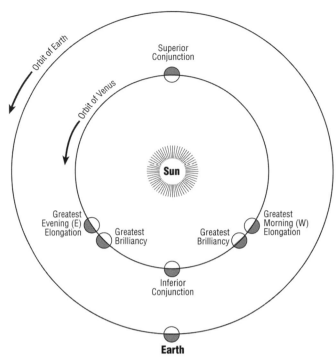

Highlights during the orbit of Venus. (Illustration by Holley Y. Bakich)

In the fifteenth century, many accurate measurements of Venus were made by the Persian astronomer Ulugh Beg (1393–1449), grandson of the Mongol conqueror Tamarlane. He built an impressive observatory in the city of Samarkand and used a sextant, 18.3 m in radius, to make detailed observations of the planets and the stars.

From 1582–1588, the Danish astronomer Tycho Brahe (1546–1601) made a large number of daytime measurements of Venus. He then compared the position of Venus to the Sun, and then that night (or in the morning) he compared Venus' position to specific stars to obtain the most accurate positions up to that time.

In December 1610, the Italian astronomer Galileo Galilei (1564–1642) became the first to observe the phases of Venus. This had been predicted by a few individuals who supported the Copernican view of the solar system. (In 1543, the Polish astronomer Nicolas Copernicus (1473–1543) had published *De Revolutionibus Orbium Coelestium*, in which he placed the Sun at the center of the solar system.) Galileo stated that Venus imitated the Moon in appearance. In addition, Galileo also saw that when Venus was nearly full it was small. That view contrasted with the large apparent size of the planet when he observed Venus as a thin crescent. This difference in apparent size, along with the phases Venus went through, was the strongest possible observational evidence for the validity of the Copernican theory.

In 1645, the Italian astronomer Francesco Fontana (1602–1656) first recorded seeing dusty shadings on the visual globe. It was not readily determined that these markings were atmospheric in origin. In fact, it was over a hundred years before Venus was scientifically shown to possess an atmosphere.

In 1666, the French astronomer Jean-Dominique Cassini(1625–1712) made the first measurements of the rotation rate of Venus. He obtained a value of 23 hours and 21 minutes. Cassini also reported sighting a moon in orbit around Venus. This "discovery" had first been reported by Fontana in 1645.

The German amateur astronomer Johann H. Schroeter (1745–1816) claimed to have seen one of the horns of a crescent Venus blunted. He attributed this to the presence of a high mountain. Sir William Herschel was an opponent of such unverified claims and discounted Schroeter's observations, although Herschel did try to verify them. The Italian astronomer Francesco Bianchini (1662–1729) made a great number of drawings of Venus, compiling them into a book published the year before his death.

Mikhael Vasilyevich Lomonosov (1711–1765), at the St. Petersburg Academy of Sciences in Russia, observed the Venus solar transit of 1761. Lomonosov carefully observed the disk of Venus as it entered and left the bright solar disk. He noticed that the dark spot that represented Venus was surrounded by a luminous halo as it entered onto and exited from the solar disk. He correctly concluded this was evidence of an atmosphere. These observations were not made public outside Russia until 1910.

It was during the nineteenth century that studies of Venus began in earnest. Many observers gave a value of nearly 24 hours for the rotation of Venus. The Italian astronomer Giovanni Schiaparelli, however, disagreed with these results. His investigations led him to believe that Venus always kept the same side toward the Sun, the same conclusion he had reached about Mercury.

In 1911, the American astronomer Vesto M. Slipher determined (by spectral analysis) that the rotation rate of Venus was much greater than one day. The rotation rate of Venus had been a major planetary question up until this study was completed. The problem arises from the nearly identical length of Venus' synodic period (the time interval between successive inferior conjunctions of Venus) and the rotational period of the planet. The difference between these two values (which are both slightly more than 243 days) is less than two hours! Since Venus is best observed near inferior conjunction, it means that the same region of Venus is being seen by observers on Earth for a large number of such events.

Observing Venus

Venus is much easier to observe than Mercury, and some observers find it quite a bit more interesting. Like Mercury, Venus goes through a pattern of inferior conjunction, greatest western elongation, superior conjunction and greatest eastern elongation. There is also a time when the Sun–Venus–Earth angle permits Venus to be seen at its brightest. This event, known as greatest brilliancy, occurs when the planet lies at an elongation of 39°, approximately 36 days before and after inferior conjunction.

Conjunctions of Venus

Inferior	Superior
(Transit) 8 Jun 2004	18 Aug 2003
14 Jan 2006	31 Mar 2005
18 Aug 2007	28 Oct 2006
28 Mar 2009	9 Jun 2008
29 Oct 2010	11 Jan 2010

The phases of Venus are interesting to observe, along with an aspect a little easier to observe with Venus than with Mercury – size change. While Mercury looks twice as big near inferior conjunction as it does at superior conjunction, Venus is more than six times as large! Daytime observations of Venus are also much easier than with Mercury, due to Venus' magnitude. In fact, it is recommended that, for serious observations, Venus be observed during the daytime, or at least in twilight, when the deleterious effects of its brilliance are diminished. Observing Venus in the early daytime sky is easy. Simply set your telescope on Venus and allow the drive to track it until the Sun rises.

There is one problem with daytime observations, especially in the summer. Solar heating of the air (and your telescope) can produce some really bad seeing conditions. Most locations report the worst daytime seeing in the afternoon. Here on the west side of El Paso, my afternoon seeing is actually better, as there is a mountain range which divides the town. During the morning, updrafts associated with the mountains really destroy the seeing, producing very bad images of Venus (and the Sun).

Venus within a day of inferior conjunction, 29 Mar 2001. (Unitron 100 mm, Pentax ME Super 35 mm camera. Kodak Ektachrome 100 film. Daytime photo by the author)

The orbit of Venus is tilted nearly 3.4° to the plane of the ecliptic. This means that, at certain inferior conjunctions, Venus may as much as 3.4° from the Sun. Several times I made it a personal challenge to observe Venus at the exact moment of inferior conjunction. Such a feat is obviously a daytime observation. If your telescope has setting circles you can attempt this. First, record the time of inferior conjunction and the positions (RA and Dec.) of the Sun and Venus at that moment. With a solar filter in place and using your widest field eyepiece, center the Sun in your field of view. Finally, offset your telescope by the appropriate amount in RA and Dec. If Venus is not immediately seen use extreme care in searching for it. Remember, the Sun is near!

Observing Tip: If your goal is only to observe Venus' phases during the daytime, use a yellow, orange or red filter (darker ones on larger apertures) to enhance contrast and eliminate the blue daytime sky.

Amateur astronomers have reported seeing an irregular terminator, dusty shadings, bright spots, cusp-caps and cusp-bands, to name the most obvious. Viewed in visible light, there are no permanent features discernible in the clouds of Venus. The atmosphere is in a continuous state of mixing, and any patterns observed quickly dissipate.

The best – well, really the only – way to see features in the

Greatest elongations of Venus

East (evening sky)	West (morning sky)
29 Mar 2004	11 Jan 2003
3 Nov 2005	17 Aug 2004
9 Apr 2007	25 Mar 2006
14 Jan 2009	29 Oct 2007
20 Aug 2010	7 Jun 2009

atmosphere of Venus is through a violet (47A) or ultraviolet (Schuler UV) filter. These filters may be purchased via the internet through Adirondack Video Astronomy. The filter page is at

http://www.astrovid.com/schuler_astro_imaging_filters.htm

Such filters do not allow much light in, however, so this advice, unfortunately, is for those who have medium to large telescopes. The most reported sighting is of an immense C- or Y-shaped feature centered on and symmetrical with the planet's equator. This is a short-lived phenomenon, but it tends to re-form often enough to perhaps be considered a "permanent" feature in the clouds of Venus.

If you're going to perform non-daytime observations of Venus, it is a good idea to limit your viewing to when Venus is at least 20° above the horizon. The air below that level is so thick that the effects of atmospheric refraction obliterate image quality.

Now measure the diameter from north to south and measure across the crescent, or gibbous, from east to west. Divide the former into the latter and you will get a fraction. This fraction is the percent phase, or phase coefficient, of the observed disk. For example: 2.5″/4″ = 0.625 or 62.5% illumination.

The ashen light

In 1643, Giovanni Battista Riccioli (1598–1671) observed a faint glow within the dark part of the disk of Venus. Through the centuries, many explanations have been given for this ashen light of Venus, but the phenomenon remains elusive. When reported, it is likened to earthshine, which is a naked-eye monthly occurrence at times of thin crescent Moons.

Whether or not the ashen light actually exists is still a question. At this writing, evidence points to "no." However, if trying to see the ashen light gets you out to your telescope, then, well, maybe it does exist.

Some observers use an occulting bar to attempt observations of the ashen light. Such a device eliminates most of the bright portion of Venus' disk. Good luck.

Venus setting sequence, 16 Sep 1986. (Pentax ME Super 35 mm camera. Kodak Ektachrome 100 film. One exposure every 5 minutes. Photo by the author)

6.7 Mars

Mars – physical data

Size	6 794.4 km
Mass	6.421×10^{23} kg
Oblateness	0.005 19
Volume	1.643×10^{11} km³
(15.8% that of Earth)	
Albedo	0.15
Density	3.94 g/cm³
Solar Irradiance	595 W/m²

In Assyria, the planet Mars was the special sign of Nergal, often called "Shedder of Blood," a god of death, misfortune, and disaster. To the Norsemen, Mars was Tyr, or Tiu, the one-handed god of war. From this god we derive the name of the third day of the week, "Tiu's day." The planet represented the god of war to the ancient Greeks as well. There, it was known as Ares. The Romans made both the mineral hematite and iron sacred to Mars, their red god of war, and the stone or metal was often used as an amulet during battle.

To the Etruscans, Mars was not originally a god of war. His earliest image was that of a sacrificed fertility god, Maris, worshiped at an ancient shrine in northern Latium. He joined with the goddess Marica and their union produced Latinus, the legendary ancestor of all Latin tribes.

More than 2500 years ago, the Babylonians were making regular observations of Mars, which they named Salbatani. A Babylonian text which research has shown dates from January 523 BC, records the observation of a heliacal setting of Mars, in the western part of the constellation Gemini. The ancient Chinese, who called Mars Huo xing, used the planet astrologically, but the following quote does indicate that they watched its motions carefully: "When Mars is retrograding in the station Ying-she, the ministers conspire and the soldiers revolt."

Aristotle (384–322 BC) taught that the Earth was the center of the universe and the Sun, Moon, and planets revolved around it, ascribing perfectly circular orbits. But simple observation showed that there were irregularities. For one, the brightnesses of the planets changed. Also, each planet did not always move through the starry background at the same velocity. Most hostile to the theory, the direction of motion of the planets changed on a regular basis. This was most pronounced in the case of Mars. The retrograde motion of the red planet seemed intensely damaging to a theory which suggested circular motions. To account for this, Aristotle and Eudoxus (c. 408– c. 353 BC) each had systems

of rotating, nested spheres upon the surfaces of which rested Mars and the other planets. This was an incredibly complex system. Eudoxus utilized 27 spheres and Aristotle's system demanded 55, 22 of which were counter-rotating.

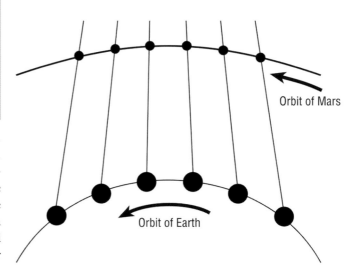

Retrograde (westward) motion for Mars (or any outer planet) occurs when the Earth passes it in its orbit around the Sun. There are also two stationary points where the planet appears to change direction. (Illustration by Holley Y. Bakich)

Two centuries later, a somewhat simpler system was championed by Hipparchus. This was the theory of epicycles. Planets were assumed to describe a circle (called an epicycle) around a point which was part of a larger circle (called a deferent) which itself was centered on the Earth. It is important to note that all motions were circular. It would have been heresy to suggest otherwise.

Two more centuries passed. In the library of Alexandria, Claudius Ptolemaeus, also known as Ptolemy (fl. 127–145), made numerous observations of Mars with which to refine the epicycle theory. Ptolemy introduced inclinations between the deferents and the ecliptic and between the epicycles and the deferents. He added deferents upon deferents and the whole system became extremely complex. However, to a rough degree of accuracy it did predict the

Mars – orbital data

Period of rotation	$1^d0^h37.4^m$
Period of revolution	$1^y320^d18.2^h$
Synodic period	$779^d22^h33.6^m$
Velocity of revolution	24.13 km/s (86 868 km/h)
Distance from the Sun	1.5237 AU 227 940 000 km
Minimum distance from Earth	0.372 AU 55 650 200 km
Maximum distance from Earth	2.68 AU 400 922 300 km
Inclination of equator to orbit	25.19°

Mars. Tan Wei Leong, of Singapore, took this sequence between 18:09–18:44 UT on 20 May 2001. (Celestron C11 and an SBIG ST7E CCD camera)

Interesting facts about Mars

The Martian rotational period (approximately 24^h37^m) is referred to as a sol. Scientists at the Jet Propulsion Laboratory in Pasadena, California, use this term to differentiate between it and an Earth "day."

Mars has a total of 1345 named features, of which 845 are craters. The largest crater on Mars, named Schiaparelli, is 461 km in diameter.

On 27 Aug 2003, Mars will reach its maximum brightness, shining almost as bright as magnitude −3.0. The most recent approach of Mars to this magnitude was on 22 Aug 1924. More recently, the red planet has twice shone at magnitude −2.9: on 7 Sep 1956 and on 12 Aug 1971. Because of changing surface features on Mars (most notably the reflective polar ice caps), opposition brightness may differ by as much as 0.3 magnitude from prediction.

The satellite which revolves around its planet closer than any other is Phobos. Phobos orbits Mars only 9377 km from the planet. This is only 2.4% the distance at which the Moon orbits the Earth.

The average atmospheric pressure on Mars, approximately 709 Pa, is equivalent to the atmospheric pressure 30 km above the surface of the Earth.

The only known incident of a meteorite killing a mammal occurred in 1911, in Nakhla, Egypt. The fall killed a dog and was witnessed by the dog's owner. The meteorite has been shown to be of Martian origin.

positions and motions of Mars and the other planets like no theory before it.

Mars was the focal point for Johannes Kepler's three laws of planetary motion. Kepler worked for more than five years to obtain an orbit of Mars whereby prediction agreed with observation. He once suggested that the Sun had a repulsive force which was variable with distance, causing a type of epicyclic orbit of the planet. But even this did not remove the error between what was expected and what was seen. Eventually, Kepler realized that his difficulties were caused by his unwillingness to forego circular orbits. As soon as he applied an elliptical orbit to Mars, the errors vanished.

Galileo first observed Mars in 1609. The following year he wrote about observations of disk and phases (full and gibbous) indicating a spherical body illuminated by the Sun. On 13 October 1659, the first sketch of Mars was made by Christiaan Huygens (1629–1695). Just 46 days later, on 28 November, Huygens recorded the first observation of a feature on Mars, almost certainly Syrtis Major. As he observed the feature on successive rotations, he arrived at an approximate 24-hour rotational period for Mars. This was refined in 1666, by the Italian-born French astronomer Jean-Dominique Cassini (1625–1712) who determined the length of the Martian day at 24^h40^m.

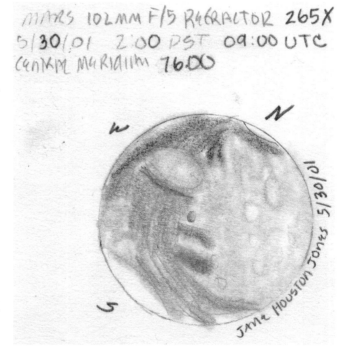

Mars. (Sketch by Jane Houston Jones of San Francisco, California)

In 1672, Jean Richer (1630–1696) was in the French colony of Cayenne making observations of Mars. At the same time, Cassini was at the Paris Observatory taking similar measurements. When compared, the first (reasonably) accurate value of the parallax of Mars was deduced, and from that, its distance. Making observations of Mars during favorable oppositions (those when Mars is near perihelion) was one of two ways that astronomers attempted to deduce the distance scale of the solar system. The other was by using transits of Venus.

Also in 1672, Huygens observed a white spot at the south pole of Mars. In 1704, Cassini's nephew, Italian astronomer Giacomo Filippo Maraldi (1665–1729) observed white spots at the poles but did not refer to them as ice caps; he noted that the south cap is not centered on the rotational pole. In 1719, Maraldi raised the possibility that the white spots were ice caps. On 25 August of that year, Mars, two days from opposition, was closest to Earth. Its brightness in the sky caused panic.

On 26–27 October 1783, Sir William Herschel (1738–1822) observed the close passage of two faint stars near Mars. He correctly concluded that Mars has a thin atmosphere, as he saw no effect on the light of those stars when they were close to the planet. During the following year, Herschel identified a 30° axial tilt for Mars; he noted the seasonal changes of the polar caps and suggested they were composed of snow and ice.

In 1809, the French amateur astronomer Honoré Flaugergues (1755–1830), working at Viviers, perceived the presence of yellow clouds, possibly an early observation of dust clouds. Four years later, Flaugergues noted variable markings on the surface of Mars, and that in the Martian spring, the polar cap shrinks rapidly; he assumed that the cap is made of layers of ice and snow and that its rapid melting proves that Mars is hotter than the Earth.

In 1840, the first global maps of Mars were created by Wilhelm Beer (1797–1850) and Johann Madler (1794–1874); they also refined Mars' rotational period to $24^h37^m22.6^s$, within 0.2^s of the current value. The rotational rate of Mars was further refined by the Dutch astronomer Friedrich Kaiser (1808–1872). During the opposition of 1862, he made a number of drawings, and even constructed a globe. Then he compared markings on his maps to similar characteristics on maps drawn by Christiaan Huygens (1629–1695) in 1666, and Robert Hooke (1635–1703) in

1667. The value Kaiser obtained for the rotational period of Mars differs from the presently accepted value by only 0.1^s.

In 1868, British astronomy popularizer Richard Anthony Proctor (1837-1888) published *The Lands and Seas of Mars, from 27 Drawings by Mr. Dawes*. This was a record of observations of Mars conducted by William Rutter Dawes (1799–1868) from 1852 to 1865. Proctor's choice of the zero meridian of Mars survives to this day. The next year, Father Secchi referred to canali, the Italian word for channels.

During the exceptional opposition of 1877, the Scottish astronomer David Gill (1843–1914) traveled to Ascension Island to measure the parallax of Mars. Gill was trying a relatively new method – measuring the position of Mars against the background of stars twice in the same night. The first observation is made early in the evening when Mars is in the east and then another is made early in the morning when Mars is in the west. Thus, a lone observer could measure the east–west component of Mars' parallax avoiding the errors inherent in using a second observer a great distance away.

It was also during the 1877 opposition of Mars that the Italian astronomer Giovanni Virginio Schiaparelli (1835–

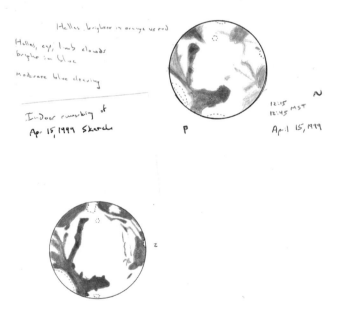

Mars. 19 Mar 1999. (250 mm Newtonian reflector at ~f/4.6. University Optics 6 mm Abbe Orthoscopic eyepiece rendered a magnification of 190x, used in integrated, orange, and blue light. Sketch by Jeff Medkeff of Sierra Vista, Arizona)

1910) caused such a stir by reporting canali. At this point, Schiaparelli also developed a nomenclature for the features observed on Mars. Finally, to cap probably the best of all "Mars years," in August, American astronomer Asaph Hall (1829–1907) discovered Deimos on 11 August and Phobos on 17 August.

In 1894, American Percival Lowell (1855–1916) built an observatory in the territory of Arizona, at Flagstaff. He made his first observations of Mars, which was the main reason for the construction of the observatory. The following year,

Mars – observational data

Brilliancy at opposition	Maximum	−2.9
	Minimum	−1.0
Angular size	Maximum	25.11″
	Minimum	13.82″

the first edition of Lowell's *Mars*, his first (and most famous) book on the red planet, was published. In it, Lowell made some wild claims about Mars, including supposed observations of canals, his mistaken translation of the word *canali*. In 1909, American astronomer George Ellery Hale (1868–1938), using the Mt. Wilson 152 cm reflector, reported "... not a trace" of canals.

During the 1920s, the atmosphere of Mars was tops in the minds of astronomers. In 1925, American astronomer Donald H. Menzel (1901–1976), studying photographs of Mars taken at different wavelengths, concluded that the air pressure on Mars was less than 6687 Pa. The following year, American astronomer Walter Sydney Adams (1876–1956) determined spectroscopically that Mars is "ultra-arid." The year after that, large temperature differences between day and night sides of Mars were measured by William Weber Coblentz (1873–1962) and Carl Otto Lampland (1873–1951); this was taken to be a sign of a very thin atmosphere. Two decades later, in 1947, using infrared spectroscopy, Dutch-born American astronomer Gerard Peter Kuiper (1905–1973) detected carbon dioxide on Mars, but no oxygen.

At this writing, the most recent data concerning Mars comes to us from the NASA Mars Pathfinder mission.

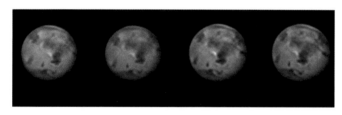

Mars. Tan Wei Leong, of Singapore, took this sequence between 15:32–16:14 UT on 28 Jun 2001. (Celestron C11 and an SBIG ST7E CCD camera)

Launched on 4 December 1996, the spacecraft touched down on the red planet on 4 July 1997. On 5 July 1997, NASA announced that it was renaming the Mars Pathfinder lander the Carl Sagan Memorial Station, in honor of the American astronomer who passed away on 20 December 1996. On 8 August 1997, the Mars Pathfinder completed its primary mission. From the data returned, scientists have concluded that surface photographs provide strong geological and geochemical evidence that fluid water was once present on the red planet.

During the first 30 days of the mission, the imager for Mars Pathfinder returned 9669 pictures of the surface. These pictures appear to confirm that a giant flood left stones, cobbles, and rocks throughout Ares Vallis, the Pathfinder landing site. In addition to finding evidence of water, the scientists confirmed that the soils are rich in iron and that suspended iron-rich dust particles permeate the Martian atmosphere.

It is known that the Martian fluvial valleys and channels are ancient features. Researchers believe that the peak of activity was about 3.5 billion years ago. Later, many valleys formed. After this period, fluvial activity became localized and episodic. Cataclysmic discharges of ground water formed the huge outflow channels during this time. This water would have ponded in the northern plains of Mars. During the recent Amazonian period, only modest fluvial activity was observed. Scientists are convinced that the water that remains on Mars today is trapped, probably as permafrost and ice beneath the Martian surface. Such valleys could not form today, because Mars is cold and dry. But Mars may have been significantly different in the past.

Bolstering their evidence for once-present water, the imaging team found evidence for a mineral known as mag-

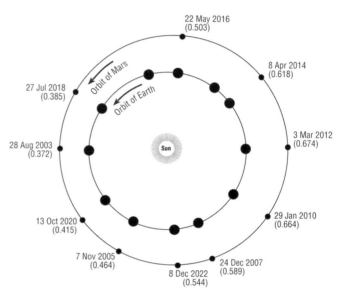

Dates of future oppositions of Mars. Numbers in parentheses are distances between the Earth and Mars in astronomical units. (Illustration by Holley Y. Bakich)

hemite — a very magnetic iron oxide. On Earth, maghemite forms in water-rich environments and could likely be formed the same way on Mars. Reddish rocks such as Barnacle Bill, Yogi and Whale Rock — pictured by the Pathfinder on the Martian surface and named by mission scientists — show evidence of extensive oxidation on their surfaces. The oxidation is possible only if water existed on the surface at some time and played an important role in the geology and geochemistry of the planet. Close examination also confirmed that the rocks have been sitting on the planet's surface for billions of years. In such a position, they endure a slow-motion sandblasting from a usually weak, dusty Martian wind.

Of course, there is, at present, no liquid water on the surface of Mars. Several theories about the disappearing water exist, such as evaporation into space, or seepage into subsurface ice deposits or liquid aquifers, or storage at the Martian poles.

Mars Pathfinder's camera also revealed that Mars' atmosphere is dustier and more dynamic than expected. Surprisingly, the scientists found wispy blue clouds, possibly composed of carbon dioxide, traveling through Mars' salmon-colored sky. White cirrus-like clouds, made of icy water vapor, also circulate throughout the thin Martian atmosphere. In such a thin atmosphere, these variations in the clouds of Mars were surprising.

Surface features

The land area of Mars is approximately equivalent to the dry land area of Earth. Due to the smaller size of the planet, the thin Martian atmosphere and the lack of erosion, large surface features on Mars tend to be more pronounced than those on Earth.

The most prominent surface features on Mars are the polar ice caps. Each of the ice caps can be subdivided into "seasonal" and "residual" caps. The southern residual ice cap is approximately 350 km in diameter and is composed of frozen CO_2. The northern residual ice cap is composed of H_2O ice, and has a diameter of 1000 km. Both seasonal ice caps are composed of frozen CO_2, which condenses directly from the Martian atmosphere when the temperature is below $-123\,°C$. In the northern hemisphere, where the winters are more severe, the extent of the ice cap may reach a latitude of $45°$, but in the southern hemisphere the extent of the seasonal ice cap never passes above a latitude of about $-55°$.

All Martian clouds are temporary phenomena. High thin clouds composed of water ice may form in mountainous

Composition of Martian atmosphere

Carbon dioxide (CO_2)	95.32%
Nitrogen (N_2)	2.7%
Argon (Ar)	1.6%
Oxygen (O_2)	0.13%
Carbon monoxide (CO)	0.07%
Water (H_2O)	0.03%
Neon (Ne)	0.000 25%
All other gases <0.0001% (1 part per million)	

regions. In addition, in low-lying areas of the planet fog is possible in the hours just prior to sunrise. Also, a stratospheric haze of CO_2 may form when crystals of dry ice condense at high altitudes.

Though not truly clouds, dust storms have been observed from time to time on Mars. These storms can be quite severe, enveloping the entire surface area of the planet for up to several months.

Observing Tip: Dust storms seem to be more active after perihelion (when the heating of Mars is greatest). Get in as much observing and imaging as you can prior to perihelion. The onset of a dust storm, and its subsequent development, is one of the most fascinating observations an observer can make.

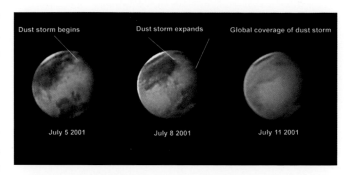

The development of a Martian dust storm 5–11 Jul 2001. (Image by Ed Grafton, Houston, Texas, using a Celestron 14 and an ST5c CCD camera)

Upcoming conjunctions and oppositions of Mars

Upcoming conjunctions	Upcoming oppositions
15 Sep 2004	27 Aug 2003
24 Oct 2006	30 Oct 2005
6 Dec 2008	24 Dec 2007
4 Sep 2011	29 Jan 2010

Early "observations" concerning the moons of Mars

The first astronomer to state that Mars had moons was the father of the laws of planetary motion, Johannes Kepler, in 1610. When trying to solve Galileo's anagram referring to Saturn's rings (which Galileo thought to be two moons close to the planet), Kepler believed that Galileo had found moons of Mars instead.

In 1643, the Capuchin monk Anton Maria Shyrl claimed to have observed two moons in orbit around Mars. We now know that such a view would be impossible with the telescopes of his time. In all likelihood, Shyrl had observed a star in the same field of view as Mars.

Of all the conjectures related to the satellites of Mars, certainly the most famous took place in the eighteenth century. 150 years prior to their actual discovery, Jonathan Swift wrote *Gulliver's Travels* (1727) in which he mentioned two small moons orbiting Mars. These satellites were known to the astronomers of Lilliput, having been observed by them for some time. Swift gives their periods of revolution as 21 and 10 hours. Swift's imagined moons were again mentioned in 1750, by Voltaire in his novel *Micromegas*. This was the story of a giant from the star Sirius who visits our solar system.

Twenty years after publication of *Gulliver's Travels*, in 1747, a German military captain named Kindermann, claimed to have observed a satellite of Mars three years earlier, on 10 July 1744. Kindermann had worked out an orbit for this "moon" and stated that its period of revolution around Mars was 59 hours 50 minutes and 6 seconds.

In 1877, Asaph Hall, an astronomer working at the US Naval Observatory, finally discovered the two small moons of Mars. He was given the honor of naming these satellites and,

The satellites of Mars

	Size (km)	Mass (kg)	Density(g/cm3)	Orbital period	Eccentricity	Inclination	Distance from planet
Phobos	26 × 18 (km)	1.8×10^{15} (kg)	1.750	$07^h39.2^m$	0.0151	1.08°	9.377×10^3
Deimos	16 × 10 (km)	1.08×10^{16} (kg)	1.900	$01^d06^h17.9^m$	0.00033	1.79°	2.3436×10^4

in deference to the mythology concerning the god of war, chose the names Phobos (fear) and Deimos (dread). The orbital periods of these two satellites are remarkably close to the periods imagined by Jonathan Swift 150 years earlier.

Observing Mars

Observing the "Red Planet" was once described to me by a friend as "two long years of waiting for four to six weeks of panicked activity." When you do have the chance to view Mars, one word says it all: filters! We'll get to this in a bit.

First, let me make a recommendation about a wonderful Windows freeware program that will help you in planning your Mars observations. Mars Previewer II was written by Leandro Rios of Argentina. It shows the view Mars presents toward Earth at any given date and time. Download it at the bottom of the internet page
http://www.astronomysight.com/as/start/books.html

Mars exhibits a wide range of sizes. Don't wait for it to reach an opposition diameter of 20″ or more before you start observing it. Quality observations can be made even when the planet's apparent diameter is under 10″. As long as the atmosphere of Mars is transparent, detail will be seen.

Mars. Tan Wei Leong, of Singapore, took this sequence between 15:14–16:59 UT on 1 Jul 2001. (Celestron C11 and an SBIG ST7E CCD camera)

There is another reason not to wait for Mars to reach its apparent maximum size, and this is much more important for Mars than for Jupiter or Saturn. There have been times when opposition occurs near the Martian perihelion, that dust storms, caused by solar heating, have raged over huge areas of the planet. If you wait until opposition to observe Mars there is a chance that you won't see very much. This situation occurred during the opposition of 2001. When

opposition occurs near the Martian aphelion, the atmosphere is essentially dust-free and the features stand out quite prominently.

The rotational rate of Mars is 37.4 minutes longer than our own. So, if you were to observe Mars at the same time each day (night), the markings would appear to gradually change by $(37.4/1477.4) \times 360° = 9.11°$ per day to the west. In a little more than five weeks, therefore, the planet would appear to be slowly rotating backwards. All the prominent features of Mars would, at some time during this period, be placed favorably on its meridian.

You can also choose to wait for Mars' rotation to bring an object into view, or onto its meridian. Since Mars rotates once every 24.623 333 hours, in one hour it will rotate $360/24.6233\ 33 = 14.62°$.

Longitudes of 20 Martian features

Approximate Longitude	Feature
0°	Sinus Meridiani
35°	Chryse Planitia
35°	Mare Acidalium
35°	Niliacus Lacus
45°	Xante
65°	Protei Regio
90°	Solis Lacus
95°	Tharsis
125°	Arcadia
145°	Nix Olympica
180°	Mare Sirenum
210°	Elysium
260°	Libya
260°	Mare Tyrrhenum
270°	Isidis
285°	Syrtis Major
290°	Hellas
295°	Zen Lacus
320°	Sinus Sabaeus
345°	Noachis

Note: R. A. Proctor's choice for zero longitude on Mars corresponds to the crater Airy-0 which may be found in Sinus Meridiani.

Use high magnification to observe Mars, right up to the limit of the seeing. There's a lot of small detail on the planet and you must give yourself the best chance to see it. Prepare to spend a lot of time at the eyepiece, waiting for moments of good seeing. When those moments arrive, concentrate your attention on a small area or a single feature. Trying to "see" the entire disk all at once or darting your gaze from one feature to another is not a profitable use of your observing time. Compare your views to a detailed map of Mars, or use Mars Previewer II as noted above.

Dust storms

Above, I indicated that dust storms were more likely to occur near Martian perihelion. This is true, but it also bears noting that dust storms can occur any time. Summertime (on Mars) dust storms are generally larger and have greater coverage. They can either be localized dust storms, associated with a desert region, or global dust storms. True global dust storms were unheard of until 1956. Since then, six have been seen, including the most recent one (at this writing) in 2001.

Observing Martian clouds

Several different types of clouds are observable in the Martian atmosphere. Some are termed seasonal clouds and are related to the seasonal heating and cooling, which causes sublimation and condensation.

There are also discrete clouds. These are generally related to an area being carried along as the planet rotates. Most of them are found in Mars' northern hemisphere during the spring and summer.

Certain discrete clouds are known as orographic clouds. These are caused by wind passing over the high peaks of Martian mountains and volcanoes and are composed of water. To view the high-altitude orographic clouds, use a blue or violet filter. For the low-altitude ones a green filter works better.

A good medium-size telescope observing challenge would be to try to observe the Syrtis Blue Cloud of Mars. This is a very famous discrete cloud associated with the Libya basin and Syrtis Major. It was first observed in 1858 by the Italian astronomer Angelo Secchi (1818–1878). Since this cloud, viewed with no filter, turns Syrtis Major a bluish color, if you use a yellow filter, the portion of Syrtis Major covered by the cloud will appear greenish.

Morning and evening clouds may also be observed. These are bright, isolated patches of surface fog seen at sunrise, (the western edge of Mars) or sunset. This sighting can also be of ground frost. The difference is that the fog usually dissipates in a few hours, while the frost may last all day. Evening clouds are generally larger and there are more of them. They tend to grow as the Martian night approaches. Telescopic views of morning and evening clouds are enhanced through the use of blue or violet filters.

Filters to try when viewing Mars

Filter	Use
#8 Light Yellow	Somewhat useful in increasing contrast between light and dark areas in the smallest of apertures; larger telescopes should ignore this filter.
#12 Yellow	Brightens desert regions, darkens bluish and brownish features; good for small apertures as it lets through a lot of light.
#15 Deep Yellow	Increases the contrast of the dark areas with the background more than the above two filters; also can be used with small telescopes.
#21 Orange	Further increases contrast between light and dark features over the #15; penetrates hazes and most clouds; allows some detection of dust clouds.
#23A Light Red	Will increase the contrast of details and assist in the identification of dust clouds slightly more than the #21.
#25A Red	Gives maximum contrast of surface features, enhances fine surface details, dust clouds boundaries, and polar cap boundaries; for use only on medium and large telescopes.
#30 Magenta	Enhances red and blue features and darkens green ones; improves polar region features and some clouds.
#38A Dark Blue	Shows atmospheric clouds, discrete white clouds and limb hazes, equatorial cloud bands, polar cloud hoods; darkens reddish features; medium–large aperture required.
#47 Violet	All the features of #38A to a greater degree; you'll need a big telescope to make good use of this filter.
#56 Light Green	Brings out limb hazes and terminator clouds and frost patches; great filter for polar caps.
#58 Green	Darkens red and blue features, enhances frost patches, surface fogs and polar caps to a greater extent than the #56, but lets only half the light of the #56 through.
#64 Blue–Green	Helps detect ice fogs and polar hazes.

Filters to try when viewing Mars (*cont.*)

Filter	Use
#80A Blue	If you have at least a medium-sized telescope, this is probably the first filter you should purchase to observe Mars; enhances all clouds and ground frost, along with polar caps.
#82A Light Blue	Clouds, frost, and polar caps for smaller apertures; allows much more light through than #80A.

Sketching Mars

When sketching Mars, be aware that subtlety is best. You probably don't want somebody asking you if your name is Percival Lowell! Start with the polar ice caps, if they are visible. Try to gauge accurately the extent in latitude of the edges of the caps. This is a starting point and will aid tremendously in positioning other features on your sketch. Next, sketch the most prominent dark features that you see. Finally, lightly fill in the finer details.

Take your time. Most sketches require approximately 15–30 minutes to complete. Of course, the more you sketch, the less time it will take. Remember that Mars only rotates 14.62° per hour. During your sketch most features of the planet will not move much.

When Mars is near opposition, you can simply use a circle to denote its disk. Several months on either side of opposition, however, Mars will show a distinct gibbous phase. Its maximum phase occurs when the planet is 89% illuminated. An 11% difference may not sound like much, but when observed at the eyepiece it is very noticeable. This is approximately the difference between a Full Moon and a Gibbous Moon three days from Full.

Imaging Mars

Although good work can be done when Mars is not at maximum apparent angular size, the best images of Mars are always obtained near opposition. Also, those observers whose latitude on Earth places Mars highest in the sky have a decided advantage. This is not to say that good work has not been done when Mars is low in the sky. It's simply one more factor for you to consider.

I want to emphasize one final point to those of you considering imaging Mars. Although individual images are nice, you might think about taking a number of them over a long period of time, say one month prior to opposition to one month past. A series of images with the same Martian longitude on its central meridian may show differences in seasonal clouds, changes in the size of the polar ice caps, or even the development of dust storms.

6.8 Jupiter

Jupiter – physical data

Size	142 984 km (polar diameter, 133 717 km)
Mass	1.9×10^{27} kg
Oblateness	0.064 81
Volume	1.531×10^{15} km^3 (1321 times that of Earth)
Albedo	0.52
Density	1.33 g/cm^3
Solar irradiance	51 W/m^2

The Babylonians, who called Jupiter Nibiru-Marduk, noticed a periodicity with regard to its motion. They calculated that 71 years equaled 65.01 synodic periods of Jupiter (essentially the round number 65) and also 5.99 sidereal periods (essentially 6). Thus, to a rather high degree of accuracy, the Babylonians could foretell the position of Jupiter on certain dates by looking at the data accumulated 71 years in the past. The Babylonians began this detailed record keeping earlier than 250 BC.

The Chinese used their observations of Jupiter, which they called Mu xing, for more than astrology. They sectioned the sky into 12 "Jupiter stations." They observed that the sidereal period of Jupiter is roughly 12 years. That is, they noticed that Jupiter took approximately 12 years to move completely through the zodiac, where it would begin the cycle anew. This value was short by only 50 days, or slightly more than one percent – certainly accurate enough for their purposes. Chinese astrologers associated each of these with one character from a sequence of 12 animals.

During summer 364 BC, the Chinese astronomer Gan De made an amazing visual observation of what was almost certainly Ganymede; this was 1974 years before Galileo (and he was using a telescope!). At the time, Jupiter was within the boundaries of the constellation Aquarius.

In 1610, the Italian astronomer Galileo Galilei (1564–1642) discovered the four large moons of Jupiter. On 15 March 1611, Galileo reported Jupiter without a visible satellite for several hours, however calculations show that at least one satellite was visible at all times. His observations demonstrate the "quality" of his telescope.

The Galilean satellites were actually named by the German-born Dutch astronomer Simon Marius (1573–1624), who also claimed to have discovered them before

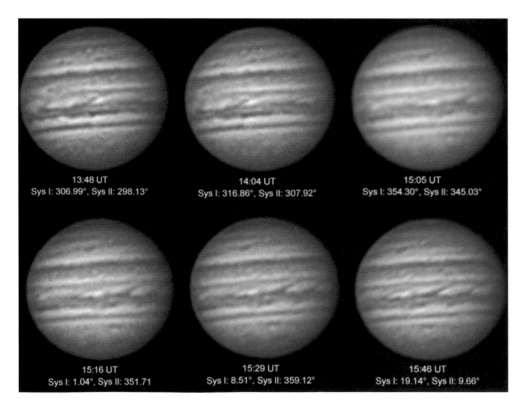

13:48 UT Sys I: 306.99°, Sys II: 298.13°	14:04 UT Sys I: 316.86°, Sys II: 307.92°	15:05 UT Sys I: 354.30°, Sys II: 345.03°
15:16 UT Sys I: 1.04°, Sys II: 351.71	15:29 UT Sys I: 8.51°, Sys II: 359.12°	15:46 UT Sys I: 19.14°, Sys II: 9.66°

Tan Wei Leong, an excellent planetary imager who lives in Singapore, took this sequence of images on 30 Dec 1999. He used a Celestron C11 and an SBIG ST7E CCD camera.

Jupiter – orbital data

Period of rotation	$0^d9^h55.5^m$
Period of revolution	$11^y315^d1.1^h$
Synodic period	$398^d21^h07.2^m$
Velocity of revolution	12.572 m/s (45 259.5 km/h)
Distance from the Sun	5.2028 AU 778 330 000 km
Minimum distance from Earth	3.934 AU 588 518 000 km
Maximum distance from Earth	6.47 AU 967 898 200 km
Inclination of equator to orbit	3.13°

Galileo. In 1664, British chemist and physician Robert Hooke (1635–1703) observed the rotation of Jupiter but did not measure it. He also discovered the feature that would come to be called the Great Red Spot. The very next year, Italian-born French astronomer Jean-Dominique Cassini (1625–1712) measured the rotational rate of Jupiter.

Jupiter by CCD and eye. (Image by Ed Grafton, sketch by Carlos E. Hernandez, both of Houston, Texas. The image was taken at 05:12 UT through a Celestron 14; the sketch made at 05:30 UT through a Celestron 8, both on 28 Nov 2000)

In 1675, the Dutch astronomer Ole Romer (1644-1710) made a number of observations of the Galilean satellites as they disappeared into the shadow of Jupiter. It had been noticed that the predicted timings of these events differed from observation based upon where Jupiter was in its orbit. Romer correctly deduced that light had a finite velocity and the travel time of the light from these eclipses was causing the errors. The value he arrived at was 225 300 km/s, a little slower than the modern value of 299 793 km/s.

In the latter half of the nineteenth century, after spectroscopy revealed the processes occurring within stars, and later, when we learned something about star formation, the question was raised as to whether Jupiter was a failed star. This was based upon the observation that Jupiter has

essentially the same chemical makeup as the Sun. As more was learned, however, astronomers realized that the mass of Jupiter was far too low to allow nuclear fusion at its core. The least massive star that can support nuclear reactions has a mass approximately 0.08 of the Sun. Jupiter's mass is approximately 0.001 of the Sun. Therefore, one would have to crush 80 Jupiters together to make even the least massive viable star.

Interesting facts about Jupiter

The largest single structure in the solar system is the magnetosphere of Jupiter. If it were visible from Earth, it would appear larger than the Full Moon.

Under favorable conditions, Jupiter can cast a visible shadow at night.

At Jupiter's equator, the speed of rotation is 45 259.5 km/h. This is more than 27 times as fast as at the Earth's equator.

Eight outer moons of Jupiter (J6–J13) are named in an interesting way. Those satellites with direct orbital motion have names ending in "a" (Elara, Himalia, Leda, Lysithea). Those with retrograde orbits have names ending in "e" (Ananke, Carme, Pasiphae, Sinope).

Jupiter's average apparent motion (against the background of stars) is approximately 5 minutes of arc per day. Thus, in a little over six days, Jupiter can move the width of the Full Moon.

The brightest satellite visible from Earth (not counting the Moon) is Ganymede, with an apparent visual magnitude of 4.4 at opposition. Io is next brightest at magnitude 4.7, then Europa at 5.1, and Callisto at 5.4.

The four large satellites of Jupiter were discovered in 1610. It was to be 282 years before the fifth satellite (Amalthea) was discovered, by Edward Emerson Barnard (1857–1923), on 9 September 1892. This was the last satellite discovered by visual means.

On 9 September 1892, American astronomer Edward Emerson Barnard (1857–1923) discovered the fifth satellite of Jupiter, now known as Amalthea. He was observing with what was then the largest telescope in the world, the 91 cm refractor at the Lick Observatory. Even so, it was a fantastic discovery considering the brilliance of Jupiter and the faintness of the satellite. Barnard has the distinction of making the last discovery of a planetary satellite with the naked eye.

The most recent data about Jupiter come from the Galileo spacecraft, composed of an orbiter and a probe, which was released from the main spacecraft on 12 July 1995, and entered the atmosphere of Jupiter on 7 December 1995. The

Jupiter. Planetary imager Tan Wei Leong, of Singapore, took this image at 15:24 UT on 19 Feb 2002. (Celestron C11 and an SBIG ST7E CCD camera)

The rotation of Jupiter. Tan Wei Leong, of Singapore, took these images on 25 Mar 2002. From left to right, the times were 13:04 UT, 14:19 UT and 14:42 UT (Celestron C11 and an SBIG ST7E CCD camera)

Jupiter's clouds

The overall appearance of Jupiter is that of a planet whose atmosphere is divided into bands of various colors that are oriented parallel to the planet's equator. These bands, of course, represent clouds. The lighter colored bands are known as "zones" and the darker colored ones are called "belts." Within the zones and belts are eddies which may produce temporary spots or streaks within or between Jupiter's cloud bands.

Jupiter's atmosphere (gas concentrations of more than 1 ppm)

Hydrogen (H_2)	82%
Helium (He)	18%
Methane (CH_4)	0.1%
Ammonia (NH_3)	0.02%
Water vapor (H_2O)	0.1%
Ethane (C_2H_6)	0.0002%

The Great Red Spot

For more than a century, humans have been observing Jupiter's Great Red Spot (GRS). In fact, it was almost certainly first observed in 1664, by the English astronomer Robert Hooke (1635–1702). The GRS is an anti-cyclonic (high-pressure) storm located 22° south of Jupiter's equator. The closest analogy would be to a terrestrial hurricane. Since it is anti-cyclonic in Jupiter's southern hemisphere, the rotation of the GRS is counterclockwise, with a period of about 6 days. For comparison, a hurricane in Earth's southern hemisphere rotates clockwise because it is a low-pressure system. The Spot itself is enormous: it has a north–south width of 14 000 km and a variable east–west width of 24 000–40 000 km, although as of this writing it has been observed to be shrinking. The clouds associated with the GRS appear to be about 8 km above neighboring cloud tops.

The Coriolis effects that are responsible for cyclones and anti-cyclones on Earth are greatly magnified on Jupiter. This is understandable when we compare the Earth's rotational period of approximately 24 hours with Jupiter's rotation, once every 10 hours, approximately. This difference alone does not account for the persistence and size of the GRS.

probe, after the most difficult planetary atmospheric entry ever attempted, detected extremely strong winds and very intense turbulence. This provided evidence that the energy source driving much of Jupiter's distinctive circulation phenomena is probably heat escaping from the deep interior of the planet. The probe also discovered an intense new radiation belt approximately 49 900 km above Jupiter's cloud tops.

As the probe entered Jupiter's atmosphere, instruments showed upper-level atmospheric density and temperatures to be much greater than expected. The high temperatures cannot be explained by current theory. Apparently, there is something unexplained occurring in the upper Jovian atmosphere, causing higher-than-predicted temperatures to be found there.

Following parachute deployment, the probe collected data throughout the descent. During that time, the probe endured severe winds, periods of intense cold and heat, and strong turbulence. The probe eventually stopped transmitting due to extreme temperature and pressure.

Due to its vast size and rapid rotation rate, it was anticipated that the Galileo probe would find strong winds. But instead of gusts of up to 300 km/h as expected, the probe recorded fairly constant winds at an amazing 530 km/h. Wind speed did not vary significantly during the probe's trip through the atmosphere, suggesting that Jupiter's winds are not caused by differences in the intensity of sunlight between the equator and the poles. The origin of Jupiter's winds appears to be the internal heat source which radiates energy up into the atmosphere from the planet's deep interior. Because of this internal heat source, Jupiter's overall atmospheric movement is dominated by a jet stream-like mechanism rather than swirling hurricane or tornado-like storms.

Jupiter on the first day of a new year. Tan Wei Leong, of Singapore, took this image at 19:08 UT on 1 Jan 2002. Note the smallness of the actual "Red Spot" within the Red Spot Hollow. (Celestron C11 and an SBIG ST7E CCD camera)

missing. Another factor adding to the persistence of the GRS is that its motion is driven by Jupiter's internal heat source. Computer simulations suggest that such large disturbances may be stable on Jupiter, and that stronger disturbances tend to absorb weaker ones, which may explain the size of the GRS.

Finally, it is not known how long the GRS will last. Serious changes in its size have taken place during the twentieth century. As of this writing, the Spot is approximately half as large as it was 100 years ago. Some observers have gone so far as to suggest that the GRS should be renamed. My opinion? Well, I am more a slave to history than some, so my one-word answer: "No!"

Galilean satellites

The Italian astronomer Galileo Galilei (1564–1642) made a number of discoveries with his early telescopes. Some contend that the most miraculous result of his exploration of the heavens was made on 7 January 1610. On that night, Galileo looked at Jupiter through his telescope and saw three little stars in a straight line, two on one side of Jupiter and one on the other. The next night, the stars were still there, but their positions were different. On 13 January 1610, Galileo noticed a fourth star.

After watching them for a number of weeks, Galileo came to the conclusion – and he could scarcely believe it – that the four "stars" were actually planets revolving around Jupiter in the same way that the Moon circles the Earth. These were the first solar system objects discovered that were invisible to the unaided eye. It was a tremendous discovery, and one which helped to establish Copernicus' ideas as the true model of the solar system.

An observation remembered – SL9

It is with fondness that I remember Comet Shoemaker–Levy 9 (SL9). It was discovered in March 1993, by Eugene and Carolyn Shoemaker and David Levy. Two months later it was recognized that SL9 was on a collision-course with Jupiter, the impact to take place in mid-July 1994.

On 7 July 1992, SL9 passed within 1.3 Jupiter radii. At that time, the comet was fragmented into 21 individual pieces. Soon after fragmentation, a great deal of dust was observed,

There are other features similar to the GRS within the clouds, but none are as large.

So the GRS is an anti-cyclone positioned within the South Equatorial Belt (SEB) of Jupiter. The GRS itself lies under a layer of ammonia-rich clouds. Since the upper atmosphere of Jupiter is moving at a different rate than this layer of clouds, an instability is caused. Furthermore, it is believed that these clouds provide the reddish color to the spot. The GRS also drifts in longitude through the SEB. The instability is also responsible for what is known as the Red Spot Hollow, a slightly larger area stretching into the South Tropical Zone: i.e., the bit it takes out of the zone to the south and, more loosely, the bits it takes out of the SEB itself. The GRS is thought to change color (or fade, in the words of some amateurs) because clouds at a higher level and of different composition are condensing within the region which lies above the spot.

Presumably the persistence of the GRS (or hollow, or whatever) is related to the fact that it never comes over land, as in the case of a hurricane on Earth. Thus, on Jupiter, the friction which would ordinarily dissipate such a structure is

Galilean satellites of Jupiter

	Size (km)	Mass (kg)	Density (g/cm³)	Orbital period	Eccentricity	Inclination	Distance from planet (km)
Io	3630	8.94×10^{22}	3.57	$01^d18^h27.6^m$	0.041	0.040°	4.216×10^5
Europa	3120	4.799×10^{22}	3.018	$03^d13^h14.6^m$	0.0101	0.470°	6.709×10^5
Ganymede	5268	1.482×10^{23}	1.936	$07^d03^h42.6^m$	0.0015	0.195°	1.07×10^6
Callisto	4806	1.076×10^{23}	1.851	$16^d16^h32.2^m$	0.007	0.281°	1.883×10^6

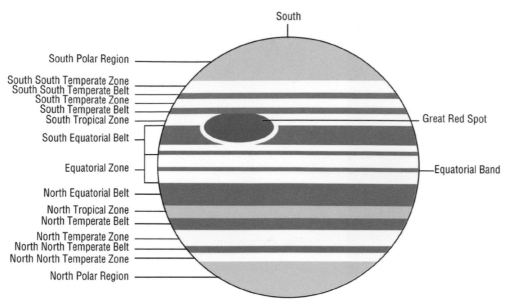

South

South Polar Region

South South Temperate Zone
South South Temperate Belt
South Temperate Zone
South Temperate Belt
South Tropical Zone

South Equatorial Belt

Equatorial Zone

North Equatorial Belt

North Tropical Zone
North Temperate Belt

North Temperate Zone
North North Temperate Belt
North North Temperate Zone

North Polar Region

Great Red Spot

Equatorial Band

Observing nomenclature for Jupiter.
(Illustration by Holley Y. Bakich)

but this diminished over time. Before impact, the comet was out of reach of even the largest amateur instruments.

The diameter of the original comet nucleus has been estimated at between 4–5 km. From calculations based upon the energy released during each impact, the minimum fragment diameter has been estimated at 350 m with the largest 1–2 km in diameter.

The impact sites were not directly observable from the Earth. However, after about 30 minutes, the fireballs had been transformed into black clouds (referred to as dark spots or plumes), easily visible at all wavelengths and observable with even small telescopes. These structures resembled "pancakes," and formed in the stratosphere of Jupiter. Each of the plumes had a diameter greater than 10 000 km. The major plume structures were still clearly visible in late September.

The high-speed easterly and westerly jets turned the dark plumes at the impact sites into striking "curly-cue" features which were visible in medium-sized telescopes. Although individual impact sites were still visible several months later despite the shearing effects of the atmosphere, the fading of Jupiter's scars had begun and after a year no trace remained.

My fondest memories of SL9 are associated with the Astronomical Society of Kansas City. For ten straight days club members brought their telescopes to Powell Observatory where the public was welcomed in throngs. I pre-

sented over 40 illustrated lectures, between which I took advantage of the views through some excellent telescopes. I will not soon forget those black patches on a super-sharp image of Jupiter through Kathy Machin's superb 250 mm Newtonian reflector. It seemed that every time Kathy looked up from her scope, I was in line!

Observing Jupiter

Next to the Sun and Moon, the celestial object with the greatest detail is Jupiter. Even small telescopes can immediately show the four Galilean satellites. They appear as bright stars on either side of Jupiter and they are generally in a straight line (although some interesting triangles and other forms are possible).

In addition to the moons, several dark stripes are easily seen on the planet. These stripes, on either side of the equator, are known as the North and South Equatorial Belts. With larger telescopes operating at greater magnifications, more belts and zones are visible.

At higher magnification, you can see that Jupiter is flattened, a result of its rapid rotation rate coupled with the fact that it is not a solid planet. Jupiter's equatorial diameter is over 9000 km larger than its polar diameter.

Watching Jupiter night after night, far from being monotonous, can be a rewarding pursuit. In addition to the changing positions of the belts and zones, the planet's rotation brings nearly all of its visible area into view in a single night around the time of opposition. There are also times when individual belts and zones become more or less prominent. Some have even been observed to disappear for extended periods of time. During each of the past two apparitions of Jupiter (at the time of this writing) pairs of large ovals have merged into one.

It is often said that Jupiter does not respond as well to high magnification as do Mars or Saturn. This may be due to the

Jupiter – observational data

Brilliancy at opposition	Maximum	−2.9
	Minimum	−2.0
Angular size	Maximum	50.11″
	Minimum	30.47″

Note: Jupiter's polar diameter is 93% of its equatorial diameter.

fact that Jupiter has too much detail and high magnification tends to muddy the water, so to speak. In practice, this is more noticeable in telescopes with large central obstructions. With quality refractors and planetary Newtonians with small central obstructions, the limiting factors are related to atmospheric conditions, not the amount of detail on Jupiter.

Systems

The planet Jupiter, being a gas giant, exhibits a difference in rotation periods over its equatorial region (System I: 9 hours 50 minutes 30.003 seconds or corresponding to approximately 10° north and south of the Jovian equator) and all latitudes north and south of the equatorial region (System II: 9 hours 55 minutes 40.632 seconds). A third (and truer) rotation period is System III (9 hours 55 minutes 29.711 seconds), which corresponds to the rotation of the interior of Jupiter, is not visible to observers but is detectable to radio astronomers.

Filters should be used by all observers as they bring out fainter detail not visualized in integrated (white) light. A blue (e.g. Wratten 38A) filter will enhance the dark reddish-brown belts over the planet. A red (e.g. Wratten 23A) will bring out the blue features (e.g. festoons) within the Equatorial Zone (EZ) of Jupiter as well as the northern and southern borders of the major belts.

Sketching Jupiter

Draw quickly! Jupiter rotates quite rapidly and some of the features you are sketching may rotate out of view if you take longer than about 20 minutes. Sketch the positions of the equatorial belts and polar regions first. Carefully estimate their widths and extent and where, in terms of latitude, they begin and end. Next, place the less apparent belts and zones as they appear. Work on one hemisphere at a time. Next, place features within the belts and zones using Jupiter's central meridian (an imaginary line from top to bottom) to help you gauge the distances. Finally, carefully shade your sketch to duplicate what you see.

Jupiter is so bright that some observers use a dim white light aimed at their sketch pad. I cannot endorse this practice because it doesn't work for me. Using a white light, I tend to lose too many of the less contrasty details on the planet. (This type of illumination sure works well for me with the Moon, however.)

One technique which some observers say works well to bring out additional detail on Jupiter is to focus the eye on a spot halfway between one of the poles and the equator. But don't look at the spot. Concentrate on seeing and sketching detail in the polar area. Repeat for the other hemisphere. Some observers, in search of more realism, go so far as to draw the outline of the planet as a slight oval rather than a perfect circle.

Upcoming conjunctions and oppositions of Jupiter

Upcoming conjunctions	Upcoming oppositions
22 Aug 2003	2 Feb 2003
21 Sep 2004	4 Mar 2004
22 Oct 2005	3 Apr 2005
22 Nov 2006	4 May 2006
23 Dec 2007	6 Jun 2007
24 Jan 2009	9 Jul 2008
28 Feb 2010	21 Sep 2010

Satellite phenomena

Four events are possible involving Jupiter and its four large moons. An eclipse of a satellite occurs when the satellite moves through Jupiter's shadow. A satellite occultation occurs when the satellite moves behind Jupiter as seen from the Earth and is hidden by the planet. Satellites always disappear into occultation at the west side of Jupiter and reappear at the east side. A satellite transit occurs when the moon moves in front of Jupiter as seen from the Earth. A satellite in transit always moves from east to west across the

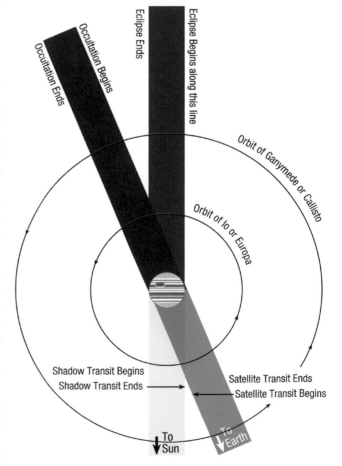

Phenomena of Jupiter's satellites before opposition. Further explanation in the text. (Illustration by Holley Y. Bakich)

face of Jupiter. The satellites themselves appear as bright dots against Jupiter's dark belts. However, when they are in front of the brighter zones, they are hard to see unless you have followed a satellite from the time it started to cross the planet's face.

A shadow transit occurs when the shadow of a satellite moves across Jupiter's disk. The shadows cast by the satellites onto the face of Jupiter as they transit the disk can be seen as small black dots in any telescope. Shadow transits also move from east to west across Jupiter.

Eclipse events are easier to observe than occultations, as they usually take place some distance from the limb of Jupiter. Occultations take place at the edge of the brilliant planet.

Prior to opposition the shadow of Jupiter is to the west of the planet as seen from the Earth. Because of this, a satellite will undergo an eclipse before it is occulted. The satellite being occulted is observed to gradually fade as it enters Jupiter's shadow. Because Io and Europa are closer to Jupiter, they each emerge from an eclipse after their occultation has started. Therefore, their reappearance from eclipse cannot be seen (nor can the disappearance into occultation). The two outer satellites, Ganymede and Callisto, are usually far enough out from Jupiter to be seen reappearing from eclipses before they are occulted. At opposition, the shadow is directly behind Jupiter. After opposition, occultations occur before eclipses, and for Io and Europa only the disappearance into occultation and the reappearance from eclipse are visible.

Regarding transits and shadow transits, prior to opposition, the shadow of a satellite falls on the disk of the planet before the satellite transit begins. After opposition, the satellite begins its transit first, with the shadow following. The shadow can be seen on the planet even after the satellite has finished its transit.

In the ten weeks before opposition, the Sun–Jupiter–Earth angle is shrinking. As that angle gets smaller, the time between an eclipse of a satellite and its occultation decreases. The same is true for the time between a shadow transit and the transit of a satellite. At opposition these times coincide. After opposition the Sun–Jupiter–Earth angle increases for about three months, and then it starts decreasing again before Jupiter moves behind the Sun.

Satellite observations

Medium-sized telescopes and good seeing will allow details to be observed on the Galilean satellites themselves. With high powers (exceeding $350\times$), distinct disks may be resolved, especially during satellite transits when the glare of the satellite is reduced due to Jupiter providing a lighter background. Ganymede is the best candidate for this. Look for the lighter shaded frost in its polar regions. With large telescopes, even the colors of these satellites may be observed.

Phenomena of Jupiter's satellites after opposition. Further explanation in the text. (Illustration by Holley Y. Bakich)

Jupiter and Ganymede (with detail). (Image by Ed Grafton, Houston, Texas, using a Celestron 14 and an ST5c CCD camera)

Observing challenge: Himalia

Edward Emerson Barnard discovered Amalthea in 1892, the last satellite discovered through visual means. Twelve years later, Himalia was discovered photographically. Himalia is actually much easier to see than Amalthea even though it is 0.7 magnitudes fainter at 14.8. This is because Himalia lies over 60 times as far from Jupiter at greatest elongation. Himalia takes 250 days to orbit Jupiter and can be nearly a degree from the giant planet at its furthest point. A confirmed observation has been made with a telescope as small as 250 mm. Your results may differ!

Filters and Jupiter

For an object as bright and diverse as Jupiter, amateurs have tried some strange techniques, far beyond the range of standard colored filters. Jeff Medkeff, of Sierra Vista, Arizona, offers the following: "In narrowband filter land, LPR filters can do dramatic things to Jupiter, but they are even more dramatic if you cut off the red tail using a blue filter in conjunction with them."

At the time of this writing, I had just received a Variable Filter System from Sirius Optics in Kirkland, Washington. Unfortunately, Jupiter was quite low in the west, but I see this filter's potential. By turning the variable control wheel, different wavelengths are allowed through. The result is amazing! It's like looking at Jupiter with every colored filter in rapid succession. At times I got too excited and moved the wheel too quickly, as it doesn't have much friction. There is also a tendency to move from one color to the next too quickly, but I think that once the novelty of the filter wears off (it really is a great accessory!), its usefulness will increase.

There are many filters which work well on Jupiter. Blue filters bring out the contrast within the bright zones and sharpen brighter cloud features. Green and blue filters will darken the belts of the planet, perhaps allowing you to see greater detail. Yellow filters will darken bluish features such as festoons that appear from time to time within the equatorial region. Red filters will brighten and enhance white spots and ovals seen in the South Temperate Belts and Zone.

6.9 Saturn

Saturn – physical data

Size	120 536 km (polar diameter 107 566 km)
Mass	5.688×10^{26} kg
Oblateness	0.1076
Volume	8.183×10^{14} km³ (785 times that of Earth)
Albedo	0.47
Density	0.69 g/cm³
Solar irradiance	15 W/m²

Saturn – orbital data

Period of rotation	$0^d 10^h 14^m$
Period of revolution	$29^y 167^d 6.7^h$
Synodic period	$378^d 02^h 09.6^m$
Velocity of revolution	9.67 km/s (34 812 km/h)
Distance from the Sun	9.5388 AU 1 429 400 000 km
Minimum distance from Earth	8.01 AU 1 198 279 000 km
Maximum distance from Earth	11.09 AU 1 659 040 000 km
Inclination of equator to orbit	25.33°

To the early Chinese, the planet Saturn was known as Tu xing. In Babylon, it was the star of the god Genna or Ninib. Ninib was primarily an agricultural deity, regulating the changes of seasons. The Babylonians did not set the planets up as equivalent to the gods. Instead, they saw them as celestial talismans of the gods' power and intentions. The planets' positions and motions, therefore, were important because they were thought to respond to the whims of the gods.

Through careful observation, the Mesopotamians saw that Saturn took almost 30 years to pass completely through the band of the zodiac, making it the slowest-moving of the (visible) planets. Because of this fact, they correctly surmised that it was the most distant.

The Greeks first called Saturn "the star of Kronos." As the planet Saturn, Kronos was recognized to be a deposed king. He had once ruled the world as the Sun rules the day. For this reason they called it the "Sun of the night" and in some ways made it the Sun's antithesis. And although it may seem strange to us, the Greeks occasionally called Saturn, Phainon, or "brilliant star." Although not nearly the brightest planet, if found in an area where there are no bright stars, Saturn can stand out quite nicely in a dark night sky.

The planet also gave its name to the last day of the week. The Romans called it "Saturnus dies," which translates as "Saturn's day." In Rome, Saturn's seven-day feast was the Saturnalia, a holiday that began on December 17 and was timed in part by the winter solstice – the day of shortest daylight in the northern hemisphere.

Beginning with Galileo's telescopic observations of Saturn, the story of Saturn is inextricably bound to its magnificent system of rings. The initial question, "What were they?" was answered by Huygens. But many others helped us to know them in detail.

The rings

Chronology

The first to observe Saturn's rings was Galileo, in 1610. He thought the rings were "handles" or moons on either side of the planet. He said: "I have observed the highest planet [Saturn] to be tripled-bodied. This is to say that to my very great amazement Saturn was seen to me to be not a single star, but three together, which almost touch each other." In 1612, Galileo was astounded when he found that the rings he first observed a couple of years earlier had disappeared; he is the first person to have observed and recorded a Saturn ring plane crossing.

It was not until 1655 that the true nature of the rings was discerned. In that year, the Dutch astronomer Christiaan Huygens (1629–1695) stated that Saturn was surrounded by "a thin, flat ring, nowhere touching, and inclined to the ecliptic." On 25 March 1655, using a refractor which provided a magnification of 50 (that he designed himself), Huygens discovered the first Saturn moon, Titan.

Four years later, in 1659, Huygens published his book, *Systema Saturnium*, in which he explained that every 14 to 15 years the Earth passes through the plane of Saturn's ring. In the same year, British physicist James Clark Maxwell stated that Saturn's ring would be broken by the tensions of attraction and centrifugal force; he suggested that Saturn's rings consisted of numerous small bodies which, like a ring of meteorites, freely circle the planet. In 1660, French astronomer Jean Chapelain (1595–1674) suggested that

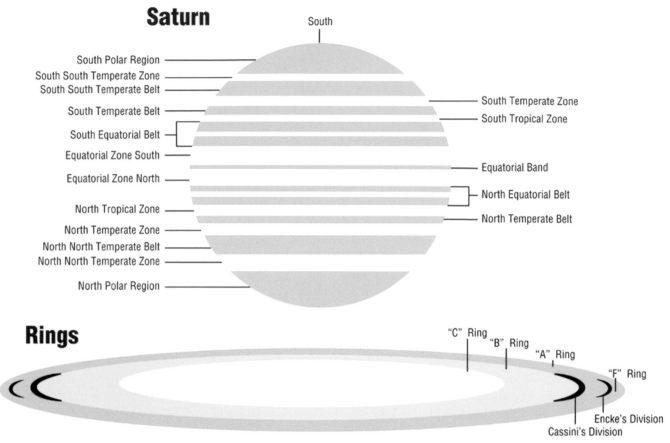

Saturn

South

- South Polar Region
- South South Temperate Zone
- South South Temperate Belt
- South Temperate Belt
- South Equatorial Belt
- Equatorial Zone South
- Equatorial Zone North
- North Tropical Zone
- North Temperate Zone
- North North Temperate Belt
- North North Temperate Zone
- North Polar Region

- South Temperate Zone
- South Tropical Zone
- Equatorial Band
- North Equatorial Belt
- North Temperate Belt

Rings

- "C" Ring
- "B" Ring
- "A" Ring
- "F" Ring
- Encke's Division
- Cassini's Division

Observing nomenclature for Saturn. (Illustration by Holley Y. Bakich)

Saturn's rings were made up of a large number of very small satellites.

In 1676, the French astronomer Jean-Dominique Cassini (1625–1712) discovered a gap between the outer (A) and the inner (B) rings. This would soon be known as the Cassini Division and it is often observed by amateur astronomers even in small telescopes. In 1684, Cassini discovered two more Saturn moons, Tethys and Dione, just prior to the ring plane crossing of 1685. It would be over 100 years until the next Saturn moon would be discovered.

In 1789, Sir William Herschel discovered two new Saturn moons, Enceladus and Mimas, during the ring plane crossing of 1789–1790. Herschel also found that Saturn was flattened at its poles, something he had suspected since 1776. He also made one of the earliest estimates of the thickness of the rings at 483 km, and reported observations of eclipses of Saturn's moons by the planet's shadow. The following year, Herschel was able to determine the rotation period of Saturn's ring to be 10 hours 32 minutes.

In 1837, the German astronomer Johann Encke noticed a dark band – later called Encke's Division – in the middle of the A ring. For amateurs, this is a much more difficult sighting and requires either (1) medium aperture and great seeing or (2) large aperture and good seeing. Just as a reference, in 1888, American astronomer James Edward Keeler (1857–1900) became the first person to clearly

Interesting facts about Saturn

From Saturn, the Sun appears only 1% as bright as it does from Earth.

The most reflective object in the solar system is Enceladus, with a geometric albedo of 1.0. This means that Enceladus reflects essentially 100% of the light striking it back into space.

Saturn's north pole is pointed toward right ascension 02^h34^m and declination +83.3° (1950.0 coordinates). This is a point within the boundaries of the constellation Cepheus. It is interesting that Saturn's north celestial pole is only 6° from that of Earth.

Thirteen of Saturn's moons have been discovered around the time of a ring plane crossing.

Saturn's average apparent motion (against the background of stars) is approximately 2 minutes of arc per day. Thus, in about 15 days, Saturn can move the width of the Full Moon.

The orientation of Saturn's rings at opposition can dramatically increase its brightness. The difference between the rings being totally open and Earth passing through the plane of the rings can be as much as 0.9 magnitude.

observe the Encke Division. In 1849, Edouard Roche (1820–1883) proposed a ring formation theory. He stated that a satellite had approached Saturn so closely that tidal forces had broken it apart.

During the mid-nineteenth century, more and more evidence was being gathered that the rings of Saturn were not solid, but made of numerous particles. In 1872, Daniel Kirkwood was able to associate the Cassini Division and Encke's Division with resonances of the four then known interior moons: Mimas, Enceladus, Tethys and Dione. In 1883, A. A. Common took the first photograph of the rings. Late in the nineteenth century, measurements showed that the inner part of the ring system rotates more rapidly than the outer part – observational evidence of non-solid rings.

Ring formation theories

With more detailed observations of Saturn's rings being made, theories of their formation seemed to know no end. Today, three prevailing theories are under consideration. Before we discuss the theories, however, a short review of Roche's limit may prove helpful.

The Roche limit is the minimum distance from the center of a planet that a satellite can maintain equilibrium, that is, without being pulled apart by tidal forces. If a planet and a moon have the same density, the Roche limit is 2.446 times the radius of the planet. Any sizeable satellite orbiting inside the Roche limit will be destroyed. The Roche limit for the Earth is 18 470 km. If our Moon somehow ventured within this distance, it would be pulled apart by tidal forces and a ring system might form around the Earth. The ring systems of the four Jovian planets are within their respective Roche limit. The approximate values are listed below:

Roche limits

Planet	Roche limit (km)
Jupiter	173 800
Saturn	148 000
Uranus	62 800
Neptune	59 500

One theory of the formation of Saturn's rings assumes that the particles are material left over from the formation of the solar system. This matter could not form into a satellite because it was located within the Roche limit for Saturn. Eventually, collisions and interactions formed the rings as we see them today.

Another theory states that at some time early in its history, one of Saturn's moons ventured too close to the planet. Because this satellite was inside the Roche limit, it was pulled apart by tidal forces. As above, after a suitable length of time the ring system formed.

The third theory is that of violent encounter. At an early time in the history of the planet and its moons, one of Saturn's satellites suffered a devastating impact, either from bombardment by meteors or through a collision with another moon. Astronomers view these three theories as equally probable.

Shepherding moons

When astronomers gained some grasp of the origin of Saturn's rings, their next question was, "What keeps them in place?" Most theoretical calculations show the rings either being pulled into Saturn, due to the planet's gravity, or dissipating in space, due to collisions among ring particles.

The leading theory which explains how the ring particles are kept in place is the theory of shepherding satellites. The gravitational attraction of Saturn's moons force the particles into rings that rotate around the planet.

Shepherding moons are satellites that orbit near a ring. Due to gravitational effects from the shepherding moon, the edges of the rings are kept sharp and distinct. If the shepherding moon was not present, then the ring material would have a tendency to spread out. In the case of two satellites orbiting on each side of a ring, the ring will be constrained on both sides into a narrow band.

The concept of shepherding moons was first proposed by Peter Goldreich and Scott Tremaine in 1979 to explain why the rings of Uranus were so narrow. Voyager 1 discovered the first Saturnian pair of shepherding moons, Prometheus and Pandora, in 1981, near the narrow F ring. Voyager 2 later found shepherding moons at Uranus in 1986. During the Saturn ring plane crossing of 22 May 1995, the Hubble Space Telescope discovered a third shepherding moon orbiting near the F ring.

Ring structure

The rings have a maximum thickness of 0.8 km, but are probably much thinner. Results from Voyager 2 indicate an average thickness of no more than about 200 m. Space probes have shown that the large rings are actually made up of a series of smaller ringlets, giving the rings a highly structured appearance. Voyager also discovered the presence of wave patterns in the rings.

Near-infrared observations from Earth have shown that the surface of the ring particles is predominately water ice. As impurities have been detected, some small amount of silicate material may be mixed in with the ring particles.

The size of most of the ring particles ranges from 1 cm to 5 m. It is likely a few kilometer-sized objects exist as well. One of Saturn's moons, Pan, is inside the Encke Division and is 20 km in diameter. Additional moonlets may yet be discovered in some of the other gaps in the rings.

Strange dark radial features up to 20 000 km in length that move about in curious patterns within the B ring have also been discovered. Since these spokes have been observed on both sides of the ring plane, they are thought to be microscopic grains that have become charged and are levitating away from the ring plane. Since the spokes are seen to rotate at the same rate as Saturn's magnetic field, it is thought that the electromagnetic forces are also at work.

The Cassini Division is the largest clear space in the ring system. Voyager 2 revealed that this space is not empty, but the majority of the ring debris in that area has been removed. The cause is believed to be the moon Mimas. The Cassini

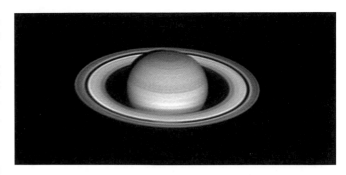

Saturn. (Image by Ed Grafton, Houston, Texas, using a Celestron 14 and an ST5c CCD camera)

The ring system of Saturn

Name	Distance from center of ring to Saturn (km)	Width (km)
D ring	71 000	7 500
C ring	83 500	17 500
Maxwell Gap	88 000	270
B ring	105 200	25 500
Cassini Division	120 000	4 200
A ring	129 500	14 600
Encke Division	133 500	325
Keeler Gap	137 000	35
F ring	140 600	30–500
G ring	170 500	8 000
E ring	250 000	300 000

Division is maintained by a 2:1 resonance with Mimas. According to Kepler's third law, any particle that orbits at the distance of the Cassini Division would orbit exactly twice every time Mimas completes one orbit. Thus a particle in the gap would feel a gravitational attraction from Mimas every other time it completed an orbit. Eventually, Mimas would pull the particle out of its orbit, leaving a gap behind.

Ring plane crossings

It takes Saturn approximately 29½ years to complete one revolution around the Sun. During this time, the angle of Saturn's rings relative to the Sun varies by 27.3°, which is the tilt of the rings to Saturn's orbital plane. Twice during Saturn's orbit, the rings are edge-on to the Sun. Since, as seen from Saturn, the Earth appears not more than 6° from the Sun, it too crosses the ring plane at roughly the same time. Due to the relative thinness of the rings compared to their distance from Earth, they seem to disappear in all but the largest telescopes.

The Earth may experience either one or three ring plane crossings (always an odd number) during any half-orbit of Saturn. If there is only one ringplane crossing then Saturn and the Earth will be on almost opposite sides of the Sun, making observations of Saturn difficult. If there are three crossings, the middle one is near opposition and the other two are near quadratures. The chance of three intersections is about 53% and the chance of one intersection is about

47%. There is an occasional case where the Earth hangs in the plane without passing through it.

When the rings of Saturn are nearly edge-on to Earth, the glare from the rings is reduced considerably, and faint objects near Saturn are easier to see. Months before and after the ring plane crossings, observations of Saturn, its rings and moons can be made from Earth which are available at no other time. Thirteen of Saturn's moons (and possibly more by the time this book is published) have been discovered around the time of a ring plane crossing.

The next ring plane crossing will be the hard-to-observe single crossing which will occur on 4 September 2009 at 10^h19^m UT. Another single crossing will take place on 23 March 2025 at 14^h38^m UT.

On 15 October 2038, the first of three passages of the Earth through the ring plane of Saturn (a triple-plane crossing event) will take place at 13^h41^m UT. The heliocentric longitude of the Earth will be 22° and that of Saturn, 170°.

Saturn – observational data

Brilliancy at opposition	Maximum	−0.3
	Minimum	+0.9
Angular size	Maximum	20.75″
	Minimum	18.44″

Upcoming conjunctions and oppositions of Saturn

Upcoming conjunctions	Upcoming oppositions
24 Jun 2003	31 Dec 2003
8 Jul 2004	13 Jan 2005
23 Jul 2005	28 Jan 2006
7 Aug 2006	10 Feb 2007
22 Aug 2007	24 Feb 2008
4 Sep 2008	8 Mar 2009
17 Sep 2009	22 Mar 2010
1 Oct 2010	

The next will be at 23^h17^m UT on 1 April 2039 (heliocentric longitude of the Earth will be 191° and that of Saturn 175°), and the final at 12^h43^m UT on 9 July 2039 (heliocentric longitude of the Earth will be 286° and that of Saturn 179°).

Saturn's atmosphere

The atmosphere of Saturn is very similar to that of Jupiter. Although the cloud bands are not as pronounced, the nomenclature used to differentiate them is the same. Light colored cloud bands are called "zones" and darker bands are known as "belts." Colors are subtle, ranging from yellow and tan to light brown, depending on the observer. Temporary whitish spots are sometimes seen. These tend to last several months on average. The maximum wind speeds, as measured by Voyager 2, are 2294 km/h.

As mentioned above, the composition of Saturn's atmosphere is similar to that of Jupiter, but with more hydrogen. That gas comprises 94% of the molecules in the atmosphere of Saturn and 75% of the mass. Helium represents 6% of the molecules and nearly 25% of the mass. Methane, ammonia and water vapor are the only other gases in quantities greater than 0.0001% (one part per million).

Satellites

Properties of ten moons of Saturn

Satellite	Size (km)	Mass (kg)	Density (g/cm³)	Orbital Period	Eccentricity	Inclination	Distance (km)
Prometheus	1528	1.4×10^{17}	0.27	$14^h42.7^m$	0.0042	0.0°	139 350
Epimetheus	138 × 110 × 110	5.05×10^{17}	0.63	$16^h40.2^m$	0.009	0.34°	151 422
Mimas	397.6	3.75×10^{19}	1.14	$22^h37.1^m$	0.0202	1.53°	185 520
Enceladus	498.2	7.3×10^{19}	1.12	$01^d08^h53.1^m$	0.0045	0.02°	238 020
Tethys	1059.8	6.22×10^{20}	1.00	$01^d21^h18.4^m$	0	1.09°	294 660
Dione	1120	1.052×10^{21}	1.44	$02^d17^h41.2^m$	0.0022	0.02°	377 400
Rhea	1530	2.31×10^{21}	1.24	$04^d12^h25.2^m$	0.001	0.35°	527 040
Titan	5150	1.3455×10^{23}	1.88	$21^d06^h38.3^m$	0.0292	0.33°	1 221 850
Janus	199 × 191 × 151	4.98×10^{18}	0.65	$16^h40.2^m$	0.007	0.14°	1 514 720
Iapetus	1436	1.59×10^{21}	1.02	$79^d07^h55.5^m$	0.0283	7.52°	3 561 300

Observing Saturn

Where to begin? A better question, probably, is where to end? So much could be said about observing this magnificent planet that it would fill a volume on its own. Up front, let me give you a word of advice: Saturn requires magnification, lots of it. Spend your time observing Saturn

on nights of better-than-average seeing. Here in El Paso, Texas, nights of great seeing are often experienced. Quite frequently, when Saturn is high in the sky, I observe it through my 100 mm refractor at magnifications in excess of 250×.

Observing the rings

First, notice the rings. (They're hard to miss!) Are you able to see the Cassini Division? That feature has been observed in apertures as small as 50 mm. After locating this familiar feature, note the relative brightnesses of the various rings. Are any textures visible? With larger apertures you may be able to pick up differences in colors among the rings. These are subjective to some extent, based on each individual's color perception, but they still should be noted carefully.

Time for a definition. You will often see references to the ansa (plural = ansae) of Saturn's ring system. An ansa is the portion of a ring that appears farthest from the disk of a planet. The word comes from the Latin word for "handle," since the earliest views of Saturn's rings suggested that the planet had two handles extending out on either side. The ansae may be designated "east" or "west" or by the way the planet is rotating: "preceding" or "following," often abbreviated p. and f.

At much higher magnification, look for the elusive Crepe ring. I am usually able to see the Crepe ring in my 100 mm telescope, but its detail is only revealed on nights where the seeing might be described as "legendary." On such nights, I can stare at it for hours. With a 200 mm (or larger) telescope on a night of good seeing and during a time when the rings are "open," that is, when their tilt is near maximum, the Crepe ring is one of Saturn's finest features.

Saturn, showing the Crepe ring. (Image by Ed Grafton, Houston, Texas, using a Celestron 14 and an ST5c CCD camera)

If the Crepe ring is visible, try for the Encke Division (often mislabeled "Encke Gap"). It is located near the outer edge of the A ring (about 4/5 of the way from the Cassini Division to this edge). This is a very narrow division and it will take a telescope of at least 200 mm aperture using high magnification to reveal it. With smaller telescopes, some observers see an albedo feature often termed the "Encke Minimum." This is a decrease in brightness of the A ring near its outer edge. Telling this feature apart from the Encke Division is a matter of position. The so-called Encke Minima (plural when visible on both sides of the ring) are located starting at the center of the A ring. The true Encke Division is much nearer the outer edge.

For what it's worth, I have no problem denoting the above mentioned albedo feature as the Encke Minimum. It is certainly easier than the "A ring albedo feature," and it has come into common usage. With my 100 mm refractor, I can view the Encke Minima when seeing permits using magnifications of 300× or over. I cannot see the Encke Division with this telescope.

With a large telescope, you might even be able to glimpse some structure within Saturn's rings. In 1977, Stephen James O'Meara became the first to observe dark radial features (which would later be called "spokes") on Saturn's rings and recorded them in a sketch. (The Voyager 1 spacecraft confirmed the existence of spokes in 1980).

Observing the disk

The disk, or globe, of Saturn has markings on it which, for the most part, are quite subtle. Changes in the appearances of belts and zones are usually gradual. Note carefully the positions of any bright or dark spots and their relative brightnesses compared to the belt or zone in which they are found. From night to night, note any changes in these features. The only way to recognize these features is by spending lots of time observing Saturn and familiarizing yourself with its markings.

Much easier to observe are the positions and shapes of the shadow of the globe on the rings or of the rings on the globe. Before each opposition, the shadow of the globe on

the rings is found on the preceding ansa. Watch the shadow shrink in size over the weeks prior to opposition. After opposition, the globe's shadow will be found on the following ansa, and will begin to grow. If the shadow of the globe is found on the inner side of the rings be sure not to confuse it with a sighting of the Crepe ring superimposed against the disk.

Sketching Saturn

Templates showing Saturn's rings at different orientations to us, as well as complete observing forms, are available from the Association of Lunar and Planetary Observers (ALPO). Find the specific forms online (labeled as "Report Form 1 . . . 6") at
http://www.lpl.arizona.edu/~rhill/alpo/satstuff/satfrms.html

The general attributes of Saturn are not difficult to sketch. In addition, there are nowhere near as many features as on Jupiter. However, it is this subtle nature and low contrast of the features that makes Saturn a more difficult subject to sketch than it first appears.

Filters

Colors on the disk of Saturn are less bright and have less contrast than those seen on Jupiter. You may be able to see a difference in the color of the rings. For example, one ansa may appear brighter than the other when examined with a red filter. If you find this to be true, switch to a blue filter and see if the brightness variation is reversed.

As for the colors themselves, the brighter zones are described as appearing off-white, slate-gray or yellowish at times. Saturn's belts exhibit bluish-gray, brown and reddish colors easily seen using red, orange or yellow. Brighter patches sometimes appear on this ringed planet and are best seen through a green filter. The rings are highlighted using light green (#56) or blue (#80A) filters.

Imaging Saturn

Compared with most celestial objects, Saturn is very bright. Unfortunately, to bring out details on an image, you need at least a medium-sized aperture, a very stable mount, excellent polar alignment, great seeing, and a method of creating an image of high magnification (refer to the "Astrophotography" and "The CCD" chapters for more information). For film astrophotography of Saturn or any other object, my absolute highest recommendation goes to the book *High Resolution Astrophotography* by Jean Dragesco (Cambridge University Press, 1995).

Observing Saturn's moons

The satellites of Saturn, though nowhere near as bright as the Galilean satellites of Jupiter, are easy to observe in medium-

The eight brightest moons of Saturn

Satellite	Magnitude
Titan	8.4
Rhea	9.8
Dione	10.2
Tethys	10.3
Iapetus	11.2
Enceladus	11.8
Mimas	13.0
Hyperion	14.3

sized telescopes. I have seen the brightest six with my 100 mm refractor. Saturn itself does not mask the satellites unless they are right next to the disk.

A very good dos-based computer program which will display the locations of Saturn's satellites may be found on the Switch to Space website. The internet address is http://www.ibiblio.org/ais/space.htm

Click on "Software" and look under "Saturn's moons positions." The name of the program is Satsat.

No filter is necessary to view the moons of Saturn. In fact, any filter would be counterproductive, as it would cut down the light from the satellites themselves. Watch for interesting alignments of the satellites and for the very rare mutual satellite events. One of these occurs when one satellite eclipses another or a shadow of one satellite falls upon another. In 1921, the first observation of a mutual satellite event, an eclipse of Rhea by Titan, was observed by L. Comrie and A. Levin.

6.10 The outer planets

Uranus

For untold thousands of years, humanity knew of only five wanderers, or planets, which traveled against the background of stars. But on 15 November 1738, Friedrich Wilhelm Herschel was born in Hanover, Germany. Forty-two years and 118 days later – on 13 March 1781 – Herschel altered the celestial landscape forever:

> On Tuesday the 13th of March, between ten and eleven in the evening, while I was examining the small stars in the neighborhood of H Geminorum, I perceived one that appeared visibly larger than the rest: being struck with its uncommon magnitude, I compared it to H Geminorum and the small star in the quartile between Auriga and Gemini, and finding it so much larger than either of them, suspected it to be a comet.
>
> *Excerpt from paper read by William Herschel 28 March 1781 to the Bath Philosophical Society.*

In a comparatively short period of time, it was realized that Herschel had not discovered a comet, but rather the first new planet since antiquity. The astronomical community was delighted. With this one observation, Herschel had doubled the size of the solar system.

The outer planets – physical data

	Uranus	**Neptune**	**Pluto**
Equatorial diameter (km)	51 118	49 572	2 320
Polar diameter (km)	49 584	48 283	2 320
Mass (kg)	5.688×10^{26}	1.024×10^{26}	1.29×10^{22}
Oblateness	0.030	0.026	0
Volume (km³)	6.995×10^{13} (67.1 times Earth)	6.379×10^{13} (58.7 times Earth)	6.545×10^{9} (15.8% of Earth)
Albedo	0.51	0.41	0.30
Density (g/cm³)	1.29	1.64	2.05
Solar irradiance (W/m²)	3.8	1.5	0.9
Maximum wind speeds (km/h)	720	2400	none

Herschel had initially assumed the new object to be a comet because it increased in apparent diameter as he increased the magnification of his telescope. His initial sighting was with an eyepiece which provided a magnification of 227×. To prove that the object was not a star, Herschel first increased the magnification to 460×, and then to 932×. In Herschel's day, new comets were sighted infrequently, but they were not once-in-a-lifetime events.

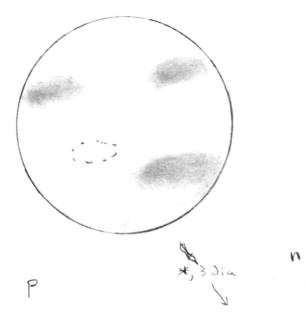

A sketch of Uranus. 09:44 UT 10 Jul 1998. (250 mm Newtonian reflector at ~f/4.6. University Optics 9 mm Abbe Orthoscopic, combined with a University Optics Klee Barlow, rendered a magnification of 354x. (Sketch by Jeff Medkeff, of Sierra Vista, Arizona)

The German astronomer Johann Elert Bode (1747-1826) was the first to suggest the name Uranus. His reasoning was mythological – Saturn was the father of Jupiter, so the next planet out should be the father of Saturn. However, the honor of naming the new planet naturally fell to Herschel. (This was more than a century before any thought was given to planetary nomenclature committees.) Herschel proposed the name Georgium Sidus (George's Star) in honor of George III, then King of England.

As may be imagined, outside of England the thought of a major member of the solar system named after a British monarch was met with less-than-enthusiastic support. Many non-British astronomers took to calling the new planet "Herschel." Some years after Herschel's death (in 1822), Bode's suggestion of Uranus as the name for the new world finally gained the total support of the astronomical community.

Major satellites of Uranus

	Size (km)	Mass (kg)	Density (g/cm³)	Orbital period	Eccentricity	Inclination	Distance from planet (km)
Miranda	470	6.59×10^{19}	1.15	$01^{d}09^{h}54.7^{m}$	0.0027	4.22°	1.298×10^{5}
Ariel	1160	1.353×10^{21}	1.56	$02^{d}12^{h}28.8^{m}$	0.0034	0.31°	1.912×10^{5}
Umbriel	1169.4	1.172×10^{21}	1.52	$04^{d}03^{h}27.4^{m}$	0.0050	0.36°	2.66×10^{5}
Titania	1577.8	3.527×10^{21}	1.70	$08^{d}16^{h}56.6^{m}$	0.0022	0.10°	4.358×10^{5}
Oberon	1522.8	3.014×10^{21}	1.64	$13^{d}11^{h}06.7^{m}$	0.0008	0.10°	5.826×10^{5}

Satellites

The above table of satellite data lists those moons which may be directly observed or imaged by amateurs.

Observing Uranus

The visible atmosphere of Uranus is generally a featureless haze. Throughout the history of observations of the planet, a number of observers have seen detail. The first such observer, the British astronomer William Buffham (1801–1871), observed two round bright spots and a bright zone in 1870. In 1883, the American astronomer Charles Augustus Young (1834–1908) reported markings, along with both polar and equatorial belts. All detail was, of course, very faint.

In April 1891, American astronomers Edward Singleton Holden (1846–1914) and James E. Keeler (1857–1900) and German astronomer John Martin Schaeberle (1853–1924), using the Lick Observatory 91 cm refractor, all saw bands. In 1924, the French astronomer Eugène Marie Antoniadi (1870–1944) saw grayish polar caps and two faint equatorial bands.

The Voyager 2 mission to Uranus, in January 1986, also showed a nearly featureless globe. The best recent detail-bearing image of Uranus was taken on 30 May 1993, at near-infrared wavelengths from the Multiple-Mirror Telescope near Mt. Hopkins, Arizona. The image showed a dark spot near the center of the disk of Uranus along with a bright region and a subtle, irregular dark band near the pole.

Viewed through a telescope, Uranus appears as a greenish disk never brighter than about sixth magnitude, and it is slightly elliptical because of its rapid rotation. The combination of a small disk and seeing effects caused by the Earth's atmosphere make it difficult to see the planet in any great detail. Observing these outer planets is difficult and their blue-greenish color begins to change to a bright blue using moderate to large aperture telescopes (300 to 600 mm).

The orbit of Uranus is tilted less than 1° from the plane of the ecliptic, so it is always found quite close to that line. Uranus' average apparent motion (against the background of stars) is approximately 42 seconds of arc per day. It takes Uranus about 44 days to move a distance equal to the width of the Full Moon.

Upcoming conjunctions and oppositions of Uranus

Upcoming conjunctions	Upcoming oppositions
17 Feb 2003	24 Aug 2003
22 Feb 2004	27 Aug 2004
25 Feb 2005	1 Sep 2005
1 Mar 2006	5 Sep 2006
5 Mar 2007	9 Sep 2007
9 Mar 2008	13 Sep 2008
13 Mar 2009	17 Sep 2009
17 Mar 2010	21 Sep 2010

Uranus and its four largest satellites. (Image by Ed Grafton, Houston, Texas, using a Celestron 14 and an ST5c CCD camera)

Neptune

In 1820, the French astronomer Alexis Bouvard (1767–1843) undertook the task of compiling positional tables for Uranus. He set out to define the entire orbit of that planet from observations made during the nearly forty years that had elapsed since its discovery in 1781.

Bouvard found that Uranus had been observed at least

The outer planets – orbital data

	Uranus	Neptune	Pluto
Period of rotation	$0^d17^h14^m$	$0^d16^h6.6^m$	$06^d09^h17.6^m$
Period of revolution	$84^y03^d15.66^h$	$164^y288^d13^h$	$248^y197^d5.5^h$
Synodic period	$369^d15^h50.4^m$	$367^d11^h45.6^m$	$366^d17^h31.2^m$
Velocity of rotation (km/h)	8971.5	9667.1	47.55
Distance from the Sun	19.1914 AU 2870990000 km	30.0611 AU 4504300000 km	39.5294 AU 5913520000 km
Inclination of equator to orbit	97.86°	28.31°	122.52°

fifteen times before its discovery by Herschel. Indeed, between 1690 and 1771, it had been recorded in catalogs as a fixed star. Knowing that earlier positions would help to refine the orbit, Bouvard welcomed the fact that the planet had been so often seen before. On the completion of his calculation, however, he met with a most extraordinary and puzzling circumstance. The orbit he deduced was found unsatisfactory to either the pre-discovery or the post-discovery set of observations, and its deviation from the older ones was striking.

Urbain Jean Joseph Leverrier (1811–1877) was a well-respected astronomer who worked at the Paris Observatory. He had attempted to reconcile irregularities in Mercury's orbit by proposing, in 1860, the existence of an intra-Mercurial planet which he eventually named Vulcan. Now his mind was turned to the orbit of Uranus. Leverrier quickly reduced the problem to a simple form: "Can the anomaly be explained by the supposed action of a foreign and hitherto unknown body on Uranus?"

After a lengthy series of calculations, Leverrier communicated the position of the object to a number of observatories around the world. At the Berlin Observatory, Johann Gottfried Galle (1812–1910) discovered the planet on the very evening of the day on which he received the letter from Leverrier indicating its place. The following is his reply to Leverrier:

Berlin, 25th September, 1846
Sir, – The Planet whose position you marked out actually exists. On the day on which your letter reached me, I found a star of the eighth magnitude, which was not recorded in the excellent map designed by Dr. Bremiker, containing the twenty-first hour of the collection published by the Royal Academy of Berlin. The observations of the succeeding day showed it to be the Planet of which we were in quest . . .
–J. G. Galle

The longitude of the new planet, as ascertained by Galle, was less than a degree from Leverrier's predicted position. Truly this was a triumph for Leverrier and celestial mechanics, for Newton and the Law of Universal Gravitation, for Galle and the Berlin Observatory, and for a modest man from Laneast,

Cornwall, England, named John Couch Adams (1819–1892).

Several years before, Adams had begun a mathematical search for the planet responsible for the irregularities in the orbit of Uranus. He completed the bulk of his research by October, 1843, and first worked out an approximate position for the trans-Uranian body in the middle of 1845.

In a letter dated 22 September 1845, James Challis, Professor of Astronomy at Cambridge University, wrote to the Astronomer Royal, George Biddel Airy (1801–1892) regarding the calculations of his former student, Adams. In the letter, he mentions that Adams would call on Airy personally. Adams did so on 22 September but Airy was in France. Adams' second call was on 21 October. He actually visited the Airy residence twice. The first time Airy was not at home, and the second time the butler refused him admittance because the Airy family was at dinner. Adams left a letter in which he provided Airy the results of his work.

In a letter dated 5 November 1845, Airy wrote to Adams in Cambridge, requesting more information about the radius vector of the hypothetical planet. Adams never replied, as he considered the request unimportant to the overall problem. The next correspondence between Adams and Airy was in September 1846, when Adams provided a detailed account of his work. It was then a week after Galle had found Leverrier's planet.

A firestorm ensued. Airy was berated publicly, privately, and professionally. The correspondence among astronomers during the latter part of 1846 is among the most heated in the history of the science. As the furor subsided and the truth came out, both Leverrier and Adams were honored for the most amazing theoretical discovery in the history of astronomy.

Observing Neptune

To the amateur astronomer with a medium or large telescope, Neptune is no problem to find, presenting a small bluish disk of about 7.7 magnitude (at opposition). If you have a small scope or are using binoculars, you will need Neptune's position and a good star chart.

Neptune lies at such a great distance that it is too dim to be visible with the naked eye. A telescope won't reveal much in the way of surface features on its tiny disk, although you may be able to see Neptune's largest moon, Triton. Basically, an amateur can't get much further than identifying the planet against the background stars. The thrill to observing Neptune is simply finding it in your telescope, where it is a small blue disk without any noticeable features at this distance in almost all telescopes.

Neptune, with an average apparent motion (against the background of stars) of only 22 seconds of arc per day, takes approximately 85 days to traverse a distance equal to the width of the Full Moon.

When it was discovered, Neptune was within the boundaries of the constellation Aquarius. Since its discovery, Neptune has not made one complete orbit around the Sun. On Wednesday, 8 June 2011, Neptune will finally complete one post-discovery orbit. Sounds to me like a good time for all of us to go out and re-discover Neptune.

Upcoming conjunctions and oppositions of Neptune

Upcoming conjunctions	Upcoming oppositions
30 Jan 2003	4 Aug 2003
2 Feb 2004	6 Aug 2004
3 Feb 2005	8 Aug 2005
6 Feb 2006	11 Aug 2006
8 Feb 2007	13 Aug 2007
11 Feb 2008	15 Aug 2008
12 Feb 2009	18 Aug 2009
15 Feb 2010	20 Aug 2010

Triton

Triton is Neptune's largest moon. It was discovered by the British astronomer William Lassell (1799–1880) scarcely a month after the discovery of Neptune, in 1846.

Triton is an oddity among moons in that its orbit is highly tilted to the plane of Neptune's equator (156.8°), and it is in a retrograde orbit. It is the only large satellite in the solar system to circle a planet in a retrograde direction. These facts have led scientists to believe that Triton formed independently of Neptune and was later captured by Neptune's gravity. If that is the case, tidal heating could have melted

Triton in its originally eccentric orbit, and the satellite might even have been liquid for as long as one billion years after its capture by Neptune. It is also the second most distant of Neptune's satellites, lying some 355 000 km from the planet. The diameter of Triton is 2 705 km.

Triton's surface is covered with a thin layer of nitrogen and methane ice. Most of the geologic structures on Triton's surface are likely to be formed of water ice, because nitrogen and methane ice are too soft to support their own weight.

An extremely thin atmosphere extends as much as 800 km above Triton's surface. Tiny nitrogen ice particles may form thin clouds a few kilometers above the surface. Triton is very bright, with an albedo of 0.60–0.95, depending on the surface composition.

The atmospheric pressure at the surface of Triton is about 1.4 Pa. That is equal to 1/70 000th the sea level atmospheric pressure on Earth. The temperature at Triton's surface is a mere 38 K, the coldest of any body yet visited in the solar system. At 800 km above the surface, the temperature is 95 K.

Observing Triton

The observation can only be carried out when certain geometric conditions are satisfied. Mainly, Triton must be separated enough from Neptune. Triton has a semi-major axis of 14.15 Neptune radius, which corresponds to separation angle of 16 arcseconds.

Pluto

The discovery of Pluto was made by Clyde William Tombaugh (1906–1997) at Lowell Observatory in Flagstaff, Arizona. The third search for Planet X began in 1927. Plans were formulated and money was secured for a new photographic telescope to better continue the hunt. The 33 cm A. Lawrence Lowell Astrographic Telescope took its first exposure in the hunt for the trans-Neptunian planet on 6 April 1929. It was an hour-long exposure centered on the star δ Cancri. The man at the telescope was Clyde Tombaugh.

Tombaugh had been corresponding with Vesto Melvin Slipher (1870–1963), Director of Lowell Observatory in 1928. Early in 1929, Slipher offered Tombaugh a position to work with the 33 cm Astrograph. Tombaugh traveled from Burdette, Kansas, to Flagstaff, Arizona, arriving at Lowell Observatory on 15 January 1929, with absolutely no idea that he would be conducting a search for a new planet.

Tombaugh exposed more than 150 plates before his epic

Major satellites of Neptune

	Size (km)	Mass (kg)	Density (g/cm³)	Orbital period	Eccentricity	Inclination	Distance from planet (km)
Triton	2 705.2	2.147×10^{22}	2.054	$05^d21^h02.7^m$ (retro.)	0.00	156.834°	3.548×10^5
Nereid	340	unknown	unknown	$360^d03^h16.11^m$	0.7512	7.23°	5.513×10^6

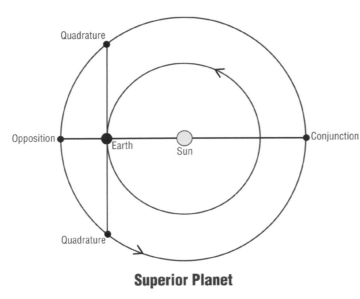

Superior Planet

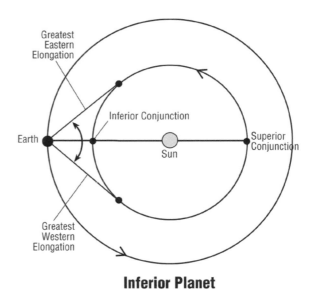

Inferior Planet

Planetary configurations. (Illustration by Holley Y. Bakich)

The outer planets and observational data

		Uranus	*Neptune*	*Pluto*
Brilliancy at opposition	Maximum	+5.65	+7.66	+13.6
	Minimum	+6.06	+7.70	+15.95
Angular size	Maximum	3.96″	2.52″	0.11″
	Minimum	3.60″	2.49″	0.065″

exposures on 23 and 29 January 1930. Even then, the discovery was not immediate. To examine the plates, Tombaugh used the observatory's blink comparator, where first one plate is illuminated and then the other. A small portion of the plates is studied with a microscope incorporated into the apparatus. The necessity of blinking other plates did not allow Tombaugh to examine the exciting pair until several weeks after they were exposed. Then, on 18 February 1930 at 4pm, Planet X was discovered on the comparator.

V. M. Slipher was a cautious man. He wanted confirmation

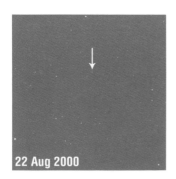

22 Aug 2000

10 Sep 2000

The motion of Pluto against the background of stars is a slow one. These two images show Pluto's motion over a nearly three-week period. (Images by Larry Robinson, Olathe, Kansas)

Interesting facts about Pluto

On 1 January 2000, Pluto had travelled 28.14% of its orbit around the Sun since its discovery. Pluto will not complete one post-discovery orbit until 8 August 2178.

Even though it takes Pluto more than 248 years to orbit the Sun, it still travels, on average, at a respectable 17 064 km/h.

Pluto and Neptune will never collide. They can never be less than about 386 million km from each other.

On average, the Sun appears 1905 times fainter on Pluto than it does on Earth.

Pluto may appear within the boundaries of 41 constellations.

The surface gravity of Pluto is only 4.1% that of the Earth.

We often think of Uranus when we imagine a planet "tipped over on its side." However, Pluto's axial tilt is more than 24° greater than that of Uranus.

On average, it takes the light of the Sun 5 1/2 hours to reach Pluto. Because of Pluto's varying distance from the Sun, this time can be as short as 4 hours 6 minutes, when Pluto is at perihelion, or as long as 6 hours 58 minutes, when it is at aphelion.

that the object identified by Tombaugh was indeed Percival Lowell's long-anticipated Planet X. More exposures were taken through all the telescopes at the observatory. A determination of the orbit was made.

Finally, on 13 March 1930, on what would have been the 75th birthday of the observatory's founder, Percival Lowell, the announcement was made. It is interesting to note that

Charon

Size (km)	Mass (kg)	Density (g/cm³)	Orbital period	Eccentricity	Inclination	Distance from Planet (km)
1172	1.7×10^{21}	1.800	$06^d09^h17.3^m$	0.00	96.56°	1.9405×10^4

this was also the date of the first planetary discovery since antiquity, by Sir William Herschel, 149 years earlier. Lowell Observatory's reputation was secure and Clyde Tombaugh – who less than two years before his discovery was a farmboy on the plains of Kansas – was assured a place in history.

Observing Pluto

As seen from the Earth, Pluto moves very slowly. Its average apparent motion (against the background of stars) is a leisurely 14 seconds of arc per day. This means that it takes Pluto roughly 130 days to travel a distance equal to the width of the Full Moon.

You can see Pluto with a 200 mm telescope, although you will need a dark site and an accurate star chart to do so. At the Anderson Mesa site of Lowell Observatory, Brian Skiff observed Pluto with a 70 mm refractor. This may be the "small aperture record" for Pluto. I know of no smaller telescopes through which it has been observed. I have seen it with both 75 mm and 100 mm refractors.

Upcoming conjunctions and oppositions of Pluto

Upcoming conjunctions	Upcoming oppositions
12 Dec 2003	9 Jun 2003
13 Dec 2004	11 Jun 2004
16 Dec 2005	14 Jun 2005
18 Dec 2006	17 Jun 2006
21 Dec 2007	19 Jun 2007
22 Dec 2008	21 Jun 2008
24 Dec 2009	23 Jun 2009
27 Dec 2010	25 Jun 2010

The easiest way to observe Pluto is on successive nights. Print-out a star chart of suitable limiting magnitude (say $v = 15$) which shows the position of Pluto. Remember to invert and/or reverse the printed image to match what you will see through your telescope/diagonal/eyepiece combination. At the eyepiece, identify the position of any object in the location of Pluto, as shown on the star chart. The following night (or, on the next clear night) find the same star field and search for Pluto. After noting the exact date and time, mark what you believe to be the new position of Pluto on your original map. Finally, find the position of Pluto which corresponds to the new date and time and compare. If the position given by your software matches the position of the mark on the original star chart, you have successfully observed Pluto.

A current reference for observing Pluto may be found on the internet at

http://www.pietro.org/Astro_C5/Articles/PlutoCurrent.htm

Charon

Observations in the late-1990s by the Hubble Space Telescope showed that Charon is bluer than Pluto. Planetary scientists interpret this as showing that these two bodies have different surface composition and structure. A bright highlight on Pluto indicates that it might have a smoothly reflecting surface layer. A detailed analysis of the image suggests that there is a bright area parallel to the equator of Pluto. Subsequent observations are needed to confirm that this feature is real. The image was taken when Charon was near its maximum elongation from Pluto, which, at the distance of the Earth, amounts to only 0.9 seconds of arc.

Well, 0.9 arcsecond is tough in a 100 mm telescope but not in one with a 600 mm aperture. Charon's magnitude of 15.7 allows it to be observed, with some difficulty, when at maximum elongation from Pluto. This occurs every 3 days 4 hours 38 minutes, approximately.

6.11 Asteroids

The sexiest celestial objects in the 1990s were asteroids, or minor planets. Sure, everyone wants to discover a comet, but few people will. A lot of people have been discovering asteroids. And some people have discovered lots of asteroids.

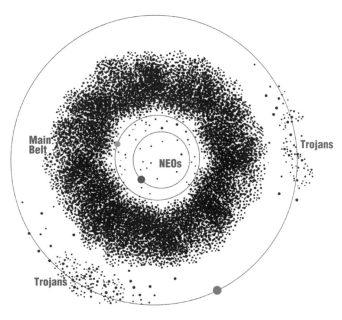

Most asteroids are confined to the region between the orbits of Mars and Jupiter. NEOs travel near the Earth. Trojans are at two of Jupiter's Lagrangian points. Not to scale. (Illustration by Holley Y. Bakich)

Asteroids are small rocky bodies which orbit the Sun. The first asteroid, Ceres, was discovered by the Italian astronomer Giuseppe Piazzi (1746–1826) on 1 January 1801. Ceres is also the largest of the asteroids. By the way, the name asteroid was bestowed upon these bodies by John Herschel. As of this writing, there are over 30 000 numbered asteroids, with about twice that many awaiting permanent designations.

Asteroid types

Main belt asteroids lie 1.8–4.0 AU from the Sun, between Mars and Jupiter. Generally, this area is referred to as the asteroid belt. Within the belt are areas called Kirkwood gaps where there are very few asteroids. The gravitational attraction of Jupiter is responsible for these gaps.

Asteroids in the main belt are divided into families, among them the Cybeles, Eos, Floras, Hildas, Hungarias, Hygiea, Koronis, Phocaea, Themis, and Veritas. Families are separated by Kirkwood gaps. One theory is that each family may have been a single body shattered long ago, leaving the pieces to occupy the same orbit as the original. The families are named after the main asteroid in them.

Near-Earth asteroids (NEAs or NEOs when called Near-Earth Objects) are found in the region close to our planet. Each has a perihelion of 1.3 AU or less. There are three subgroups of NEAs, as shown in the following table.

Near-Earth Asteroid subgroups

Subgroup	Semi-major axis	Perihelion
Aten	<1.0 AU	>0.983 AU
Apollo	>1.0 AU	<1.017 AU
Amor	>1.0 AU	$1.017 \leqslant \chi \leqslant 1.3$ AU

As is evident, the Aten and Apollo asteroids cross the orbit of the Earth while the Amor asteroids do not reach Earth's orbit but do cross that of Mars.

The Trojan asteroids are located along the orbital path of Jupiter, at stable Lagrangian points. The French mathematician Joseph Louis Lagrange (1736–1813) suggested that when objects in the same plane form an equilateral triangle with the Sun they can share the same orbit without catching up with or colliding with each other. The two largest groups of asteroids are at the so-called L4 and L5 points on Jupiter's orbit, at an angle of 60° to both the Sun and Jupiter. Their orbits are stable, with L4 leading and L5 trailing Jupiter.

Trans-Neptunian Objects (TNOs) are, as their name indicates, located past the orbit of Neptune. They are often

Asteroid types by chemical composition

Type	Percent of total	Albedo	Color	Properties
C	75	0.03–0.06	Grayish	Rich in carbon
S	17	0.10–0.22	Greenish to reddish	Metal-rich silicates; brighter than C-type
M	8	0.10–0.18	Reddish	Mainly iron and nickel; the brightest type

grouped with asteroids but are small, icy bodies more like comets. Their range of distances is 30 AU (Neptune) to further than 50 AU. Observations show that TNOs are mostly confined within a thick band around the ecliptic. This region is generally referred to as the Kuiper Belt and often objects there are known as Kuiper Belt Objects (KBOs).

Centaurs are asteroids which have orbits between Saturn and Uranus. These objects may, in fact, be more like comets or KBOs than asteroids. However, because of their distance, Centaurs do not get sufficiently close to the Sun to heat up to show cometary activity.

There are also a dozen or so other rarer types of asteroids included in the classification scheme with their own letter designations associated with clustering of brightness and color or spectral properties.

Asteroid designations

Upon discovery, an asteroid is given a temporary designation consisting of six characters. The first four characters are the year of the discovery, the next character is a letter indicating the half-month in which the discovery occurred (the letters I and Z are unused), and the last character is another letter numbering from 1–25 the asteroids discovered in that half-month (this time only I is unused). For example, the asteroid 2000 MB would be the second asteroid discovered in the first half of July 2000. If there are more than 25 asteroid discoveries in a half-month then the last character reverts to A but with the addition of a number. Should the asteroid be studied long enough for its orbit to be determined it will be given a permanent number. At this point, the discoverer is allowed to name the asteroid. A citation is submitted to the Minor Planet Center.

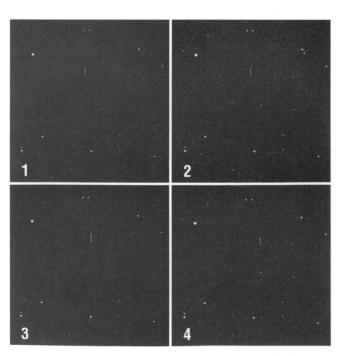

Four images of an asteroid, showing its apparent motion through the stars. (Images by Larry Robinson, Olathe, Kansas)

When an asteroid moves from its temporary designation to being named, it is assigned a permanent number. This number is simply the next in a sequential list and is usually listed when the asteroid's name is used, as in 1 Ceres or 18873 Larryrobinson.

Observing asteroids

Although asteroids are essentially point sources and appear as stars in amateur telescopes, many are bright enough to be observed even through instruments as small as the author's 100 mm f/15 refractor. The key to recognizing an asteroid visually is to observe it several times. One method is to make a sketch of the star field at two different times. The second method is to mark a printed star chart with the position of the possible asteroid and then return to the field at least an hour later. Observe asteroids visually near opposition when they are brightest and when their relative motion is greatest.

The Astronomical League offers two certificates related to their Asteroid Observing Club. Find the description for this club on the internet at http://www.astroleague.org/al/obsclubs/asteroid/astrclub.html

Asteroid astrometry

Astrometry is the positional measurement of any celestial object. This is what is done to determine orbits of comets, asteroids . . . even planets. Three positions are necessary to determine an orbit. The greater the number of accurate positions measured for any asteroid, the more precisely its orbit can be determined. An encouraging number of amateur astronomers are now involved in this endeavour and they have submitted literally millions of positional measurements, and, as a side benefit, they have discovered thousands of asteroids as a result.

Start doing astrometry

Here are some points to consider if you would also like to try your hand at asteroid astrometry.

- Choose good-quality equipment. It doesn't have to be the biggest or the best, but the telescope/CCD combination must be able to provide quality images. A driven telescope with a 200 mm aperture is sufficient, but, as in most cases, the bigger the better. A CCD coupled to this size telescope will record asteroids as faint as about eighteenth magnitude.
- Match the CCD to the focal ratio of your telescope to obtain an image scale of approximately 2 arcseconds per pixel. Such a scale provides a good limiting magnitude for your system as well as allowing good position measurement.
- The longitude, latitude, and altitude of your observing site must be measured to an accuracy of better than one arcminute. Any GPS receiver will provide this information.

The central times of all your images are also necessary and it must be accurate to one second. Many CCDs log the time of exposures by interfacing with a computer clock. Be certain that the central time of the exposure is recorded, rather than the start. If not, a correction factor will need to be applied. Try your hand at asteroid recovery first, perhaps even with some of the brighter, numbered asteroids. This is still important work and will allow your system, software, and blinking and measuring procedures a good test.

- Begin by taking three equally-spaced images of the same fields. Some experienced astrometrists take only two but when you're first learning your system, three images helps you to identify spurious images caused by hot pixels, etc. Allow between 15 and 20 minutes between exposures.
- Since asteroids are solar system objects, more will be found in the region near the ecliptic. An asteroid's motion is also more apparent when it is near opposition. On average, a main belt asteroid moves about 0.5 arcsecond per minute at opposition.
- Become familiar with the Minor Planet Center (MPC) and their database resources for asteroid identification and prioritized search lists. The MPC has a tiny, but efficient, staff, so they are not able to answer basic questions.
- Automate as much of the process as possible. Blinking images is fun, at the beginning. As your observations increase, however, you will find that your time is better spent elsewhere. Selection of appropriate software that can align images, recognize potential asteroids, measure positions, even prepare properly-formatted reports to the MPC, can save you a lot of time.
- Become very familiar with the reporting procedure. If you have internet access, start here:
  http://cfa-www.harvard.edu/iau/info/TechInfo.html
- Consider yourself part of a team. The main purpose is to collect data and further asteroid-based science. Share with those who ask every detail of your process and don't be afraid to ask questions – lots of questions. Astrometry is not a competition.
- Join the MPML (see later).

Asteroid photometry

While many amateurs are doing astrometry (possibly for the thrill involved in discovery), few are involved in photometry. This is also important work and is just as easily accomplished by amateur astronomers. Obtaining light curves for asteroids establishes their rotational rates and can aid in determining their compositions.

Collecting data for asteroid light curves takes an enormous amount of time, and is a great project for amateur astronomers. What astronomers are looking for are numbers, plain and simple. And having done some photometry myself (back in the days of photomultipliers, before CCDs), I can tell you that watching a light curve being built up point-by-point can be an addictive process.

The big asteroid surveys

The goal of professional observatories is a complete survey of all asteroids within our solar system down to a relatively small size. Astronomers wish to identify potentially life-threatening NEAs. These surveys use telescopes larger than the average amateur uses, as well as very sensitive and expensive CCDs. The automated searches are usually performed by scanning a large region of the sky several times each night. Objects identified to have moved from one image to the next are noted. The astronomer then determines if the object is a typical main belt asteroid, or something different, such as an NEA.

Asteroid surveys

Survey	Nightly area covered	Times photographed
Lowell Observatory Near-Earth Object Survey	600 degrees²	3
Catalina Sky Survey	417 degrees²	3
Lincoln Near Earth Asteroid Research program	1200 degrees²	5
Spacewatch	24.6 degrees²	3

One note on the above surveys. The Spacewatch survey has a different strategy. Their procedure searches a smaller area but to a much fainter limiting magnitude.

There are four reasons these surveys discover far more objects than amateurs do: (1) because of the carefully selected locations of their observations, they have more clear nights; (2) they are automated and maximize their imaging time; (3) their optical systems are very efficient regarding the light collected; and (4) they have high-grade software.

Mailing list resource: MPML

The Minor Planet Mailing List is an e-mail service subscribed and contributed to by professional and amateur astronomers with an interest in asteroids.

It was started in June, 1998 in order to facilitate better communication within the minor planet community. The specific focus of the list is to aid asteroid observers worldwide, but it has been known to branch out into other, related fields. This is a popular list and many of the posts are technical. To subscribe to this mailing list, visit http://groups.yahoo.com

In the search box, type in MPML and follow the directions.

6.12 Comets

Comets are wonderful. Most appear unannounced, seemingly out of nowhere. Some become bright. A few rise to the level of "spectacle." Amateur astronomers live for comets like that, and we've had a few lately. In centuries past, comets were thought by some to be the finger of God, pointing at the Earth, warning us all about some dire catastrophe to come. Today, we know a bit more about these "hairy stars."

The great Comet Hyakutake. (150 mm f/2.5 lens piggybacked, 20 minutes, Scotchchrome 400, 07:30 UT, 23 Mar 1996, Sierra Vista, Arizona. Image by David Healy)

Comets are small, irregularly-shaped bodies composed of a mixture of dust grains and frozen gases. Most have highly elliptical orbits that bring them close to the Sun and then take them deep into space, far beyond the orbit of Pluto.

Comets themselves are diverse and very dynamic, but they all develop a surrounding cloud of thinly distributed material, called a coma, that grows in size and brightness as the comet approaches the Sun. A small nucleus (usually less than 10 km in diameter) remains invisible within the cloud of the coma. Together, the coma and the nucleus make up the head of the comet.

When far from the Sun, the nucleus is very cold and its material is frozen solid. A great description was provided back in the 1950s by the American astronomer Fred L. Whipple, who first referred to comets as "dirty snowballs." When a comet approaches within a few AU of the Sun, the surface of the nucleus begins to warm and the ice begins to evaporate. The gas molecules boil off and carry small solid particles with them, forming the comet's coma.

When the nucleus is frozen, the comet is very faint, being seen only by reflected sunlight. When a coma develops, however, released dust reflects still more sunlight, and gas in the coma absorbs ultraviolet radiation and begins to fluoresce. At about 5 AU from the Sun, fluorescence contributes more to the comet's brightness than reflected light. Also, as the comet absorbs ultraviolet light, chemical processes release hydrogen, which forms a large envelope around the coma. Such envelopes cannot be observed from Earth as their light is absorbed by our atmosphere, but they have been detected by spacecraft.

As comets approach the Sun they develop tails of luminous material that extend from the head. Some of these tails are enormous, stretching for tens of millions of kilometers. The solar wind accelerates materials away from the comet at a range of velocities according to the size and mass of the material. This can create two types of tails, called dust and ion. The dust tail of a comet contains some mass and so is accelerated slowly and tends to be curved. The ion tail is much less massive, and so is accelerated as a nearly straight line extending in a direction directly opposite the Sun.

Observing comets

Comets will reward the patient observer. Actually, the good ones reward even the impatient observer. For bright comets, start with a detailed naked-eye observation from a dark site. What is the altitude of the comet? In which constellation(s) does it lie? Is the comet seen against the background of the Milky Way? Note the apparent size of the coma. Next, try to determine the full extent of the tail. Take some time with this. How wide is the tail (both near the coma and at the end of the tail)? Can you see both a dust tail and an ion tail? How do they differ? Try to estimate the comet's overall magnitude (see

Raymond Shubinski of Prestonsburg, Kentucky, took this photo of Comet West on the morning of 3 Mar 1976, in East Lansing, Michigan.

25 brightest comets since 1950

Name	Designation	Visual magnitude
Ikeya-Seki	C/1965 S1	−7
West	C/1975 V1	−3
Hale-Bopp	C/1995 O1	−0.8
Arend-Roland	C/1956 R1	−0.5
Hyakutake	C/1996 B2	0
Bennett	C/1969 Y1	0
SOHO	C/1998 J1	0.5
Mrkos	C/1957 P1	1
Seki-Lines	C/1962 C1	1
IRAS-Araki-Alcock	C/1983 H1	1.7
Halley	1P/1982 U1	2.4
Kohoutek	C/1973 E1	2.5
Ikeya	C/1964 N1	2.7
Aarseth-Brewington	C/1989 W1	2.8
Ikeya	C/1963 A1	2.8
Ikeya-Zhang	C/2002 C1	2.9
LINEAR	C/2001 A2	3.0
Burnham	C/1959 Y1	3.5
Tago-Sato-Kosaka	C/1969 T1	3.5
Bradfield	C/1980 Y1	3.5
Wilson-Hubbard	C/1961 O1	3.5
Mrkos	C/1955 L1	3.5
Levy	C/1990 K1	3.6
Kobayashi-Berger-Milon	C/1975 N1	3.7
Bradfield	C/1974 C1	3.9

below for more on this). What is the comet's degree of condensation (again, see below)? Finally, note the color of all parts of the comet, especially the tail. If you choose to sketch the comet, be sure to note the direction north on your sketch.

Comet Hale-Bopp. (Image by Adam Block/NOAO/AURA/NSF)

When the naked-eye observation is complete to your satisfaction, you should next observe the comet with binoculars. For most bright comets, using binoculars will provide the best views. They offer some magnification and darken the sky background a little while still providing a fairly wide field of view. With binoculars, try to observe the further extent of the tail. How much further from the comet's head can you observe the last wisps of the tail? Also, determine at this point the width of the comet, especially the tail. If both a dust tail and an ion tail are visible, note any increased definition in either and any colors seen. Also note the separation of the two tails and the shape of each.

Finally, a telescopic observation of even the brightest comets is required. This will allow you to examine the coma in detail. Be prepared to use a variety of magnifications and do take your time. Note any irregularity in shape or brightness. Does the coma look elliptical (and to what degree) at all magnifications? The nucleus of any comet is too small to resolve, however the pseudonucleus may be seen as a very condensed or star-like area. Examine the pseudonucleus closely. Does the brightness of the coma vary with distance from it, and if so, how? On a number of occasions even observers with medium-sized telescopes have seen fragmenting within the coma. If you're lucky enough to see this, the comet will appear to have several pseudonuclei. Oh, and when observing at high power, be aware of the possibility of jets. These features will appear as lines or angular rays, generally in the direction of the Sun (since they are caused by outgassing due to solar heating on the sunward side of the comet).

Degree of condensation

One of the details observers should note about comets is the degree of condensation (DC). This is an indicator of how much the surface brightness of the coma increases toward the center of the coma. In general, DC = 0 indicates totally diffuse and DC = 9 means "stellar." As the DC increases, the coma size usually decreases and becomes more sharply defined. A totally diffuse comet, with no brightening toward the center, is rated DC = 0. With DC = 3–5, there is a distinct brightening. By DC = 7 you have a steep overall gradient and by DC = 8 the coma is very small, dense, and intense with fairly well-defined boundaries. With DC = 9 the comet looks like a soft star or a planet in bad seeing.

Estimating magnitude

I'm a huge fan of nineteenth century astronomy. Observers during that century, however, were not the greatest for determining the brightnesses of comets, and magnitude estimates done on the same night varied greatly. Making an estimate of a comet's brightness is not a simple thing to do. One of the mistakes amateur astronomers make is to underestimate the total brightness of the coma. They concentrate too much on the stellar central portion of the coma which is often seen. Also, from a less-than-dark site, such as an urban setting, the outer coma is often lost to sight.

There are a number of ways to estimate the visual magnitude of a comet. The easiest and, unfortunately, the least accurate, is to directly compare the brightness of the comet with nearby stars. Try to find a star slightly fainter than the comet and one slightly brighter. Also, the stars should be at the same rough altitude as the comet to take into account the extinction caused by our atmosphere. This is the only method you can use naked-eye. It will also work (for fainter comets) using binoculars or a telescope.

Another, slightly more accurate, method involves defocus-

Comet LINEAR 2001 A2. (Image taken 13 Jul 2001 by Robert Kuberek of Valencia, California)

ing the images of nearby stars to the same size as the normal, in-focus view of the comet's coma. Brightnesses are then compared. This technique works better for diffuse comets. Comets with a high DC usually have a large brightness difference between the outer coma and the central region. That makes comparison with out-of-focus stars difficult, as those images are much more uniform.

Yet another way to determine the magnitude of a comet involves a huge defocusing of the images of the comet and surrounding stars. This is done to such an extent that the out-of-focus circles can be directly compared. Some observers have taken this a step further by continuing to defocus the images until objects start to disappear into the sky background. The brightnesses of objects disappearing are noted, especially those that disappear just prior to, and just after, the comet.

Atmospheric extinction

In the magnitude determination of a comet, you are not finished until you take into account the extinction of its brightness caused by our atmosphere. In the table opposite, z refers to zenith distance (the altitude may be found by subtracting z from 90°). For example, if your elevation is 1 km above sea level and the comet is 10° off the horizon, the extinction is 1.16 magnitudes. In other words, the comet appears 1.16 magnitudes fainter than it would if you were not observing through our atmosphere.

Photographing a comet

If there's a bright comet in the sky, get out your 35 mm camera and shoot, Shoot, SHOOT! Even if you are a

Atmospheric extinction at sea level and different elevations above sea level

z	sea level	500m	1km	2km	3km
1	0.28	0.24	0.21	0.16	0.13
10	0.29	0.24	0.21	0.16	0.13
20	0.30	0.25	0.22	0.17	0.14
30	0.32	0.28	0.24	0.19	0.15
40	0.37	0.31	0.27	0.21	0.17
45	0.40	0.34	0.29	0.23	0.19
50	0.44	0.37	0.32	0.25	0.21
55	0.49	0.42	0.36	0.28	0.23
60	0.56	0.48	0.41	0.32	0.26
65	0.64	0.54	0.47	0.37	0.30
70	0.82	0.70	0.60	0.47	0.39
75	1.08	0.92	0.79	0.62	0.51
80	1.59	1.34	1.16	0.91	0.74
85	2.91	2.46	2.13	1.66	1.36
89	7.38	6.26	5.40	4.22	3.46

Comet Hyakutake. Note the handle of the Big Dipper involved in the tail of the comet. (50 mm f/2.5, 3 minute exposure. Photo by Steve Coe, Phoenix, Arizona)

beginning astrophotographer you may be surprised by what you record. I mentioned earlier that I am a devotee of nineteenth century astronomy. The best images of comets from that era, with few exceptions, were painstakingly made at professional observatories by people who did astrophotography for a living. Your images could easily rival theirs, even with a simple setup. Think of it this way: you already have a major advantage. The images you take will be in color!

Discovering your comet

Knowledge of the sky, familiarity with star charts or astronomy software, patience, and lots and lots of time at the eyepiece . . . these are the fundamental requirements of comet hunting. While it is true that most comets are found by the large surveys, it is also true that amateurs can and do find them. My favorite comet of all time, Comet Hyakutake, was found by Japanese amateur astronomer Yuji Hyakutake with binoculars. Sadly, Mr. Hyakutake passed away just before this chapter was being written. He was a humble man and will be missed.

Doug Snyder of Palominas, Arizona, talks about his comet discovery.

What led to the discovery, using my trusty 20 inch f/5 Obsession, mainly in use at 149✕ (Nagler Type IV,

17 mm eyepiece) was determination (or lunacy?) and only about 70 hours of star hopping! Yep, it was intermittent star hopping, and the darn thing kept moving on us, but I finally caught up with it. Along the way, I bagged a good number of objects that don't move around nightly, and that was rewarding in itself.

On Sunday night (11 Mar 2002) and into Monday

Comet Ikeya-Zhang. (5-minute exposure at f/2.5. Kodak Elite Chrome 200, pushed 1 stop. Nikon F2 with a 105 mm Nikkor lens. Photo by Ulrich Beinert of Kronberg, Germany)

Comet photography tips

- For tripod-mounted shots, use fast film and bracket your exposures time-wise. Set the aperture on your lens wide open. Always use a cable release.
- Some of the most pleasing shots of comets show foreground objects such as buildings, an observer with binoculars or telescope or even a body of water in which the comet's image is reflected. If the comet is low in the sky (and many bright ones never get too far from the Sun), you may consider a more "artsy" type of image.
- For piggyback shots try a somewhat slower speed film with finer grain structure to record more cometary detail. Stopping the lens down will produce better star images. Always use a cable release.
- Don't be afraid to experiment with colored filters that screw onto your camera lens. A light blue filter will accentuate the bluish ion tail. A light or dark yellow or even an orange filter (if the comet is bright) may bring out unseen detail in the dust tail.
- Always plan to record more of the comet tail than you can see with your naked eye. Frame the comet in the shot appropriately.
- Remember that a comet is an extended object, not a point source. The limiting magnitude you will see on your image will depend on the f/ratio of the lens, and not simply the size of the lens used (as is the case with star images).

morning, I was observing mainly to the south (quite a southern horizon from here in southeastern Arizona), and it was turning out to be a tremendously good night with seeing and transparency almost, but not quite, at max. I would've turned in around 3 a.m., but I just couldn't get my eye unstuck from the eyepiece, doing east-to-west scans and checking objects off in the latest edition of *Uranometria*.

Once I saw that summer Milky Way rising (so dog-gone bright!), I knew I had to check out some early summer objects, because once our summer monsoons start up here, you can almost forget about observing during those months. I trained the scope on Cygnus for awhile, and while not looking for any object in particular, I was sure amazed (once again) at the number and wealth of objects in that constellation. But I had to move on – down through part of Lyra and then into Aquila, the Eagle (mainly because it was mostly at eye level, and I didn't have to climb the ladder!). I hadn't been in Aquila long, scanning that part of the Milky Way when I came across that smudge that didn't belong there. But I had to check anyway – nothing was supposed to be at that position, although there are few small and faint planetaries and some not well known open clusters in that region. I'm using the Sky Commander DSCs on the scope, so that has helped to identify many objects (but certainly not all), and let me keep track of where I'm at.

While checking various object resources, and going back to the eyepiece, I noticed that this son-of-a-gun was moving! When I first came across it, I noted that it was evenly split between two very faint stars that were several arcminutes apart, and now it was covering one of them. That's when the shakes started and my CD player was cranking out *Some Enchanted Evening* (I know, I'm kind of retro). Well, since morning twilight was not too far away by that time, I stayed with it as long as I could, noting its position as best I could and checking up on other comets' positions that are all over the sky (most of them way too faint to be seen visually). I finally made it into the house to hesitantly send an e-mail to CBAT (Central Bureau for Astronomical Telegrams). Hesitant because I was still not 100% sure, and this would have been my first discovery report to them. Well, that part is now history, but there is more to be unfolded about this comet and I am really counting my lucky "stars" that I just happened on the right spot at the right time.

One little bit of irony in the initial report that I sent in – you don't think they get a good number of false reports, but are very diplomatic about telling folks the bad news? Well, when I entered the "Subject" line in the e-mail, my brain said "Constellation – Aquila", but my fingers typed "Constellation – Aquarius!" Fortunately, I had correctly entered the Aquila coordinates of the object in the body of the text. So when the gentleman at CBAT first read the report, he was probably thinking, "oh, another crank report! – reporting it in Aquarius, but giving Aquila coordinates." But I was lucky enough to have caught my mistake before too much damage was done, and re-sent the report before I heard from CBAT (although I did hear from them about the mix-up – very diplomatically, of course).

I mentioned earlier about *Some Enchanted Evening*; very appropriately, here are the first lyrics:

> *Some enchanted evening, you may see a stranger, you may see a stranger across a crowded room,*
> *an' somehow you know, you know even then,*
> *that somewhere you'll see her again and again!*

A different route to a comet discovery was taken by Brian Skiff, who works at Lowell Observatory in Flagstaff, Arizona. His recent comet find (as of this writing) was P/2001 R6.

I think this is simply another case where the fundamental comet-hunting rules apply: (1) you won't find them if you're not out there observing/taking data; and (2) pay attention. Actually it was Bruce Koehn's software that found it, but the pipeline isn't so automated that rule 2 doesn't apply.

The LONEOS reductions operate by showing strings of "moving object" detections from sets of four images of each field. We run a Vaisala orbit on everything, which in this incarnation calculates a suite of about two dozen orbits with varying assumptions about the current distance from Earth. If a main-belt orbit is among the solutions, objects are flagged as such, and usually we simply skip inspecting them. If a detection doesn't have main-belt motion, the reduction program stops and requires the observer to decide if this is a false detection or something real and interesting.

R6 presently is near opposition but a fair bit off the ecliptic. Most real non-main-belters are Hungarias or Mars-crossers. At the moment there is also a modest portion of Jupiter L5 Trojans, which move at about half the main-belt speed. When the non-main-belt detection listing for the comet came up, I noticed the motion was too fast for a Trojan, but not main-belt speed.

Bruce's program has an option that sends the date/position to the Minor Planet Center's mpchecker to look for known objects. Since this is not the first time such a circumstance has happened to me, about the time I sent that request, I thought, "It's a comet." I then started loading the images into a display window, and as they came up it was obviously cometary, so I mainly waited for the mpchecker report to say which NEAT (Near-Earth Asteroid Tracking) comet it was (thinking "weren't there like four southern NEAT

comets recently?"). Of course it wasn't, and it also didn't appear on the NEOCP (Near-Earth Object Confirmation Page), so it was duly reported. Again because of the location and daily motion, as I measured the comet size and orientation, I thought "another faint stinker Jupiter-family comet," a statistically-likely guess on my part, and correct in this instance. Maybe the next one will be brighter.

How to report a discovery

The Central Bureau for Astronomical Telegrams (CBAT) is the organization responsible for designating new comets and coordinating and distributing information on discoveries. They have guidelines for the reporting of potential comet discoveries. The following text is from their website http://cfa-www.harvard.edu/iau/cbat.html

"Discovering" a comet means different things depending on whether one is a very experienced observer or is a beginning observer. For every real new comet discovery, the Central Bureau for Astronomical Telegrams (CBAT) gets perhaps five reports of "discoveries" that do not pan out. And in most of these unconfirmed or erroneous discovery reports, the observers declare "NEW COMET" or "COMET

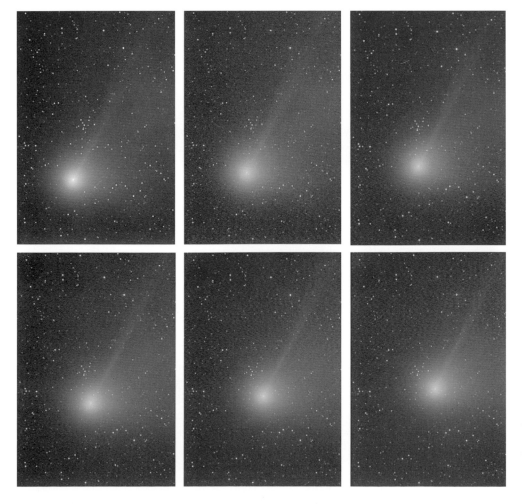

Robert Kuberek of Valencia, California, made a wonderful animation of the motion of Comet Ikeya-Zhang (2002 C1). Here are six equally-spaced frames. Note the comet's motion against the background of stars.

Comet Hyakutake. (Photo by Mark Cunningham, Craig, Colorado)

DISCOVERY," even though they have only seen the possible object once (with no detectable motion), or even though they only have a single photograph on one night with a suspicious-looking object.

If you think that you might have come across a new comet, there is a check-list that you should go through before proceeding further.

Can you be absolutely sure that the image is real? This is a very common problem, with both visual and photographic observers finding "ghost images" caused by nearby bright objects (stars, planets, etc.) and then reporting them as possible comets. Beware that even very experienced professional astronomers are sometimes fooled by ghost images!

Is there motion? If so, determine how much the object moves in a measured amount of time. If there is no definitely-detectable motion, be very skeptical. Be sure that there are no close asterisms (a big problem with visual observing) or no galaxies at the position in question. Many observers reporting "discoveries" do not have access to good atlases or catalogues and do not notice that their "comet" is in fact an entity far beyond our own solar system! Do not rely solely on Atlas

2000.0 or the Skalnate Pleso Atlas, for example; photographic atlases are far preferable to drawn or computer-generated atlases (both of which often have glaring omissions).

As accurate a position as you can measure will be needed if you still think that you have a comet. A position accurate to at least $1'$ in declination and 0.1 minute of time in right ascension is not too difficult using most available atlases with grid overlays, and such accuracy is often very necessary to prevent observers from wasting much time in searching for an object.

The equinox should also be stated (2000.0 or 1950.0 being the most-widely used).

Just as important as the position is recording the time of observation, which should be reported in Universal Time (not local time). If you have a photograph or especially a CCD, you really should provide accurate positions from all of your images, to 0.01 second of time for the RA and to $0.1''$ for the declination.

A reasonable description is also useful, including an estimation of the object's total magnitude and a note concerning its size, amount of diffuseness, possible tail information, and degree of central condensation.

When reporting information, give your full name, mailing address, and a telephone number where you can be reached. Also give information such as your observing location and instrument used for detecting the object (including aperture, type of telescope, and, in the case of photography, what kind of film and exposure times are involved). Specify what sources you have checked to rule out alternative explanations (including known comets and deep-sky objects).

Always note that a CONFIRMATION observation on a second night is always recommended before reporting the object to the professional community; at the least, observe the comet with different instruments if possible over as much time on a single night as possible, and try obtaining multiple CCD or photographic exposures – reporting all such observations/images available. This has always been a standard policy of the prolific comet discoverer William Bradfield of Australia (who has all 17 of "his" comets named solely for him, with no other names of other discoverers shared). Telescope time is so scarce and valuable at professional observatories that it becomes a real problem when lots of false alarms are searched for.

Check the International Comet Quarterly Comet Information website for explanations for any of the terms or concepts discussed on this page.

Now, if you're still convinced that what have found is really a comet, it would be helpful if you could check one of the several readily-available publications listing ephemeredes of already-known comets. At any given time, there are usually at least 2 or 3 (often more) comets visible in the night sky which are within the visual range of an 8-inch reflector. Good sources for the positions of known comets include the *International Astronomical Union (IAU) Circulars*, the *Minor Planet Circulars*, the annual *Comet Handbook of the International Comet Quarterly*, and the annual *Handbook of the British Astronomical Association*. The Central Bureau for Astronomical Telegrams also has a computer service which can be reached via modem through telephone lines, or via SPAN or Internet, whereby messages can be left, search ephemeredes can be computed (to see if a possible comet is already known), and *IAU Circulars* can be read.

It is a good idea to contact a local observatory (or one or more experienced amateur astronomers) privately (not publicly) to ask for confirmation before reporting it to the CBAT. But do not post discovery information on any Internet venue (whether discussion groups or websites), because many inexperienced people will potentially be exposed to this information – leading to potential wild-goose chases and even perhaps eliminate the chance that a comet will be named for you (to ensure that discovery rights rest with you, if you find a comet, you must contact the Central Bureau first). The proper procedure for getting information to the Bureau is either:
- via a message left in the CBATs Computer Service; or
- via the CBAT discovery form; or
- via e-mail messages to both marsden@cfa.harvard.edu and dgreen@cfa.harvard.edu
- The generic e-mail address cbat@cfa.harvard.edu may be used instead.

6.13 Meteors and meteor showers

Grab a reclining chair, a cold (or warm) drink and no optical equipment. Doesn't sound like astronomy? Well, it is...it's meteor observing. Beginning amateur astronomers observe meteors a lot. It's easy. It's fun. I've noticed that as a general rule, the more one gets "into" our hobby the less meteor observing is done. That may be a natural result of specialization, but, in my opinion, it's too bad nonetheless.

Meteor. The word is derived from the Greek *meteo*, meaning "of the air," and refers to a time when meteors were thought to be atmospheric phenomena. And, to clear up something every amateur astronomer needs to know, the object is a meteoroid when in space, a meteor when seen in our atmosphere and a meteorite if it survives its fiery plunge and is found on the Earth.

Meteors are (generally) small pieces of rock which the Earth, in its orbit around the Sun, "sweeps up" into its atmosphere. Because of the great speeds involved, as the meteor passes through our atmosphere it causes a column of gas to become incandescent. This column of glowing gas is what we see as the meteor. Very little of the overall light output is caused by the piece of rock itself burning up.

Individual meteors can be observed at any time of night. The chances that you will see one increase during events called meteor showers, discussed below. Meteors not associated with any shower are called sporadic meteors. On any given night, up to eight sporadics can be seen per hour by a good observer at a dark site.

Meteor watchers are especially fond of fireballs and bolides. A fireball, by loose definition, is any meteor bright enough to cast a shadow. The International Meteor Organization (IMO) defines "fireball" as a meteor equal to or brighter than magnitude −3. A bolide may be a fireball, but it is defined as an exploding meteor.

Meteor showers

Sporadic meteors can be impressive but most observers tend to focus their attention on meteor showers. Meteor showers originate with comets. As the comet orbits near the Sun, volatiles are boiled off its surface, carrying with them small particles of dust. These particles become meteoroids and, due to the orbits of the comets, distribute themselves along the comets' paths in what are called streams. If, in its orbit around the Sun, the Earth intersects one of these streams, a meteor shower will occur. If you happen to be so lucky as to see a display with a rate of 1000 meteors/hour or above, it is termed a meteor storm. At the other end of the scale, any shower with an observed rate of 10 or less meteors/hour is known as a minor meteor shower.

Because the orbits of the Earth and the particle streams are relatively unchanging, we see the same shower on the same date year after year. Also, the way the stream intersects the Earth's orbit, and the direction the particles are traveling, determine the average speed of the particles of the showers. For example, the average speed of a Leonid meteor is 71 km/s, whereas an average Geminid travels half as fast, 35 km/s.

Meteor showers are named for the constellations in which their radiants are found. The radiant is the point on the sky to which all the paths of the meteors in a shower may be traced back. Another way to look at this is that the radiant is the point toward which the Earth is heading in its journey around the Sun. This is like running in a rain or snow shower where the drops of rain (or flakes of snow) seem to be coming directly at you. Meteor showers are generally better after local midnight when your location is facing towards the direction the Earth is traveling in space.

Thus, we have meteor shower names like the Leonids, whose radiant lies in the constellation Leo; or the Perseids, with the radiant in Perseus; or the Quadrantids . . . wait a minute! Actually, the Quadrantids meteor shower's radiant lies in the constellation Bootes, in an area of sky once known as the constellation Quadrans Muralis, the Mural Quadrant. That constellation is now extinct and is no longer used. Bayer designations may be used to distinguish between two showers in the same constellation, as Leonids vs. Delta Leonids. The constellation name is always included however. Example: Alpha Aurigids is correct, Capellids is not.

If you are a beginning meteor observer, concentrate on the most active showers, those with rates of 20 meteors/hour or

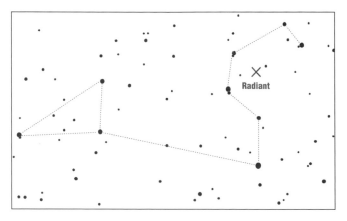

The radiant of the Leonid meteor shower. (Image from *TheSky*, courtesy Software Bisque)

Best annual meteor showers

Name	Active	Maximum	Radiant	Pop. Index	ZHR
Quadrantids	1 – 5 Jan	3 Jan	$12^h47^m +49$	2.1	120
Delta Cancrids	1 – 24 Jan	17 Jan	$08^h40^m +20$	3.0	5
Alpha Centaurids	28 Jan – 21 Feb	7 Feb	$14^h00^m -59$	2.0	8
Delta Leonids	15 Feb – 10 Mar	24 Feb	$11^h12^m +16$	3.0	12
Gamma Normids	25 Feb – 22 Mar	13 Mar	$16^h36^m -51$	2.4	8
Virginids	25 Jan – 15 Apr	26 Mar	$13^h00^m -04$	3.0	5
Lyrids	16 – 25 Apr	22 Apr	$18^h00^m +34$	2.9	15
Pi Puppids	15 – 28 Apr	23 Apr	$07^h20^m -45$	2.0	10
Eta Aquarids	19 Apr – 28 May	5 May	$22^h32^m -01$	2.7	60
Sagittarids	15 Apr – 15 Jul	11 Jun	$16^h28^m -22$	2.5	15
Epsilon Pegasids	7 – 13 Jul	10 Jul	$22^h40^m +15$	3.0	8
July Phoenicids	10 – 16 Jul	14 Jul	$02^h08^m -48$	3.0	20
Piscis Austrinids	15 Jul – 10 Aug	28 Jul	$22^h44^m -30$	3.2	5
S. Delta Aquarids	12 Jul – 19 Aug	28 Jul	$22^h36^m -16$	3.2	25
Alpha Capricornids	3 Jul – 15 Aug	30 Jul	$20^h28^m -10$	2.5	15
S. Iota Aquarids	25 Jul – 15 Aug	4 Aug	$22^h16^m -15$	2.0	8
N. Delta Aquarids	15 Jul – 25 Aug	8 Aug	$22^h20^m -05$	3.4	10
Perseids	17 Jul – 24 Aug	12 Aug	$03^h04^m +57$	2.6	95
Kappa Cygnids	3 – 25 Aug	17 Aug	$19^h04^m +59$	3.0	3
N. Iota Aquarids	11 – 31 Aug	19 Aug	$21^h48^m -06$	3.2	10
Alpha Aurigids	25 Aug – 5 Sep	31 Aug	$05^h36^m +42$	2.5	20
Delta Aurigids	5 – 10 Sep	8 Sep	$04^h00^m +47$	3.0	6
Piscids	1 – 30 Sep	19 Sep	$00^h20^m -01$	3.0	7
Draconids	6 – 10 Oct	9 Oct	$17^h28^m +54$	2.6	2
Epsilon Geminids	14 – 27 Oct	18 Oct	$06^h48^m +27$	3.0	3
Orionids	2 Oct – 7 Nov	21 Oct	$06^h20^m +16$	2.9	25
S. Taurids	1 Oct – 25 Nov	5 Nov	$03^h20^m +13$	2.3	5
N. Taurids	1 Oct – 25 Nov	12 Nov	$03^h52^m +22$	2.3	5
Leonids	14 – 21 Nov	17 Nov	$10^h12^m +22$	2.5	25
Alpha Monocerotids	15 – 25 Nov	21 Nov	$07^h20^m +03$	2.4	3
Chi Orionids	26 Nov – 15 Dec	2 Dec	$05^h 28^m +23$	3.0	3
Phoenicids	28 Nov – 9 Dec	6 Dec	$01^h12^m -53$	2.8	100
Puppids	1 – 15 Dec	7 Dec	$08^h12^m -45$	2.9	10
Dec. Monocerotids	27 Nov – 17 Dec	8 Dec	$06^h 40^m +08$	3.0	3

Best annual meteor showers (*cont.*)

Name	Active	Maximum	Radiant	Pop. Index	ZHR
Sigma Hydrids	3 – 15 Dec	11 Dec	08^h 28^m +02	3.0	3
Geminids	7 – 17 Dec	13 Dec	07^h 28^m +33	2.6	100
Coma Berenicids	12 Dec – 23 Jan	19 Dec	11^h 40^m +25	3.0	6
Ursids	17 – 26 Dec	22 Dec	14^h 28^m +76	3.0	15

above. More meteors will give you more practice in recording data and to become familiar with the appearances and brightnesses of meteors overall. The table above provides a list of the best annual showers. Explanations of Pop. Index and ZHR follow.

Population index

To arrive at a calculation for zenithal hourly rate ZHR (which we will shortly do), we must know the population index for a particular shower. This is an estimate of the ratio of the number of meteors by magnitude. For example, if a particular shower has a population index of 3.0, there will be three times as many second magnitude meteors than first magnitude, three times as many third magnitude meteors as second magnitude, and so on. When a population index is not given, or cannot be found for a particular shower, use the value 2.5. For sporadic meteors, a value of 3.0 is usually assumed for the population index.

The population index also provides us a rough number of meteors that we are not seeing from a particular site. For example, if the population index is 3.0 and the limiting magnitude of a site is 4.0 and we observe four meteors per hour of magnitude 4, we are missing approximately 12 meteors per hour of magnitude 5. The higher the population index, the more meteors we will miss due to a light-polluted site, or one with a high percentage of clouds.

Astronomers also note that a small value for the population index indicates an older meteor particle stream. This is due to small meteoroids (which would drive up the value of the population index) being stripped out of the stream over time, through various processes. In such cases where the population index is small, the shower generally produces an average meteor magnitude which is brighter than showers with high population indices.

Zenithal hourly rate

We now come to the most misunderstood concept involved in observing meteor showers, the zenithal hourly rate (ZHR). I know experienced observers who cannot under-stand the significance of the ZHR, and have thus never bothered to calculate it for any of their observations. I will now simply state the value of the ZHR: the ZHR is the only thing which allows me, in El Paso, Texas, the ability to directly compare a meteor shower observation with an observer in King's Lynn, Norfolk, UK, or with an observer outside of Kansas City, Missouri, or any other observer who saw the same shower, no matter their location. It standardizes disparate observations.

The ZHR also gives an observer a gauge as to the number of meteors which could be seen if conditions were ideal: perfect sky conditions and the radiant at the zenith. Don't underestimate the value of a radiant high in the sky. The following table gives the number of meteors you could expect for a shower with ZHR = 100 and if the sky is clear and has a limiting visual magnitude of 6.5. Thus, the numbers are dependant only on the elevation of the radiant.

Expected number of meteors by radiant elevation

Elevation (degrees)	90	70	50	40	30	20	10
Number of meteors	100	94	77	64	50	34	17

There are several methods to finding the ZHR, some more and some less complicated. Here's a relatively simple formula that I like:

$$ZHR = (HR \times r^{(6.5 - LM)}) / \sin a$$

where HR = observed hourly rate, r = population index, LM = limiting magnitude, and a = altitude of radiant above the horizon in degrees.

For those observers lucky enough to observe from a location with a limiting magnitude fainter than 6.5, the formula becomes:

$$ZHR = (HR \times r^{(1 - (LM - 6.5))}) / \sin a$$

Meteor colors

Different meteors can be composed of different compounds. When these compounds are superheated, as in the flight of a

Obscuration factor

Dedicated meteor observers often report the percentage and length of cloud cover during a session. Usually, cloud percentages are taken every 15 minutes or so and recorded, either vocally or written down. After the observation is over, the percentages listed on your report form can be converted into a correction factor F by using the equation:

$$F = \frac{1}{1 - k}$$

where $k = \dfrac{\text{percent blockage} \times \text{minutes}}{\text{total observing period}}$

Example: Leonid meteor shower. Observing period 08:30–10:00 UT (90 minutes). During this time the sky was covered with 15% clouds for 15 minutes.

$$k = \frac{0.15 \times 15 \text{ minutes}}{90 \text{ minutes}} = 0.025$$

When calculating k, be sure to include all of your individual cloud cover estimates in the percent × minutes value.

Returning to the original equation gives us a value for F of 1.03.

This correction factor should be listed in your report. Note that when there are no clouds, F equals 1.00. This should also be listed for each observing period on your report.

Due to several factors, observations when cloud cover exceeds 20% are not recommended. Instead, take a break or wait for another night when the clouds do not prevent observations.

meteor through our atmosphere, they glow with different colors. That's simple enough. But what causes color changes in meteors seen during a shower? Presumably, the particles are made of the same stuff.

A question was posed some time back to dust expert Amara Graps of the Max-Planck-Institüt für Kernphysik about color changes in Leonid meteors. This is her reply:

> I brought up your question to our dust group in our dust meeting this morning, and while none of us knows for sure, it seems more likely to us that the different colors are caused by particles of different sizes moving through layers in our Earth's atmosphere. The ionized trail of green would be due to oxygen, for example, and then the color changes as the particles move deeper. These are the same colors you see in the aurora.
>
> Since the particles are all moving at the same speed (70km/s), the main differences between particles in the Leonids are their sizes, and the Leonids have a steep size distribution. The bigger particles are probably the ones moving deeper into the Earth's atmosphere. That's our best guess answer to your question.

Observing meteor showers

- Look about two-thirds of the way from the horizon to the zenith. Do not place the radiant in the center of your field of view.
- Pick a night near New Moon and observe from the darkest site possible. Mark the position of the radiant on a star chart and, when darkness falls at the site, find that position in the sky and memorize it.
- If you observe in a group with other people every

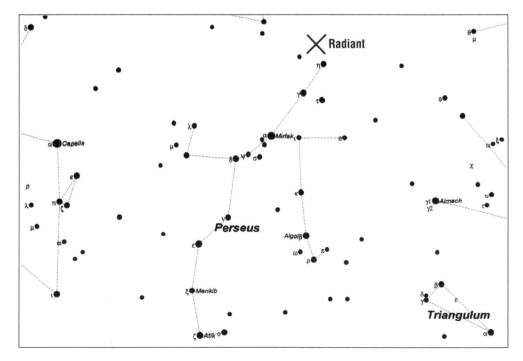

Usually the most reliable of the meteor showers during the year, the Perseids peak around 12 August. This illustration shows the radiant at the peak of the shower. (Image from TheSky, courtesy Software Bisque)

observer has to observe independently from the others. Never try to combine data from different observers. Each observer must keep their own notes and fill in their own report forms.

- Always note the limiting magnitude of the site. This is defined as the magnitude of the faintest star near the zenith that the observer can detect. The limiting magnitude is a quantity different for each observer. It not only defines the sky conditions but also the quality of the observer's eyes.
- Shower meteors appear faster the larger their distance from the radiant (up to 90°) and the higher their elevation above the horizon. Near the radiant or near the horizon shower meteors generally seem slower.
- Avoid talking and especially avoid listening to music. When you get tired take a break: walk around or eat something. If you do become too tired, stop observing and get some sleep. Even a short nap will help. Take a nap in the late afternoon or evening prior to the session if at all possible.
- Most observers follow the motion of the sky, and thus become more familiar with certain regions. If the position of your chosen field becomes unfavorable due to clouds or its elevation getting low, choose another field. Be sure to record the center of the new field in your notes.
- Record your observations on a tape recorder or by writing onto paper. Do not take your eyes off the sky. (This takes practice if you are using paper.) Your notes should include:
  - o UT of beginning and ending of the meteor watch with all breaks noted
  - o Limiting magnitude and any changes to it during your watch
  - o Details of any cloud cover present
  - o The RA and Dec. of the center of the field of view
  - o Details of all meteors observed (magnitude, shower meteor or sporadic, train, color)

Unless you are going to photograph meteors or a fireball appears, the time of individual meteors is of little interest. Note the time in your notes approximately every 30 minutes. So-called burst rates, high numbers of meteors over very short periods of time, have recently been reported by some observers. They are worthless in the calculation of ZHR but may be useful in determining maximum meteor activity if the length of the peak of the shower is very short.

- Do not look at constellations. Rather, try to see only a random collection of stars. More difficult than this, try not to concentrate on the stars. You want to watch the field, not the points of light.

Photographing meteors

Refer to the "Astrophotography" chapter for general information on cameras, film, etc. Meteor photography is usually done unguided (see the section on "Tripod-mounted shots"), but can be guided. Here are some specific recommendations regarding photographing meteors.

- Keep an accurate record of exposure times and durations and also the coordinates of the center of the image. You should be able to identify at least six stars on the image.
- For tripod-mounted shots, keep the exposures between 5 and 15 minutes to reduce star trailing. If there is an exceptional rate of meteoric activity, keep all exposures to 5 minutes.
- Aim the camera between 50°–70° above the horizon so

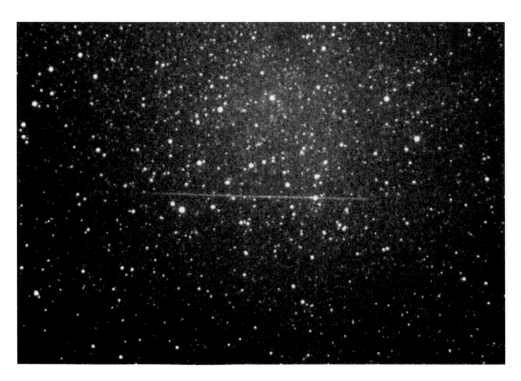

A Perseid meteor. (Kodak Elite Chrome 200, pushed 1 stop. Nikon F2 with a 50 mm Nikkor lens. Photo by Ulrich Beinert of Kronberg, Germany)

that the largest portion of sky is within the field of view. If possible, also aim your camera 30°–40° from the radiant itself. Most meteors don't become visible until they have traveled some distance from the radiant.

- Align the long axis of your film plane so that a meteor coming from the radiant toward the center will be perpendicular to it. If the long axis is parallel to that line, you will image fewer meteors.
- Keep a careful record of your images. If you have prints made, be sure to make them at least 5 inches × 7 inches in size. Include all edges of the negative in your print and make the print lighter than normal to catch fainter meteors.
- Some meteor photographers have made rotating shutters that are placed in front of the camera lens. These relatively simple devices resemble a two-bladed fan. When a bright meteor passes through the field of view, the meteor image will be "chopped." From this, it is possible to compute the meteor's duration within ten thousandths of a second. Simply divide the shutter revolutions per second (times two shutter blades) into the number of meteor breaks. Most motors used for this purpose have rpm values around 1000.

Observing meteors with optical aid

Do you desire an observing project that will set you apart? One that almost nobody else is doing? Well, then, how about observing meteors with telescopes or binoculars? This involves observation of meteor events well below the limit of either photography or naked-eye observation and can cover a size range of meteor particles recorded by professional scientists using radar techniques.

The restricted field of view of even wide-angle telescopes and binoculars means that these observations are much more accurate than results from naked-eye work. Initially, your counting rates will be low but they will steadily improve with experience. The restricted and magnified field of view allows the paths of meteors to be determined more accurately than visually. This lets us investigate the properties of meteor radiants, detect minor showers more easily, and even find new showers.

There is no single best telescope or binocular for telescopic observing of meteors. However, the instrument should have a low power and a wide apparent field of view.

To put that into numbers, the magnification should be in the range of 1.4–2.0 times the aperture in centimeters. So for example, a 7 × 50 pair of binoculars has a magnification of 1.4 times the aperture in centimeters and a 10 × 50 has a magnification twice the aperture.

The apparent field of view is governed by the eyepiece design. You can derive it from the product of the magnification and the true field of view. For example, a 10 × 50 binocular, with a 6° true field, has an apparent field of 60°. A wide field of view will encompass more of the sky,

and hence you will see more meteors. The recommended range is 45° to 70°, with 50° to 60° being preferred.

As the apparent field of view enlarges, the average plotting accuracy goes down. So ultra-wide fields (>65°) are best for determining rates, and hence deriving the time of maximum for a shower, whereas for field sizes of around 50° rates are still reasonable and accurate positional data can be obtained. Given the choice between the two, you should err on the side of the smaller apparent field as it offers more flexibility and science.

Below 50° the loss of sky coverage starts to become important. If rates become too low, boredom and loss of concentration can soon set in.

Binocular vision is the natural way to look, and since comfort is a critical consideration for the telescopic observer, a binocular is preferred to a (monocular) telescope. Aperture is less critical, and International Meteor Organization (IMO) observers' apertures range from 40 mm to 300 mm, though most are in the range of 50–80 mm. The intermediate apertures (50–80 mm) seem to work best. The quality of the optics can make a big difference to the performance. Remember that you will be observing for long periods and considerations like accurate collimation and pinpoint images will reduce strain.

The IMO has a Telescopic Commission which has produced several sets of charts suitable for plotting telescopic meteors. Each set has its own limiting magnitude, field size, and orientation, and each is geared towards popular binocular and telescope specifications. See their website at http://www.imo.net

Meteors by radio

If you're really hooked on meteors, there is a way to count them during Full Moon, when there are clouds – even in the daytime! Monitoring meteor activity by radio got its start right after World War II. When a meteoroid enters the atmosphere, it produces an ionized column of gas molecules. This ionized gas can reflect radio signals between a transmitter and a distant receiver. The frequency range that this occurs at is 40–150 MHz, with the optimum at 40–70 MHz. The common FM band (88–108 MHz) is frequently used for detection.

The best FM radio to use is one that is digital and that has a shielded cable connection where the antenna plugs in. The digital type of radio is easier to set on the desired frequency. A Yagi-style FM radio antenna is also required. Unfortunately, which brand of antenna to use depends on trial and error. It is best to try a lower-cost antenna first.

Try all the frequencies one at a time. Note the frequencies where there is no music or talking. All that should be audible is static. You may have to turn the antenna all the way around to find this static. Do not direct the antenna very high, because in most places the antenna may receive unwanted, continuous reception. If you're lucky, you'll find several or more frequencies where nothing but static is heard.

If you can't find a frequency, don't fret. The internet can help. A site I really like is called Radio-Locator, at http://www.radio-locator.com/cgi-bin/home

I select "FM" and enter TX (Texas) and receive 711 stations! "NM" (New Mexico) gives me 226 more. "United Kingdom" lists 140 stations. Ideally, choose a station that transmits over 30kw and is located about 450 to 750km away.

Ready to "observe" yet? Simply listen to the static. When a meteor passes by it may produce a signal. The radio can detect meteors which correspond to a visual magnitude of 8 or 9. Most produce very short signals on the order of 1/4 second. The signals are actually small segments of what the station is transmitting. Some sound like thumps and chirps but longer signals will register as pieces of music or talking. These are usually very sudden, loud and clear, and will begin and end abruptly. Aircraft can interfere when they fly nearby, but they usually produce signals which are gradual before growing in volume.

Once you find a good station, use its frequency exclusively with the antenna always pointed in the same direction and at the same elevation. This way, day-to-day monitoring establishes a reliable pattern that can be compared.

Some observers tape record all sessions, adding verbal comments along the way. Others simply jot down a mark on a piece of paper. If the signal lasts longer than about a second, note this separately.

When you're out at night for your next meteor event, try to observe by radio as well as visually. Sometimes, it's possible to observe a simultaneous event.

Fireball detection network

A fireball is loosely defined as any meteor bright enough to cast a shadow. Most meteor organizations clarify this by stating that a fireball is any meteor with a visual magnitude of −3.0 or greater.

Some meteor showers produce more fireballs than others. Also, it is noted that about three times as many fireballs are observed near the time of the vernal equinox as near the autumnal equinox. In addition, a higher quantity of fireballs will appear around 18^h local time than 06^h local time. The reason is related to the speed of the meteor.

Slower-traveling meteoroids penetrate to deeper levels of our atmosphere and produce brighter meteors. If a meteoroid of the same size, but traveling faster, enters the atmosphere, it will break apart due to friction caused by its high rate of speed, producing a fainter meteor.

Earlier in this chapter, we discussed the reason that meteors are more common after midnight. Likewise, in the early evening, meteoroids must "catch up" to the Earth. This

Dick Spalding of Sandia National Laboratories (the originator of the Sandia Meteor Detection Network) holding an all-sky camera. (Photo courtesy Sandia National Laboratories)

Camera and all-sky mirror for the El Paso, Texas, station of the Sandia Bolide Detection Network. (Photo courtesy Jim Gamble, El Paso, Texas)

tends to produce slower, brighter meteors, some of which may be bright enough to be classified as fireballs.

All the major meteor organizations have fireball report forms and observers are encouraged to submit reports when brilliant meteors are seen. There is also a relatively new program which uses a video camera coupled to a wide-

Close-up of the camera assembly (cover off) for the El Paso, Texas, station of the Sandia Bolide Detection Network. (Photo courtesy Jim Gamble, El Paso, Texas)

angle mirror to monitor the sky during nighttime hours. It is known as the Meteor Detection Network.

The first system of this type was the Southern Saskatchewan Fireball Array (SSFA) constituting three all-sky, video camera systems located in the southernmost prairie region of Saskatchewan, Canada, at Regina, Moose Jaw and Laird.

The camera systems were designed and supplied by Sandia National Laboratories, New Mexico. Each system consists of a 45 cm diameter spherical mirror combined with a centrally mounted, and downward looking video camera. The systems afford all-sky monitoring (except for local buildings and obstructions). The video image is recorded onto standard VHS videotape, and, as of this writing, all tapes are manually reviewed when a report is received.

The limiting magnitude of the camera systems has been evaluated by planet image detection and through iridium flare observations. The system limiting magnitude is approximately −1.

Since the initial setup, other cameras have gone online. At least three such systems are operating in the El Paso, Texas area. Jim Gamble, amateur astronomer and meteorologist, installed one such system on the roof of his home:

I'm happy to announce that after many months of negotiations and weeks of setup and testing, The El Paso station of the Sandia Labs Fireball Detection Network is up and running. This system is solely for the purpose of collecting video records of fireballs and bolides, the information from which is submitted to Sandia Labs near Albuquerque, NM, for analysis to determine trajectory, speed, angle of attack, altitude etc. or possible "fall" events.

I and Dr. Robert Liefeld of New Mexico State University, who operates a camera there, will be notified by Sandia when fireballs are detected by either their radiometers or an automated video scanner module called "Video Sentinel." But since neither exist here, public visual reports will be critical to the success of this undertaking.

Lunar meteorite impacts

During several recent meteor showers, calls have gone out for observations of impact events caused by meteorites on the surface of the Moon. A number of observations have been submitted. These would certainly classify as high-level obser-

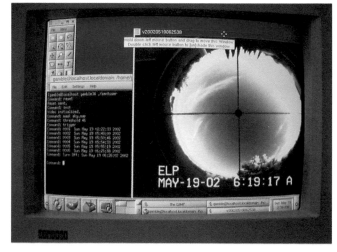

Video monitor of the El Paso, Texas, station of the Sandia Bolide Detection Network. (Photo courtesy Jim Gamble, El Paso, Texas)

NGC 2264: The Cone Nebula Region in Monoceros. (Mosaic of three images through IMG1024 and a Santa Barbara Instrument Group ST8 camera, 12.5" Ritchey-Chretien, f/7.5. Image by Robert Gendler, Avon, Connecticut)

vations by amateurs but there seem to be a number of problems.

Jeff Medkeff of Sierra Vista, Arizona, has looked into the claims of those who have observed lunar meteoric impact. His conclusions have led to the strong possibility that many presumed impact flash observations are instrumental. In Jeff's own words, "I'm not necessarily skeptical that impact flashes can be observed — but I do think that a lot of controls could be applied. The impact flash hypothesis seems somewhat uncritically accepted." Here are his reasons, edited for brevity.

(1) Lunar impactors produce localized temperatures in or above the 8000 °C range and pressures of several million atmospheres. The highest temperatures occur below the point of formation of the shock wave (thus below the surface), and prior to excavation. Additionally, there is the probability of significant penetration of the target by the impactor prior to achieving peak temperatures. This suggests that the flash would be severely extincted by lunar surface and impactor material which would obscure the line of sight from Earth. The problem is, basically, that the substantial producer of light is buried underground.

I believe that modeling suggests that to have a bright excavation would require the formation of a crater easily visible from Earth. My understanding is that squirting of liquefied rock early in the impact has a similar problem — to liquefy or vaporize enough rock to produce the observed intensities requires the formation of a substantial crater.

In short, there are some problems explaining how so much light is formed by impacts that demonstrably form craters <0.5 km in size. I'd also

be interested to see how much of this light must be in the near-infrared.

(2) A sample of presumed meteor-stream impact flashes and their brightness distributions should be found to correlate with observed bolides. As far as I know, no such observed population study has been done with the recent observations.

(3) The astrometric question should be more vigorously pursued. I am aware of at least one case in which time-correlated flashes occurred at separate stations to within a quarter second or better (the limits of my ability to measure the correlation). But it was not possible to tell the position on the Moon of these flashes in either tape. A flash occurring on the lunar surface, whether impact or endogenic, should be observable at the same selenographic position and at correlated times by dispersed stations.

(4) Considering the number of possible impact flashes reported, and the short amount of time spent observing the Moon for these events, we can infer that if the flashes are at the lunar surface, they are very common; therefore a large population of time- and positionally-correlated events should be easily discovered.

(5) Another statistical control could be applied by observing a patch of sky adjacent to the Moon and counting the number of "detections" in this location down to a specified magnitude. Perhaps observers could select their fields so that the field is 50% sky and 50% Moon areally.

(6) Observations should include some kind of photometry which can be used to constrain excavation profiles and insure the flashes are below the threshold of producing an observable crater.

CHAPTER SEVEN

What to observe: deep-sky objects

7.1 Double stars

Astronomers estimate that 60% of all the stars in space are double or multiple stars. These stars appear as one to the unaided eye but many may be resolved into pairs with the help of a telescope. I don't think there's an amateur astronomer I know who does not enjoy observing double stars. It's fun, easy, rewarding, it doesn't take a lot of aperture or a complicated setup, you can observe from within a city, and there are challenging objects for every size telescope.

Types of double stars

- *Optical binaries.* Sometimes called "line-of-sight" doubles, these stars are not physically related. They only look like a double star because of the chance lineup they have with the Earth. I think of these as "optical illusion" binaries.
- *Visual binaries.* These are the type of double stars we observe. All visual binary systems are discernible by telescopic observation.
- *Eclipsing binaries.* These stars show one or two dips in magnitude because one of the stars is passing in front of and/or behind the other star.
- *Spectroscopic binaries.* These pairs are distinguished by their Doppler shifts. The spectrum of such a pair shows alternating blue shifted and red shifted lines as the components orbit toward and away from us.
- *Spectrum binaries.* Spectrum binaries are stars whose spectra have lines from two different stars. For example, no star has the ionized helium lines of a B dwarf and also the metallic lines indicative of a K giant. It just doesn't

happen, so there must be two stars.
- *Astrometric binaries.* Stars which seem to be single but whose motion through space shows them to be co-orbiting with an invisible companion are known as astrometric binaries. This is the same science which is now finding planets around other stars.

Theories of double star formation

- *Fission.* One massive object forms. In this theoretical case it is spinning so rapidly that it splits to create a double star. Pairs formed in this way would essentially have their surfaces touching. This would not explain the great majority of double star systems.
- *Gravitational Capture.* An unlikely scenario which states that, due to the relative closeness of stars within a recently-formed cluster, double star systems would form as stars were gravitationally "caught" by other stars.
- *Common Condensation.* The least "tricky" of the three, this is the theory most astronomers think is correct. It says, simply, that stars forming in a nebula were close enough to gravitationally interact.

Observing double stars

In addition to where the double star is in the sky and the brightness of each of the components, there are two quantities with which a double star observer should be familiar. The first is the separation of the pair. This number is

The Double-Double ε Lyrae, 14 Jul 2001. (Imaged by Arpad Kovacsy with a Nikon CoolPix 950 through a Celestron CR-150 HD 6" refractor).

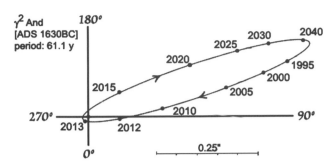

γ² Andromedae. At times impossible to see in any telescope; at other times easy. (Binary orbit drawn and supplied by Richard Dibon-Smith, Toronto, Canada. (http://www.dibonsmith.com/orbits.htm))

given in arcseconds and it is simply the distance separating the two stars. The second quantity is the position angle. This is the angle, measured from north through east, of the fainter of the pair (the companion or secondary) from the brighter (the primary). For instance, if the companion is due north of the primary, its position angle is 0°. If it is due east, 90°. If midway between south and west, 225°.

When the stars are the same magnitude, you'll just have to

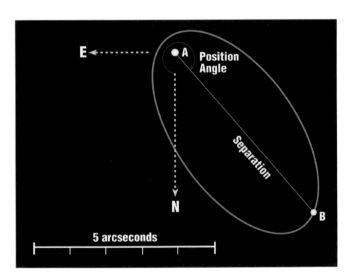

Position angle, measured in degrees, and separation, measured in arcseconds, are two important quantities in double star observations. (Illustration by Holley Y. Bakich).

pick one to be the primary. You can check a catalog at some point and if your value of the position angle differs from what you find in the catalog by 180°, you'll know that you should have picked the other star.

I asked Brian Mason, the Project Manager of the Washington Double Star Program, how the primary is determined when both stars are of the same magnitude. Here is his reply.

When known the primary is the most massive component – which is usually the brightest. However, when components are about the same brightness they are probably close to the same mass. In that event it is arbitrarily assigned or the assignment is made such that the position angle is less than 180°.

To determine the directions in your field of view, just let the stars drift through for a while. If your telescope has a motor drive, turn it off for this check. The stars will enter the field of view from the east and exit to the west. Determine the longest path for the stars you see drifting through the field. This is your east–west line. The north–south line is perpendicular to it, and to find it turn the drive back on, center a reasonably bright star, release the declination lock on your drive and move the telescope by hand so that the objective is slowly moving toward Polaris, or north. As you observe the field of view, the bright star will be heading out, toward the south. Reverse this if you are located in the southern hemisphere. The technique is only slightly more complicated if you own a telescope on an altazimuth mount. To move the telescope "north," you have to adjust the altitude and azimuth motions simultaneously.

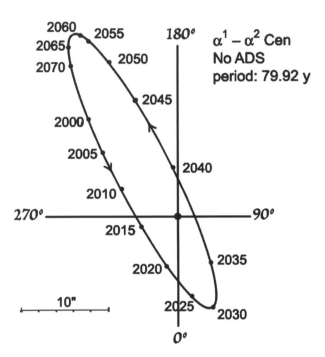

The third brightest star in the sky and the closest system to our own is α Centauri. (Binary orbit drawn and supplied by Richard Dibon-Smith, Toronto, Canada. (http://www.dibonsmith.com/orbits.htm))

A consideration that will influence which double stars you observe is the size of your telescope. As we know, the resolution of a telescope only depends on its size. Double star observers often refer to a rule of thumb called the Dawes limit. The formula for the Dawes limit is

$$r = 4.56/D$$

where r is the separation (in arcseconds) of the closest resolvable double star and D is the diameter of the objective (in inches)

or

$$r = 114/D$$

where D is the diameter of the objective (in millimeters).

The Dawes limit is certainly only a guideline. (In fact, Dawes had been dead a long time before this limit was ascribed to him.) Using this formula, my 4-inch, f/15 Unitron refractor should, at best, split a double star with a separation of 1.14 arcseconds. In May of 2000, on a night where the seeing could only be described as "legendary," I was able to attain a clean separation between a pair of double stars only 0.9 arcseconds apart. This observation was from my backyard in El Paso with six other people, three of whom are seasoned observers.

As I have often stated, it is difficult to convey how truly excellent the sky conditions of the desert southwest are. On a night such as the one described above, I believe any properly collimated and cooled (very important) quality telescope can perform near, at, or possibly beyond, the Dawes limit.

There is another way to calculate the theoretical resolving power of a telescope which is sometimes used by amateurs. It is known as the Rayleigh criterion. According to the rule of resolution, the central ring of one stellar image should fall on the first dark ring of the other. The formula is

$$r = 1.22\lambda/d$$

where r is the angular resolving power (in radians), d is the diameter of the objective lens (in centimeters), and, λ is the wavelength of the light (usually taken as 560 nm).

ζ Bootis, 25 Jun 2001. (Imaged by Arpad Kovacsy with a Nikon CoolPix 950 through a Celestron CR-150 HD 6" refractor)

When asked about resolving the components of a double star system, Brian Skiff, who does his observing from Flagstaff, Arizona, had this to say.

> The impression I got when I did a large double-star survey about 12 years ago was that once you reach the Rayleigh limit, where the Airy disks are just touching, you're indeed at the place where you can say things are "resolved. Beyond (closer than) that, the disks simply become more and more overlapped. For nearly-equal pairs you first get a figure-8 shaped object, then what some people call a "breadloaf," then an oval, then circular. The breadloaf stage comes at around half the Rayleigh separation. Although not "resolved," one can say with high certainty that the pair is recognized as

double. A couple years ago I viewed ζ Boo (about 0.9″) with my 70 mm Pronto, and got the position angle right in the "breadloaf" image. Pairs like π Aql (around 1.4″) are not more than ordinarily difficult (Dawes limit for the Pronto is 1.65″). In the nineteenth century Barnard discovered pairs with his 6-inch refractor between 0.4″ and 0.5″, which I still find amazing.

Colored double stars

Colorful double stars are a joy to behold. It does take some time at the telescope to train your eye to see colors but the payoff is a big one. And the closeness of double stars often helps in color identification. The contrast between two or even more stars in close proximity to each other brings out subtle color tones that would normally be lost if each were viewed separately.

Most amateur astronomers agree that the best double stars to observe (and certainly the best to show to the public at star parties) are those whose components have contrasting colors. Who among us can look at Albireo (β Cyg) and not be amazed? The contrast between the colors of gold and sapphire blue never fail to delight. The main colors amateur astronomers see when observing double stars are various shades of red, orange, yellow, white, and blue. Greens are sometimes seen, as are grays. Be certain to note specific colors in your observing log. If you choose to sketch the pair, add the colors later. Trying to distinguish different colors with a red flashlight is a real pain.

Albireo (β Cyg), 15 Jul 2001. (Imaged by Arpad Kovacsy with a Nikon CoolPix 950 through a Celestron CR-150 HD 6" refractor)

Some of the enjoyment in amateur astronomy is in sharing observations with friends. You will find, however, that color perception at the eyepiece is about as personal and subjective as any phase of our hobby. Colors that you see apply to your eyes, period. Consider the following example. Once, at an observing session, my friend Steve Coe of the Saguaro Astronomy Club of Phoenix, Arizona, mentioned

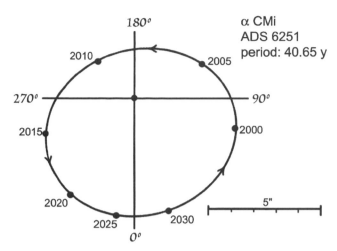

Procyon. (Binary orbit drawn and supplied by Richard Dibon-Smith, Toronto, Canada. (http://www.dibonsmith.com/orbits.htm))

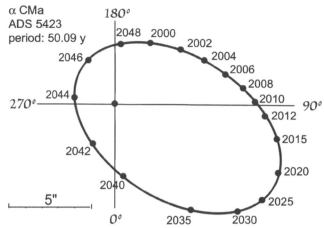

Sirius. (Binary orbit drawn and supplied by Richard Dibon-Smith, Toronto, Canada. (http://www.dibonsmith.com/orbits.htm))

Sirius B, imaged, perhaps for the first time, by an amateur astronomer. 21 Jan 2002. (Imaged by Arpad Kovacsy with a Nikon CoolPix 950 through an AP 155 EDT refractor)

that he saw 107 Aqr as white and light green. His friend Gerry Rattley immediately stepped to the eyepiece and after a moment asked, "Which star are you calling green, the orange one?"

Difficult double stars

Double stars with extremely close separation require two things, high magnification and great seeing (to allow you to use high magnification). I remember reading a very simplified rule of thumb:

$$x = 250/s$$

where x is the magnification necessary to split a double star and s is the separation in arcseconds.

With this as a guide, 125× is necessary to resolve a double star with a 2″ separation, 250× for 1″ and 500× for 0.5″. Your results will vary depending on the conditions, but this can be a starting point, at least.

Resolving double stars of similar magnitudes is rather straightforward. It is more difficult when the companion differs from the primary by several magnitudes. In some cases, a brightness difference of up to ten magnitudes may be found in a double star system.

The great test – observing Sirius B

Ask a double star observer this question: if you could positively split just one double star, which would it be? The answer, almost unanimously I believe, would be "Sirius." The discovery that Sirius was double is credited to Alvin G. Clark in 1862, who was testing an 18-inch lens that had just been installed at the Dearborn Observatory in Illinois.

Sirius A-B has an orbital period of just over 50 years. The separation of the pair varies from about 2″ to nearly 10″. Sirius A-B are now widening and are next scheduled for maximum separation in 2025. But you don't have to wait until then.

I have observed Sirius B several times in the past few years, in telescopes as small as 100 mm. Speaking of which, I don't know of a recorded observation with as small a scope. It may not be a record but I am very proud to have done it. Here is an excerpt from my astronomy log of my most recent (as of this writing) observation of Sirius B.

26 January 2002. A confirmed observation of Sirius B through my 100 mm f/15 Unitron achromat. Tonight, the seeing was legendary so I thought I'd try my luck at observing Sirius B. Several *amastro* subscribers had recently asked if I had attempted it in my 100 mm, which I had not. I had been monitoring the sky for the past week. It had been clear every night, but the seeing had not been nearly good enough. Tonight, however, the seeing was superb. We had just returned from supper at some friends' house. I immediately went out back to the observatory and, as I was headed out the door, I told my wife, Holley, that I would come and get her if I needed a confirming observation. I started at 300× (5 mm Vixen Lanthanum eyepiece) and after about 15 minutes possibly glimpsed it, but wasn't really certain. Dropping to 250×, after about 5 minutes I saw what I thought was B, so into the house I went. Holley asked if I had seen it. I said three words: "Come and look." We walked out together, she put her eye to the eyepiece and after about 3 seconds said, "Is

it at 7 o'clock?" I was actually pretty stunned and after a long pause told her that that's where I thought I'd seen it. I didn't tell her it took me longer than 3 seconds. Oh, for young eyes! This is about 2 years and a month after I'd split it in a friend's Celestron C11, also from the backyard. The time of Holley's confirming observation was 05:04 UT on 27 Jan 2002. I seem to remember it being much more difficult back in the 80s, even though it was wider then. I think the combination of a so-so 6-inch reflector and non-southwestern US skies made it a much more challenging affair. Note: the Moon tonight was nearly 94% illuminated, proof positive that a dark sky is not necessary for this observation.

After receiving numerous e-mails from observers asking how they could repeat our observation, I list below the important considerations for observing Sirius B. In fact, this list should be referred to for any difficult double star.

The following list of double stars was compiled by Brian Skiff at Lowell Observatory in Flagstaff, Arizona. As his explanation indicates, this list is to help amateur astronomers determine how good the seeing is on any particular night. But it is also a list of some mighty fine double stars which will test amateur telescopes of all sizes, so I include it here.

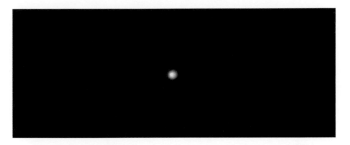

ε Bootis, 15 Jul 2001. (Imaged by Arpad Kovacsy with a Nikon CoolPix 950 through a Celestron CR-150 HD 6" refractor)

Double stars for seeing estimates

Brian Skiff, Flagstaff, Arizona

This is a first revision of a list of double stars that should be useful for estimating "seeing" (image size) in moderate-aperture amateur telescopes. Most of the stars are drawn from the list Chris Luginbuhl and I prepared for our *Observing Handbook*, with additional objects found as a result of the double-star viewing survey I did roughly ten years ago. A few special cases are included for pairs that appear near deep-sky objects.

The first column shows the ADS number; an asterisk indicates a note at the bottom of the table. The ADS

Top ten tips for observing Sirius B

(1) Get to a place and time of terrific seeing. It doesn't have to be pitch black. Just think where the Moon was for the above observation. If you have any indicators as to the quality of the seeing at your site, check them first. From my backyard, I check Saturn. If the Crepe Ring has detail and I can see the Encke Minima using 300× with my 4-inch, I know the seeing is great. Your check will differ.

(2) Let Sirius get to (or quite near) the meridian. You may as well give yourself every advantage, so look through the least amount of air.

(3) Check the position angle carefully. Know your telescope/diagonal setup so that you know where to expect Sirius B to be. If you're not familiar with position angles, look up the values of some nearby doubles. Rigel is one you can use.

(4) Be familiar with 5 arcseconds of separation. The separation between Rigel A and B is approximately 9 arcseconds. This is not quite double that of Sirius A-B. Note: Sirius B will be quite close to A, well within the diffraction ring pattern of a quality refractor.

(5) Remember, you're observing stars with approximately ten magnitudes of brightness difference. This means that A is approximately 10 000 times brighter than B.

(6) Crank up the power. I used 300× and then 250× to observe the Pup.

(7) Focus several times. Don't be satisfied with the first focus setting and adjust as needed over the course of your observation. It is probably a mistake to focus on Sirius.

Sirius B, imaged by an amateur astronomer. 22 Jan 2002. (Imaged by Arpad Kovacsy with a Nikon CoolPix 950 through an AP 155 EDT refractor)

The brightness of the star produces a very extended image which is flat through a fairly long movement of focus. Focus on one of the stars in the background, or at least check your focus against one of these. (Thanks to Jeff Medkeff for the second part of this one and also for #10.)

(8) Expect to spend some time observing the star. Not everyone has eyes like my wife. Don't rush what will be a great observation.

(9) If possible, have another observer there to confirm your observation.

(10) Don't use eyepieces that produce internal reflections of Sirius. A cheaply-coated eyepiece is a horrible idea (I checked), unless it is of a design that throws the reflections well out of focus (such that you wouldn't notice them anyway). Reflections will erode the contrast of the final view, which is exactly what you need desperately to preserve.

(Aitken Double Star) catalogue is the double-star equivalent of the NGC/IC, since it contains essentially all the pairs one can readily expect to resolve without special techniques, and is the preferred name for pairs found up to about 1930. Next comes the RA/Dec. to 0.1′ accuracy. The V magnitudes are the best that I could find without a lot of work; careful scrutiny of the Hipparcos data might yield more consistent data for some pairs. I tried to include only pairs with magnitude differences less than 1.0, but there are several exceptions. Next come the separations, in arcseconds. Separations closer than 1 arcsec are given to two decimals. Finally a common name (mostly Bayer/Flamsteed names or HD numbers) is shown in the last column.

Okay, that's how I built the list. My intention is that as part of just about any sort of visual observing, be it hunting down Himalia or mere deep-sky observing, it is useful to know what the seeing is on an absolute scale. I found that deep-sky viewing, where limiting magnitude threshold is important, required seeing no worse than about 1.5″, at least with my 15 cm refractor. The old myth that deep-sky doesn't require good seeing, is, well, a myth, and like all myths, is false. So just as one judges transparency by limiting magnitude estimates, one would like to know the seeing as well.

When I was doing a lot of double-star viewing with various apertures up to the Lowell 60 cm Clark, I found that the best pairs for estimating the size of the images scaled fairly well with aperture. Pairs whose components were between magnitude 5 and 7 were about ideal for the 15 cm (6-inch) refractor. A terrific pair of this description available in winter months is

ADS 6263, the brighter of the two pairs immediately east of Procyon (really easy to find!). At a current separation of 0.9″ (slowly closing), this is just right for seeing checks for this aperture.

With larger apertures one needs to go to progressively fainter in order to avoid problems with scattered light in the eye (called "irradiation" by old-time visual double-star observers). Thus at 10 inches, you need to choose pairs roughly a magnitude fainter, another magnitude fainter at 16 inches, etc. Obviously my list is tailored for telescopes up to only 12 inches aperture or so. And if you get a lot of good seeing, at larger apertures you'll also need much closer pairs to get accurate estimates. However, I suspect many folks will need those wider pairs as often as not.

Without some objective measurement, it is difficult to say what constitutes "resolution" or the actual image diameter. From Flagstaff with my refractor, usually the Airy disk is well defined, but moves bodily by some small amount. Thus my seeing estimate includes that image motion in addition to the image size beyond the Airy disk. Thus for example on a pair like π Aquilae, the total shift over an interval of many seconds might amount to roughly their 1.4″ separation, even though the Airy disks (0.85″ across in the 6-inch) are well-defined. Larger apertures will usually not show so much bulk motion, but instead simple blurring since there are several seeing cells in front of the aperture at any moment. Here you need to decide where the edge of the image is. So long as you do this consistently (having test pairs of similar magnitude all around the sky is one way of gaining consistency), then with a certain aperture your estimates will become pretty reliable with practice.

Some double stars for estimating seeing

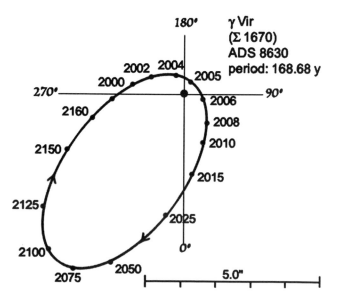

γ Virginis, one of the most famous double stars in the sky. (Binary orbit drawn and supplied by Richard Dibon-Smith, Toronto, Canada (http://www.dibonsmith.com/orbits.htm))

ADS#	RA h m s	Dec. Deg (°)	Dec. Arcmin (′)	Visual magnitudes (v = variable)	Sep Arcsec (″)	PA Deg (°)	Common name
111	0 09 21	−27	59.3	6.1,6.2	1.4	261	kappa-1 Scl
746	0 54 35	+19	11.3	6.2,6.9	0.5	198	66 Psc
755	0 54 58	+23	37.7	6.0,6.4	0.85	305	36 And
1081	1 19 48	−00	30.5	6.3,7.0	1.5	17	42 Cet
1254	1 36 03	+07	38.7	7.3,7.3v	1.6	55	HD 9817
1487	1 52 03	+10	48.6	7.8,7.8	3.3	201	HD 11386
1507	1 53 32	+19	17.6	4.8,4.8	7.5	0	gamma Ari
1579	1 59 19	+24	49.7	8.0,8.3v	1.2	277	HD 12101
1630BC	2 03 55	+42	19.9	5.5,6.3	0.5	104	gamma And BC
1654	2 06 15	+25	06.5	8.0,8.5v	1.9	161	HD 12824
1953	2 34 07	−05	38.1	7.8,8.0	3.6	346	HD 15994
2042	2 41 07	+18	48.0	7.7,8.0	3.3	119	HD 16694
2253	2 58 53	+21	37.1	7.5,7.5	0.51	265	HD 18484

Some double stars for estimating seeing (*cont.*)

ADS#	RA h m s	Dec. Deg (°)	Arcmin (')	Visual magnitudes (v = variable)	Sep Arcsec (")	PA Deg (°)	Common name
2257	2 59 13	+21	20.4	5.2,5.5	1.4	209	epsilon Ari
2616	3 34 27	+24	27.9	6.6,6.7	0.66	359	7 Tau
3297	4 33 33	+18	01.0	7.0,7.1	3.0	277	HD 28867
3734	5 10 18	+37	18.1	6.7,7.0	1.6	222	HD 33203
3799	5 13 32	+01	58.1	6.9,7.1	0.57	236	HD 33883
4490	5 54 22	+18	54.0	8.0,8.0	3.0	149	HD 39588
4728*	6 08 30	+13	58.3	7.4,8.0	2.5	109	HD 41943
4991	6 22 50	+17	34.4	7.3,8.3	2.1	18	HD 44496
5400	6 46 14	+59	26.5	5.4,6.0	1.8	74	12 Lyn
5436	6 48 12	+55	42.3	6.3,6.3	4.6	257	HD 48766/7
5447	6 47 23	+18	11.6	6.8,7.0	0.34	218	HD 49059
5871	7 12 49	+27	13.5	7.2,7.2	1.2	315	HD 55130
5996	7 20 31	+00	24.2	7.4,7.8	0.65	173	HD 57275
6180	7 34 32	+12	18.3	8.4,9.1	1.7	99	HD 60355
6258*	7 39 47	+05	16.4	9.2,9.3	0.83	186	HD 61502
6263*	7 40 07	+05	13.9	6.6,6.9	0.9	168	HD 61563
6425	7 52 42	+03	23.0	7.0,7.5v	0.97	13	HD 64165
6650AB	8 12 13	+17	38.9	5.6,6.0	0.67	103	zeta Cnc AB
7071	8 54 15	+30	34.7	6.0,6.5	1.5	313	57 Cnc
7093	8 55 30	−07	58.3	6.7,6.9	3.9	3	17 Hya
7286	9 18 26	+35	21.8	6.4,6.7	1.8	49	HD 80024
7307	9 20 59	+38	11.3	6.7,7.0	1.02	280	HD 80441
7390	9 28 27	+09	03.4	5.9,6.5	0.52	74	omega Leo
7555	9 52 30	−08	06.3	5.6,6.1	0.55	64	gamma Sex
7565*	9 55 03	+68	56.4	10.6,10.6	8.9	273	BD+69 541
7566*	9 55 04	+68	54.1	9.5,9.5	2.1	112	HD 85458
7704	10 16 16	+17	44.4	7.2,7.5	1.4	180	HD 88987
7936	10 49 17	−04	01.4	7.0,7.8	2.4	14	40 Sex
8043	11 03 59	+03	38.3	7.5,7.6	1.2	301	HD 95899
8446	12 10 47	+39	53.5	7.3,8.0	0.29	195	HD 105824
8575	12 30 34	+09	42.9	8.1,8.4	1.3	244	HD 108875
8708	12 56 27	−00	57.3	7.2,7.6v	0.97	98	HD 112398
8864	13 20 42	+02	56.5	6.7,7.4	1.1	177	HD 115995
8972	13 37 35	−07	52.3	7.5,7.5	2.7	40	81 Vir
9053	13 54 58	−08	03.5	6.6,7.5	3.5	97	HD 121325
9060*	13 56 20	+05	17.4	8.4,8.9	1.03	111	HD 121605
9174	14 13 55	+29	06.3	7.5,7.6	0.61	92	HD 124587
9343	14 41 09	+13	43.7	4.5,4.6	0.81	301	zeta Boo
9406	14 49 41	+48	43.2	6.2,6.9	2.8	45	39 Boo
9425	14 53 23	+15	42.3	6.9,7.5	1.2	167	HD 131473
9578	15 18 20	+26	50.4	7.3,7.4	1.6	258	HD 136176
9617	15 23 12	+30	17.3	5.6,5.9	0.8	56	eta CrB
9757	15 42 45	+26	17.7	4.1,5.5	0.7	116	gamma CrB
9969	16 13 18	+13	31.6	7.4,7.5	4.1	353	49 Ser
10049	16 25 35	−23	26.8	5.3,6.0	3.1	339	rho Oph
10075	16 28 53	+18	24.8	7.7,7.9	1.9	126	HD 148653
10087	16 30 55	+01	59.0	4.2,5.2	1.4	26	lambda Oph

ADS#	RA h m s	Dec. Deg (°)	Arcmin (')	Visual magnitudes (v = variable)	Sep Arcsec (")	PA Deg (°)	Common name
10345	17 05 20	+54	28.2	5.7,5.7	2.2	23	mu Dra
10374	17 10 23	−15	43.5	3.0,3.5	0.49	243	eta Oph
10650	17 35 50	+00	59.8	7.5,7.5	3.0	78	HD 159660
10850*	17 51 58	+15	19.6	6.8,7.1	0.81	349	HD 162734
10905	17 56 24	+18	19.6	7.0,7.0	2.6	292	HD 163640
11005	18 03 05	−08	10.8	5.2,5.9	1.7	281	tau Oph
11123	18 10 09	+16	28.6	6.4,7.3	1.2	220	HD 166479/80
11479	18 35 30	+23	36.3	6.4,6.7	0.7	7	HD 171745
11483	18 35 53	+16	58.5	6.9,7.1	1.7	157	HD 171746
11635AB	18 44 20	+39	40.2	5.0,6.1	2.5	352	epsilon-1 Lyr
11635CD	18 44 23	+39	36.8	5.2,5.5	2.3	84	epsilon-2 Lyr
11640*	18 45 28	+05	30.0	6.4,6.7	2.5	119	HD 173495
11869	18 57 08	+26	05.8	8.0,8.3	0.7	74	HD 176005
12239	19 15 57	+27	27.4	6.9,7.3	0.85	158	HD 180553
12808	19 42 34	+11	49.6	5.7,6.6	0.41	77	chi Aql
12962	19 48 42	+11	49.0	6.3,6.8	1.4	107	pi Aql
13277	20 02 01	+24	56.3	5.7,6.0	0.82	122	16 Vul
13465*	20 10 37	+34	51.7	9.2,9.4	4.1	172	HD 191833
13506	20 12 35	+00	52.0	6.9,7.2	2.7	206	HD 191984
14296	20 47 25	+36	29.4	4.9,6.1	0.85	11	lambda Cyg
14360	20 51 26	−05	37.6	6.4,7.4	0.96	18	4 Aqr
14499	20 59 05	+04	17.7	6.0,6.4	0.85	286	epsilon Equ
14556	21 02 13	+07	10.8	7.1,7.1	2.9	215	2 Equ
14573	21 03 03	+01	31.9	6.7,7.3	1.3	120	HD 200375
14715	21 10 59	+09	33.0	7.8,8.0	2.8	80	HD 201686
15176	21 39 32	−00	03.1	7.1,7.7	0.53	270	24 Aqr
15562	22 02 26	−16	57.9	7.1v,7.2	3.8	246	29 Aqr=DX Aqr
15639	22 07 07	+00	34.2	7.6,8.0	2.5	97	HD 209965
15934	22 26 34	−16	44.5	6.2,6.4	2.2	350	53 Aqr
15971	22 28 50	00	01.2	4.3,4.5	1.9	191	zeta Aqr
16214	22 43 04	+47	10.1	6.2,7.0	0.5	303	HD 215242
16579	23 12 00	−11	56.0	7.2,7.2	3.5	278	HD 218928
16836	23 33 57	+31	19.5	6.0,6.0	0.53	90	72 Peg
17149	23 59 29	+33	43.4	6.5,6.7	1.9	326	HD 224635/6

*Notes

ADS 4728 in NGC 2169
ADS 6258 the first mag. 9 star east of Procyon
ADS 6263 the first mag. 6 star east of Procyon
ADS 7565 on the southwest side of M81
ADS 7566 on the southwest side of M81
ADS 9060 near galaxy NGC 5363
ADS 10850 nearby mag. 10 star also double
ADS 11640 Tweedledee & Tweedledum; each component is ~0.2" pair
ADS 13465 near open cluster IC 1310

7.2 Variable stars

Stars whose brightness changes over a period of time are called variable stars. There are two main types of variable stars. They are known as intrinsic and extrinsic. Within each type are two classes. Pulsating variables and eruptive variables comprise the intrinsic type. These depend on physical changes within the star.

Eclipsing binaries and rotational variables are the two classes within the extrinsic type. Eclipsing binaries drop in brightness when one component is eclipsed by the other. Rotating stars may expose different areas to us as they spin. A large number of sunspots on one side of the star can make a great difference in its brightness.

Light curve

A plot of a star's brightness over time is known as a light curve. This is an especially valuable tool when it comes to variable stars.

History

In the middle of the nineteenth century, the German astronomer Friedrich Wilhelm August Argelander (1799–1875) and the American astronomer Benjamin Apthorp Gould (1824–1896) were calling upon amateurs worldwide to submit observations of variable stars. Gould was gathering variable star observations from amateur astronomers across the United States and publishing them in the *Astronomical Journal*, which he edited.

By the mid-1880s, Harvard College Observatory Director, Edward C. Pickering (1846–1919) and American astronomer William Tyler Olcott (1873–1936) began providing variable star observers with sets of charts which had the variable star and its comparison stars marked directly on them. 25 years later, Pickering began to enter the magnitudes of comparison stars directly onto the photographic charts.

In December 1911, spurred on by publications of and communications between Olcott with other amateurs involved in variable star observation, the first observations of the group which would become the American Association of Variable Star Observers (AAVSO) appeared in *Popular Astronomy*. In April 1914, the first meeting of the group occurred in New York City. Finally, in October 1918, the AAVSO was incorporated.

For much more on the early history of variable star observing, I cannot recommend more highly the book *Starlight Nights* by Leslie C. Peltier (Harper and Row, New York, 1965, reprinted 1999). Peltier (1900-1980) was an

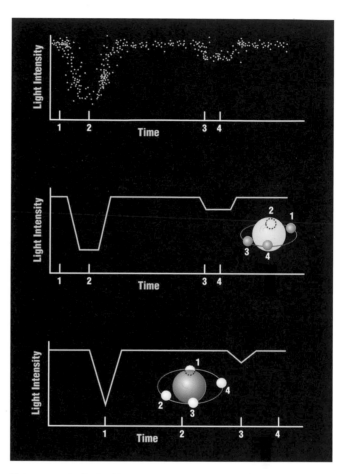

Light curves of eclipsing binary stars. Top: Raw data points. Middle: Curve based upon the raw data points. This is a binary system with a total eclipse. Bottom: The eclipse is partial. (Illustration by Holley Y. Bakich)

indefatigable observer and this book is a tribute to the pursuit of amateur astronomy.

Variable star designations

In 1862, Argelander assigned the upper case Roman letters R–Z to the initial variable stars within each constellation. After these nine letters were used came double upper case letters (RR–RZ, SS–SZ . . . AA–AZ, BB–BZ . . . QQ, QZ). Since no "J" was used to begin these pairs, a total of 334 stars could be denoted in this fashion. Thus, R CrB is the first variable star in the constellation of Corona Borealis and QZ Ori is the 334th variable star in Orion. If even more are needed, the single upper case letter V plus numbers after 334 are used. Therefore, Plaskett's Star is designated as V640 Mon, the 640th variable star in the constellation of Monoceros.

Novae are designated initially by constellation and year, as for example, Nova Cygni 1975. When the brightness of the event diminishes, such objects are assigned a variable star designation. Thus, the above object is now known as V1500 Cyg.

The AAVSO also employs a numerical system for variable stars based on the Harvard Designation. Thus, the variable R Leporis has the additional AAVSO designation of 0455-14. These numbers are the star's approximate right ascension (04^h55^m) and declination ($-14°$) for the year 1900.

Types of variable stars

Within these major classifications there are many sub-classes. The following is a short list, alphabetized, of the most familiar classifications of variable stars. Note that many variable star types are named after the first star of that class discovered, which is then known as the prototype.

Algol type

- Prototype: β Per
- Period: 2.87 days
- Magnitude range: 2.1 – 3.4

The classic eclipsing binary, Algol is actually a partially eclipsing binary star. A total eclipse occurs with many stars, one of which is U Sge. The two stars in the system are far enough apart so that most of the time their combined light is constant. Algol-type variables are fun to observe and compare to other stars. Algol itself can be compared to nearby γ Andromedae. At brightest, Algol is just (0.15 magnitude) brighter than γ And. At faintest, it is more than a magnitude fainter.

β Lyrae type

- Prototype: β Lyr
- Period: 12.91 days
- Magnitude range: 3.3 – 4.4

This system's stars are so close that they are either touching or nearly so. Each of their shapes are so distorted by the influence of the other star that they look like eggs. There is no point on their light curve when their light output is steady. When observing the prototype, β Lyr, compare it to nearby γ Lyr, only two degrees away. At brightest β is equal to γ in brightness and a magnitude fainter when dimmest.

Carbon stars

Very red and very cool stars that get their name from a high abundance of carbon which builds up on their surface. These are among the reddest stars known and great fun to observe for that reason. Examples are R Lep (Hind's Crimson Star) and μ Cep (Herschel's Garnet Star).

Brian Skiff, of Flagstaff, Arizona, has compiled a list of the brightest "very red" stars. Red stars are of general interest and several of the brightest are good for public viewing sessions. All stars listed have V magnitudes brighter than 8.0, making the list a manageable one, and an easy observing

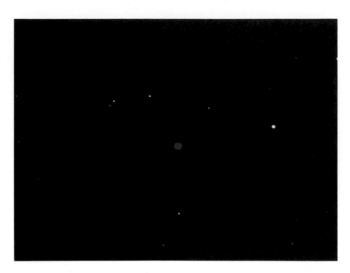

Hind's Crimson Star, R Leporis. (Image by Ed Grafton, Houston, Texas, using a Celestron 14 and an ST5c CCD camera)

project with a small telescope. (Note: most of these stars are carbon stars, but not all).

Brian received help from Australian variable star guru Fraser Farrell. The list was also added to by David Frew of Lane Cove, NSW, Australia.

The brightest very red stars
23 Oct 2001

Designation	RA (2000) h m s			Dec. (2000) Deg (°)	Arcmin (')	Arcsec (")	Visual Magnitude	B–V
R Scl	01	26	58.1	−32	32	35	5.8	3.9
TW Hor	03	12	33.2	−57	19	18	5.7	2.3
R Dor	04	36	45.6	−62	04	38	5.5	1.6
R Lep	04	59	36.3	−14	48	23	7.7	5.7
W Ori	05	05	23.7	+01	10	39	6.2	3.5
CE Tau	05	32	12.8	+18	35	39	4.4	2.1
W Pic	05	43	13.8	−46	27	14	7.8	4.8
Y Tau	05	45	39.4	+20	41	42	7.0	3.0
BL Ori	06	25	28.2	+14	43	19	6.2	2.4
UU Aur	06	36	32.8	+38	26	44	5.3	2.6
NP Pup	06	54	26.7	−42	21	56	6.3	2.3
W CMa	07	08	03.4	−11	55	24	6.9	2.5
X Cnc	08	55	22.9	+17	13	53	6.6	3.4
Y Hya	09	51	03.7	−23	01	02	6.6	3.8
X Vel	09	55	26.1	−41	35	13	7.2	4.3
AB Ant	10	11	53.8	−35	19	29	6.7	2.3
U Ant	10	35	12.8	−39	33	45	5.4	2.9
U Hya	10	37	33.3	−13	23	04	4.8	2.8

The brightest very red stars (*cont.*)

Designation	RA (2000) h	m	s	Dec. (2000) Deg (°)	Arcmin (')	Arcsec (")	Visual Magnitude	B–V
VY UMa	10	45	04.0	+67	24	41	6.0	2.4
V Hya	10	51	37.3	–21	15	00	6.8	5.5
SS Vir	12	25	19.9	+00	47	54	6.6	4.2
Y CVn	12	45	07.8	+45	26	25	4.9	2.5
RY Dra	12	56	25.9	+65	59	40	6.4	3.3
UY Cen	13	16	31.8	–44	42	16	6.9	2.8
V766 Cen	13	47	10.9	–62	35	23	6.5	2.0
X TrA	15	14	19.2	–70	04	46	5.8	3.6
TW Oph	17	29	43.7	–19	28	23	7.9	4.8
V Pav	17	43	18.9	–57	43	26	6.7	4.2
T Lyr	18	32	20.1	+36	59	56	8.5	5.5
V450 Sct	18	32	43.3	–14	51	56	5.5	2.0
S Sct	18	50	20.0	–07	54	27	7.5	3.1
V Aql	19	04	24.2	–05	41	05	7.5	4.2
V1942 Sgr	19	19	09.6	–15	54	30	6.9	2.3
UX Dra	19	21	35.5	+76	33	35	5.9	2.9
AQ Sgr	19	34	19.0	–16	22	27	7.3	3.4
RT Cap	20	17	06.5	–21	19	04	7.4	4.0
T Ind	21	20	09.5	–45	01	19	6.0	2.4
Y Pav	21	24	16.8	–69	44	02	6.4	2.8
V460 Cyg	21	42	01.1	+35	30	37	6.1	2.5
mu Cep	21	43	30.5	+58	46	48	4.1	2.3
pi-1 Gru	22	22	44.2	–45	56	53	6.6	2.0
RW Cep	22	23	07.0	+55	57	48	6.7	2.3
TX Psc	23	46	23.5	+03	29	13	5.0	2.6

Cepheid type

- Prototype: δ Cep
- Period: 5.37 days
- Magnitude range: 3.5 – 4.4

The "measuring stick" of astronomy, delta Cepheids (usually shortened to Cepheids) have an amazing characteristic. Their periods are directly related to their luminosities. So, once a Cepheid's period is measured its absolute magnitude can be calculated. Comparing that quantity with its apparent magnitude allows the star's distance to be determined. Physically, these are young stars of several solar masses and about 10 000 times the luminosity of the Sun. Cepheids pulsate in a regular way. This happens as helium in the outer layers becomes ionized, making it more opaque. In this state, the helium can absorb more energy and the star expands, with an increase in luminosity. As it expands, however, the helium cools and de-ionizes, causing the star to shrink.

δ Scuti type

- Prototype: δ Sct
- Period: 04^h39^m
- Magnitude range: 4.6 – 4.8

Visual observations of this type of variable are very difficult, as most vary in brightness by less than a tenth of a magnitude. Good light curve work may be done with CCD.

Flare stars

- Prototype: UV Cet
- Period: 10 hours
- Magnitude range: 7.5 – 12.1

Flare stars are among the fastest-acting of all variables. They are faint, red stars that experience intense outbursts from small areas on their surfaces. Typically they brighten by several magnitudes in only a few seconds. Then, during the next half hour or so, they return to minimum brightness.

Irregular and semi-regular long term variables

Unpredictability is the hallmark of these variables, some of which are very well-known stars. Their variability is generally under two magnitudes and the time spans between changes can be quite long. Examples are α Ori (Betelgeuse), α Her (Ras Algethi) and μ Cep (Herschel's Garnet Star).

Mira type

- Prototype: o Cet
- Period: 333.8 days
- Magnitude range: 2.9 – 7.3

Mira-type variables are red giants with fairly long periods of up to three years. Keeping track of this type of star can be rewarding, however, as the brightest examples have a wide range of magnitudes. Mira-type variables were some of the first types discovered and their designations show that, in most of the northern constellations, they were the first variable seen (by the letter R designated). Noting their magnitude ranges in the following table, it is easy to see why.

Mira-type variables

Star	Maximum magnitude	Minimum magnitude
R And	5.6	14.9
R Aql	5.5	12.0
R Ari	7.4	13.7
R Boo	6.2	13.1
R Cae	6.7	13.7
R Cam	6.9	14.4
R Cap	9.4	14.9
R Cas	4.7	13.5
R Cet	7.2	14.0
R CMi	7.3	11.6

Mira-type variables (*cont.*)

Star	Maximum magnitude	Minimum magnitude
R CMi	7.3	11.6
R Cnc	6.0	11.8
R Col	7.8	15.0
R Crv	6.7	14.4
R CVn	6.5	12.9
R Cyg	6.5	14.4
R Del	7.6	13.8
R Dra	6.9	13.2
R Equ	8.7	15.0
R For	7.5	13.0
R Gem	6.0	14.0
R Her	8.2	15.0
R Hya	4.0	10.9
R Lac	8.5	14.8
R Leo	4.4	11.3
R Lep	5.5	11.7
R Lib	9.8	15.9
R LMi	6.3	13.2
R Lup	9.4	14.0
R Lyn	7.2	14.3
R Mic	8.3	13.8
R Oph	7.0	13.8
R Ori	9.1	13.4
R Peg	7.1	13.8
R Per	8.1	14.8
R PsA	8.5	14.7
R Psc	7.1	14.8
R Sco	9.8	15.5
R Ser	5.7	14.4
R Sgr	6.7	12.8
R Tau	7.6	15.8
R Tri	5.4	12.6
R UMa	6.7	13.7
R UMi	8.5	11.5
R Vir	6.1	12.1
R Vul	7.4	14.3

Check out the brightest of these and see a star become invisible (and then reappear).

R Coronae Borealis type
• Prototype: R CrB
• Period: irregular
• Magnitude range: 5.7 – 14.8

These are rare, luminous, hydrogen-poor, carbon-rich, variables that spend most of their time at maximum light, occasionally fading as much as nine magnitudes at irregular intervals. They then slowly recover to their maximum brightness after a few months to a year. Members of this group have F to K and R spectral types.

RR Lyrae type
• Prototype: RR Lyr
• Period: 13^h36^m
• Magnitude range: 7.1 – 8.1

The members of this type of variable are all very similar. Their luminosities are essentially the same, around 60 times the luminosity of the Sun. Because of this they are good distance indicators, similar to Cepheids, although with much higher luminosities, Cepheids can be seen at much greater distances. Many RR Lyrae stars are found in older globular clusters. Observe the prototype over the course of an entire night. Depending on your location, either the full range of variability, or nearly so, may be seen.

RV Tauri type
• Prototype: R Sct
• Period: 140 days
• Magnitude range: 4.5 – 8.2

Stars of this type are luminous yellow giants. The minimum of brightness has two separate levels, alternating between a shallow minimum and a deep one. Their individual periods are defined as the length of time between two deep minima. Some, like the prototype, are worthy of amateur observation.

SU Ursae Majoris type
• Prototype: SU UMa
• Period: irregular
• Magnitude range: 11.1 – 14.8

Similar to the U Gem type except have short orbital periods of less than two hours, and have two distinct outbursts that are both short (duration one to two days, faint and more frequent) and long (duration ten to twenty days, bright and less frequent)

Symbiotic stars
• Prototype: Z And
• Period: irregular
• Magnitude range: 8.0 – 12.4

This type of variable is composed of interactive binaries with periods typically longer than one year. Some of them show semi-periodic outbursts that look like novae in slow-motion. The pair itself is a red giant and a hot sub-dwarf or white dwarf. A nebula due to mass loss from the stars surrounds the pair and is ionized by the hot star. A very interesting type of variable.

U Geminorum type

- Prototype: U Gem
- Period: 105.2 days
- Magnitude range: 8.9 – 14.9

Similar to symbiotic stars. A star of this type is sometimes referred to as a dwarf nova. The sudden outburst may last from one to three weeks and the magnitude gain is typically 4-5 magnitudes.

W UMa type

- Prototype: W UMa
- Period: 4 hours
- Magnitude range: 7.6 – 8.4

This type of variable is similar to the β Lyr type, meaning that it is composed of contact binaries. The difference is that the stars of the W UMa type are cooler and less massive.

W Virginis type

- Prototype: W Vir
- Period: 17.3 days
- Magnitude range: 9.5 – 10.8

Very similar to Cepheid variable stars in that they have a period–luminosity relationship. W Vir variables are older and less luminous (by about 2 magnitudes) than the Cepheids.

Z Camelopardalis type

- Prototype: Z Cam
- Period: 22 days
- Magnitude range: 10.0 – 14.5

Similar to the U Gem type, but not as numerous. Z Cam stars are more variable than U Gem stars and sometimes enter a period called a standstill. This may last several brightness cycles. When a Z Cam variable is in standstill, its magnitude is approximately one third of the way from maximum to minimum. Standstills are unpredictable.

A personal statement

For over a century, amateur astronomers have carefully measured, recorded and submitted observations of variable stars. Much of what we know of the light curves of these stars has come from dedicated visual observations made by observers who put eyes to eyepieces under some harsh conditions. That era has now passed.

With the proliferation of CCD technology, variable stars may be measured to an accuracy which could only be dreamed about by amateurs using visual means. Medium-sized telescopes equipped with CCDs can typically reach sixteenth magnitude. Add to such a system a GoTo drive and software running a script containing a night's list of variables and one can scarcely imagine the amount of data that could be collected – all while the amateur is either observing with a different telescope or asleep!

7.3 Supernovae

The greatest explosion in the universe is a supernova. This occurs when a star, much more massive than the Sun, reaches the final stages of its evolution. An instability occurs within the core of the star causing a massive collapse of the core and a tremendous outrush of energy.

The last supernova seen in the Milky Way galaxy was in 1604. Known as "Kepler's Star," this object appeared in the constellation Ophiuchus and was observed by the great German astronomer Johannes Kepler. 32 years earlier, in 1572, the Danish astronomer Tycho Brahe had observed a supernova in the constellation Cassiopeia.

In 1054, Chinese observers recorded, within the boundaries of the constellation Taurus, an observation of the most famous supernova of all. When amateur astronomers point their telescopes to this area of the sky they see a violently twisted cloud of gas called the Crab Nebula. This was such a powerful explosion that material from this object is rushing outward at an amazing 1400 km/s.

The brightest supernova since Kepler's was discovered on 23 February 1987, in the Large Magellanic Cloud. This supernova was an easy naked-eye object throughout 1987 in the southern hemisphere. This supernova was named SN 1987A. Supernovae, when they are discovered, are designated by the year in which they are discovered plus a letter of the alphabet, indicating the order in which they were discovered within that year.

All supernovae discovered since 1604 have been extragalactic supernovae. Between 17 August 1885 and 1 January 2002, 2103 supernovae have been cataloged. This works out to only 18 per year overall, but that number is skewed by the rarity of discovery in the early years. In the years 1997–2001, the total number of supernovae discovered was 996 (nearly half of all supernovae discoveries). The average for that period is 199 per year.

Supernovae in the Milky Way

Year	Constellation	RA h	m	Dec. Deg (°)	Arcmin (′)	Magnitude
185 AD	Cen	14	43.1	−62	28	−2
393/396	Sco	17	14	−39	84	−3
1006 Apr 30	Lup	15	02.8	−41	57	−9
1054 Jul 4	Tau	05	34.5	+22	01	−6
1181	Cas	02	05.6	+64	49	−1
1572 Nov 6	Cas	00	25.3	+64	09	−4
1604 Oct 9	Oph	17	30.6	−21	29	−3
1667	Cas	23	23.4	+58	50	+6

Types of supernovae

Supernovae fall into two very different types. Type Ia supernovae result from mass transfer inside a binary system consisting of a white dwarf star and an evolving giant star. Type II and Type Ib supernovae are, in general, single massive stars which, astronomically speaking, live very short lives. The two can be told apart spectroscopically as Type Ia supernovae do not show any hydrogen in their spectra, whereas Type II and Type Ib supernovae show the presence of hydrogen quite strongly.

Type Ia

The birthplace of a Type Ia supernova is an old, evolved binary system in which at least one component is a white

M1 (Celestron 14 @ f/11, 1 hour on Kodak PPE-400 color negative film (non-hypered), 12–13 Dec 1996, Sierra Vista, Arizona. Image by David Healy)

Orion. (Piggyback shot, 50 mm lens, Fuji Provia, 5-minute exposure. The faint, reddish glow of the probable supernova remnant known as Barnard's Loop may be seen. Photo by the author)

elements. When the core of the star becomes iron, so to speak, it is no longer a source of energy production and can no longer resist the crushing force of gravity. It collapses under its own weight.

The Eta Carina Nebula (reversed). At some future time, the massive star within this nebula will become a supernova. (Celestron f/1.5 Schmidt Camera, 10 minutes on Kodak Vericolor 1000 Film, Arequipa, Peru, 6–7 Apr 1986. Image by David Healy, Sierra Vista, Arizona)

dwarf star and the other a much larger star, usually thought to be a giant. A white dwarf has an upper mass limit of about 1.4 times that of the Sun. This is known as the Chandrasekhar limit, after the astronomer who determined it. Over a great deal of time the system loses some of its angular momentum and the stars begin to draw closer. At some point mass transfer occurs from the companion star to the white dwarf.

The mass transferred from the giant star eventually increases the mass of the white dwarf to a value much higher than the Chandrasekhar limit. At this point the whole star collapses and nuclear fusion of carbon and oxygen to nickel provides enough energy to utterly destroy the star.

Type II

Stars more massive than about eight solar masses become Type II and Type Ib supernovae. These are huge stars which evolve very quickly, on the order of tens of millions of years or less – a short time as far as the cosmos is concerned.

The temperature in the cores of these stars is so high that nuclear fusion moves far beyond the Sun's process of making helium out of hydrogen. I won't go into all the details here, but what ends up happening is that iron nuclei are formed. The iron nucleus is the most stable nucleus of all and no stellar core temperature can cause it to fuse into any heavier

The forces here are so unimaginably huge that atomic particles are crushed together and a neutron star forms from the once iron core. (Note: it is believed that the most massive stars form black holes, objects even denser than neutron stars.) That part of what once was a star is now stable but during the neutron star formation a huge amount of energy is released which rips through the outer layers of the star. The original massive star dies as a supernova, with only the newly-formed neutron star surviving this huge explosion. Interestingly, the point from the original collapse of the iron core to supernova and neutron star formation takes less time than it took you to read this description!

The aftermath of a supernova

The explosions of supernovae have been called the most significant events in the universe. The tremendous energy that is released affects nearby clouds of gas, setting up shock waves in them. The shock waves don't need to be too powerful, just enough to get stuff moving around a little. If the gas within these clouds was originally fairly dense, then the shock wave will begin the process of star formation.

These explosions also synthesize elements heavier than iron. This is possible due to the amount of energy in the explosion. These new elements are then distributed into the surrounding gas clouds, a process known as enrichment of the interstellar medium. It is for this reason that later generations of stars contain more heavy elements than early generations. So mercury, lead, gold, iodine, and many other elements that surround us were formed billions of years ago in one or more supernova explosions. And one of these explosions may have started the process of the Sun's formation. It appears that supernovae are more important than we may have initially thought.

Observing supernovae

The mere fact that nearly half of all recorded supernovae have been discovered in a five-year period indicates that something has changed. The change is that supernova searches are now conducted with CCD cameras coupled to telescopes. Visual searches for supernovae are a thing of the past, although an occasional visual discovery is still reported.

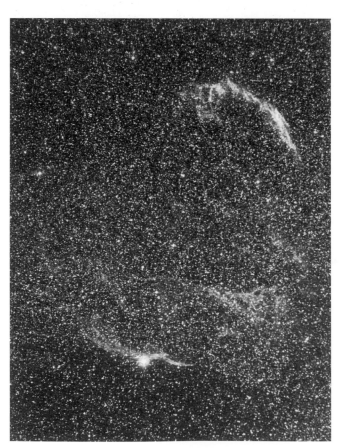

The Cygnus Loop, a huge supernova remnant. (8" Celestron Schmidt camera, f/1.5, 10 minutes on TP 2415 film, hypersensitized 9.5 days in forming gas @ 30° C, 12–13 Sep 1985, Naco, Arizona. Image by David Healy, Sierra Vista, Arizona)

In addition to providing an electronic image of the galaxy being searched, CCDs are much more sensitive than the eye. Typically, on a 200 mm telescope, a gain of more than three magnitudes can be realized.

> **Observing Note:** Supernovae are more often found in bigger, brighter and older galaxies. Spiral galaxies statistically produce more supernovae than do elliptical galaxies.

If you believe you have found a supernova, compare your image with as many reference images as you can find. One online resource that you can check is the Sloan Digital Sky Survey at http://skyserver.fnal.gov/en/

Check for movement by taking several images at least one hour apart to confirm that what you are seeing is not an asteroid, a hot pixel or a variable star. Refer to the International Supernovae Network for additional information and guidance on verifying supernovae. They can be found online at http://www.supernovae.net/isn.htm

Finally, when you are as certain as you can be, report your findings to the IAU by sending an e-mail to the Central Bureau for Astronomical Telegrams (CBAT). Find their website at http://cfa-www.harvard.edu/iau/cbat.html

Please be absolutely certain of your facts. They remember those who send in false alarms.

The following information should be included in a discovery report e-mail:

- Your name
- Your address
- Your e-mail address
- Your phone number
- Date of observation (UT)
- Time of observation (UT)
- Observation method (naked-eye, visual telescopic, photographic, CCD, etc)
- Instrumentation (aperture size, f/ratio, etc.)
- Exposure times (type of film or CCD, length of exposure, etc.)
- Observation site (name, city, state/province/country)
- Longitude, latitude, and elevation above sea level, if known

The above information should be in all reports. For supernovae specifically, also include the following:

- Identity and position of the host galaxy, noting the equinox
- Magnitude of the suspect
- Precise position for the suspect (mandatory for CCD observations)
- Visual observers should give the offset (in seconds of arc) north/south and east/west of the galaxy nucleus

In addition to the above, note the following:

- Avoid ambiguous terms such as "left," "right," "above,"and "below."
- Observations on a second night showing that the object has not moved should also be submitted.
- Indicate how you have determined that the object is new

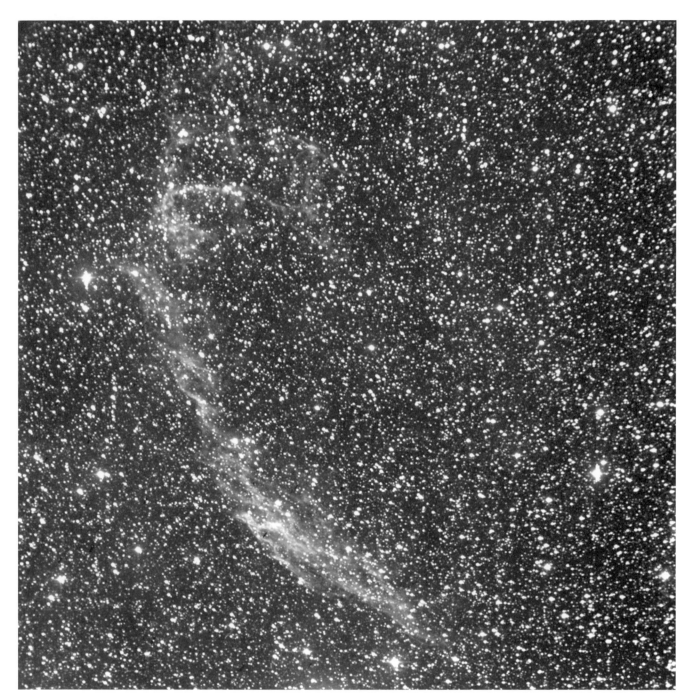

The Network Nebula. (20 minutes at f/7 with TK1024 (FLI) CCD camera and 150 mm AstroPhysics. 13 July 2001. Image by Robert Kuberek of Valencia, California)

(e.g., comparison to named atlases, observation the previous night, etc.) and whether you have checked for known supernovae and/or minor planets in the area.

- If you are a new contributor to the CBAT, please provide some background information regarding your observing experience.

The Discovery Report Form can be found online at http://cfa-www.harvard.edu/iau/DiscoveryForm.html

7.4 Occultations

Occultations occur when the Moon, a planet or its satellite, or an asteroid pass in front of a star. Such events have helped astronomers learn much about the shape of the lunar limb, about extinction caused by the upper layers of planetary atmospheres, and about the shape of asteroids. And never forget that on 10 March 1977, the rings of Uranus were discovered during an occultation of a ninth magnitude star.

Grazing occultations

The most valuable of all occultations are known as grazing occultations. This is when the solar system body's edge just barely covers (or doesn't cover) the star. For lunar occultations, grazes are even more valuable as recent spacecraft (notably Clementine) did not fully map the Moon's polar regions.

The edges of the shadow paths of occultations are known as the northern limit and southern limit. Within a kilometer or two of either limit of a lunar occultation's predicted path, you might see the star wink off and on several times as it passes behind hills and valleys near the Moon's poles. Observers spaced across this graze zone will time different sequences of events, which can later be analyzed to map the profile of the lunar limb.

Planetary grazes are of less value. Decades ago, these events were watched with care from large observatories. Once the data were compiled, general characteristics of a planetary atmosphere were deduced. Since the advent of planet-exploring spacecraft, these types of occultations have become less worthy of larger facilities.

More important are observations of occultations by asteroids. These objects are small, their paths are narrow, and occultations of even reasonably bright stars by asteroids are infrequent.

Important occultation tips

- Know where you are going! Visit the graze site days in advance as well as the meeting place that you will meet at right before the graze, and after if it applies. If you are unfamiliar with the area, be sure to take a detailed map with you.
- Check your equipment the day prior to the event. Carry spare batteries and blank tapes. Also, don't forget the information related to the actual graze: time, duration, area of the Moon where the graze will occur, etc.
- Be certain that you have a way to record the time. A short wave receiver capable of picking up several time signal

Occultation of Venus. (26 Dec 1978, Manhasset, NY. Celestron 8, 40 mm eyepiece projection (f/25), 1 second on Kodachrome 64. Image by David Healy, Sierra Vista, Arizona)

stations is perfect.
- Know the star field and identify, without doubt, the star that will be occulted. Check this several times prior to the event. Print out a detailed map oriented to your telescope/eyepiece field of view.
- If you are going to be observing an occultation by the Moon or a bright planet, the appearance of these objects may change the star field. Some stars visible without the

Moon or planet may not be seen. Be aware of this possibility. Also, if the occultation is by an asteroid or by Pluto, you may not see the occulting body prior to the event. The star will simply disappear. Great concentration is required for such events as there will be no bright body creeping up on the star to provide a warning. The same care must be taken when a star reappears from behind the Moon.

- Arrive early. You probably know how long setting up your equipment will take. Add at least 30 minutes to this, just in case. Plan on beginning your observations at least 10 minutes prior to the actual occultation. For planetary occultations, begin observing 30 minutes prior. The reason for this is that you may just catch an occultation of the star by one of the planet's moons.

- Record the position of your telescope. I suggest you use a GPS receiver. The International Occultation Timing Association (IOTA) needs 16 meter accuracy in both latitude and longitude. In arcseconds, this is equivalent to 0.5″ for latitude, and 0.5″ / (cos (lat)) for longitude. Since 1 May 2000, GPS Selective Availability has been turned off. This was a great change. Typical scatter over a 24-hour period when Selective Availability was active caused 95% of the points to fall within a radius of 45.0 m. This was simply not good enough for serious occultation work. However, once Selective Availability was disconnected, 95% of the points fell within a radius of 6.3 m.

- If you do not have a GPS receiver, quite precise measurements of your observing location can be attained in the "old-fashioned" way. Take careful measurements on a topographic map of the area. If possible, choose the most detailed maps with the best scale. Within the US, a good example is a 7.5 by 7.5 minute (1:24 000) scale USGS topographic map using a fine scale (such as millimeter or 30 to the inch) and reading to one-tenth of the divisions; using a magnifying glass helps. Note that if you are using maps of other countries with a 1:50 000 scale you will need to be a little more careful to achieve the necessary accuracy.

- Record all other relevant data. Knowing the cloud cover, wind speed and direction, temperature, transparency, and seeing (plus any unusual circumstances) will help you to better evaluate the occultation later.

- If observing the occultation visually, attempt to time the event to the nearest 0.2 second. If videotaping the occultation, strive for a timing accuracy of 0.03 second.

- Plan out how you will record disappearance and reappearance. Some observers call out a word such as "Now!" or the letters "D" and "R" signifying the two events. Others use a "clicker" able to put out a single, loud CLICK! (Note: the old clicker that I had actually put out two clicks, the latter when you relaxed the tension on it. I merely ignored the second click.)

- Always fill in a report form. This is to be done whether or not you have a positive sighting.

An occultation of Venus by the Moon. (26 Dec 1979, Manhasset, NY. Celestron 8, f/5, 1/60 second, Kodachrome 64. Image by David Healy, Sierra Vista, Arizona)

Videotaping occultations

The best way to observe an occultation is on video. Whether you attach your video camera directly to your telescope or hand-hold it up to the eyepiece for occultations of brighter stars, a video will provide a way to make much more accurate timings.

If you are going to aim a camcorder down the eyepiece of a telescope, start with your telescope's lowest-power, widest-field eyepiece. Use your camcorder's zoom to achieve focus. Try for the best illumination and field size. A little vibration from hand-holding is acceptable. Playing the video back in slow-motion or, better yet, single-frame mode will give you the most accurate timing. You simply watch the star until it disappears. This will allow you to time the occultation to 0.03 (1/30) second.

If your video camera has a quality zoom lens, you may not even need a telescope for lunar occultations of bright stars. Second magnitude stars have been captured disappearing on the bright limb and reappearances of stars as faint as fourth magnitude have been videotaped at the Moon's dark limb.

In the US, a great service is offered by Walt "Rob" Robinson (a long-time friend of mine), who will compute occultation predictions to eighth magnitude for any location in North America. Send Rob your accurate latitude, longitude, and altitude, along with either an e-mail address or a long, self-addressed, stamped envelope. His mailing address is 515 W. Kump, Bonner Springs, KS, 66012, USA. He can also be reached by e-mail at: webmaster@lunar-occultations.com. Alternately, check out his excellent website, on the internet at http://www.lunar-occultations.com

Occultation predictions for Europe are available from Hans-Joachim Bode, Bartold-Knaust-Strasse 8, D-30459 Hanover, Germany. A great European website regarding occultations is

http://astro1.physik.uni-siegen.de/uastro/occult/

7.5 Nebulae

The word nebula is Latin and it means cloud. So, when we talk about a nebula (plural = nebulae), we are referring to a cloud of gas and dust in space. Several types of nebulae exist. I have chosen to break them down into three categories: bright, dark, and planetary.

Bright nebulae

Bright (or diffuse) nebulae are frequently places of star formation. If stars have begun to form, some of them may be hot enough that their radiation excites the gas of the nebula, causing it to shine. This type of bright nebula is known as an emission nebula. The process by which the gas (hydrogen) is excited is called ionization. Neutral, un-ionized, hydrogen is designated HI. Ionized hydrogen is represented as HII. So emission nebulae are often referred to as HII regions. (This is an example of a singly-ionized element. Double and triple

NGC 6559, an emission nebula in Sagittarius. (12.5″ Ritchey-Chretien, f/7.5, total exposure of 82 minutes. Image by Robert Gendler, Avon, Connecticut)

The fabulous Eagle Nebula, M16. (Image by Adam Block/NOAO/AURA/NSF, using a 0.4 m Meade LX200 telescope)

ionizations are also possible; for example, an OIII filter allows light from doubly-ionized oxygen to pass through.)

If the stars are not hot enough to cause ionization, their light is reflected by the dust and can be seen as a reflection nebula. Most of these nebulae look blue for the same reason our daytime sky appears blue – the light is being scattered throughout the nebula. A lot of nebulae have both emission and reflection components, as some of the gas may be too far from the stars to be ionized by their radiation.

You could probably guess which was the first bright nebula discovered. Yes, it was M42, the Orion Nebula. In fact, this object was discovered about a year after Galileo first turned his telescopes to the stars. The discoverer was Nicolas Claude Fabri de Peiresc (1580–1637). The first reflection nebula discovered was M78, also in Orion, by Pierre Méchain (1744–1804) in 1780. Other diffuse nebulae in Messier's catalog are M8, M16, M17, M20, and M43. M16 is actually a double object – a star cluster (NGC 6611) and a famous nebula known as the Eagle Nebula (IC 4703). M45, the Pleiades, also contains a diffuse reflection nebula.

When you are using your eye, the Eagle isn't much to look at. The Swan, however, is a sight to behold. It resembles a checkmark in a small telescope. Both are gaseous nebulae, enormous clouds of glowing hydrogen gas. In clouds like these, stars are condensing out of primordial hydrogen as

you read these lines. A similar cloud gave birth to our own Sun and some of the bright, nearby stars that can be seen with the unaided eye.

Dark nebulae

Dark nebulae are clouds of dust and cold gas which can be seen because they obscure the light coming from stars or bright nebulae behind them. The shapes of dark nebulae are among the strangest in the sky. Some dark nebulae, for example, the Horsehead Nebula in Orion, are small and difficult to see even with large telescopes. Others, such as the Coal Sack in Crux, are large and easy to see with the naked eye.

The darkness of these objects is due to dust grains within the clouds and to the presence of cold hydrogen molecules. Very cold. The internal temperature of these clouds is only about 10 K. The largest of the dark nebulae, called molecular clouds, are huge areas of star formation. The clouds must be cold or the gas will not condense into stars. Luckily, the energy of collapse is radiated away, mainly by the dust grains within. At some point, gravity overwhelms the opposing forces in a section of the cloud and a star forms.

The first catalog of dark nebulae was prepared by the American astronomer Edward Emerson Barnard (1857–

Table of selected bright nebulae

RA (J2000) h m	Dec. (J2000) Deg(°)	Diameter Arcmin (')	Area degree^2	Name
18 04	−24 20	45 × 30	0.289	NGC 6523
18 02	−23 00	20 × 20	0.085	NGC 6514
18 10	−23 59	15 × 10	0.039	NGC 6559
18 11	−23 44	20 × 5	0.025	IC 1274
18 17	−19 44	4 × 3	0.003	NGC 6590
18 17	−19 39	4 × 3	0.003	NGC 6589
18 21	−15 59	40 × 30	0.137	NGC 6618
18 19	−13 49	120 × 25	0.268	NGC 6611
18 18	−11 59	60 × 30	0.345	S 54
18 34	−04 58	2 × 2	0.001	S 61
19 02	+02 09	2 × 1	0.001	S 71
19 11	+16 50	1 × 1	0.001	S 80
19 55	+27 17	1 × 1	0.001	S 93
19 55	+29 18	1 × 1	0.001	S 95
20 00	+35 18	20 × 15	0.042	C 173
20 13	+38 19	20 × 10	0.024	NGC 6888
20 28	+40 00	45 × 20	0.175	IC 1318B
20 17	+41 49	45 × 25	0.211	IC 1318A
20 25	+42 19	3 × 3	0.001	NGC 6914
20 26	+42 23	3 × 3	0.001	NGC 6914
20 34	+45 40	13 × 13	0.028	C 181
20 51	+44 21	25 × 10	0.036	IC 5067
20 56	+47 24	7 × 7	0.007	IC 5076
21 54	+47 14	10 × 10	0.018	IC 5146
21 02	+68 12	10 × 8	0.018	NGC 7023
21 42	+66 04	2 × 2	0.001	NGC 7129
22 47	+58 01	25 × 20	0.092	NGC 7380
23 15	+59 56	3 × 3	0.001	S 157
23 14	+61 29	8 × 7	0.011	NGC 7538
23 20	+61 11	15 × 8	0.025	NGC 7635
00 53	+56 36	35 × 30	0.148	NGC 281
00 59	+60 56	10 × 3	0.007	IC 63
01 30	+58 22	10 × 3	0.004	S 188
02 26	+61 59	12 × 12	0.028	IC 1795
04 03	+51 19	6 × 9	0.011	NGC 1491
04 41	+50 26	5 × 5	0.004	NGC 1624

RA (J2000) h m	Dec. (J2000) Deg(°)	Diameter Arcmin (')	Area degree^2	Name
04 01	+36 38	160 × 40	0.982	NGC 1499
03 46	+24 09	60 × 40	0.834	NGC 1432
05 13	+37 23	2 × 2	0.002	S 228
05 31	+34 12	4 × 4	0.002	NGC 1931
05 35	+22 02	8 × 4	0.003	NGC 1952
06 10	+20 29	40 × 30	0.229	NGC 2175
05 46	+09 03	3 × 2	0.002	C 59
06 33	+10 21	2 × 2	0.001	NGC 2247
06 33	+10 10	2 × 2	0.001	NGC 2245
06 41	+09 54	10 × 7	0.014	NGC 2264
05 06	−03 21	2 × 2	0.002	NGC 1788
06 39	+08 43	2 × 1	0.001	NGC 2261
05 47	+00 01	10 × 10	0.007	NGC 2064
05 41	−01 31	8 × 5	0.011	IC 431
05 42	−01 34	10 × 10	0.021	IC 432
06 33	+04 58	80 × 60	1.042	NGC 2238
05 42	−02 19	90 × 30	0.504	IC 434
05 42	−02 19	10 × 8	0.010	NGC 2023
05 35	−05 28	90 × 60	0.923	NGC 1976
05 36	−06 44	2 × 2	0.002	NGC 1999
06 07	−06 23	2 × 2	0.001	NGC 2170
06 09	−06 21	3 × 2	0.002	NGC 2182
07 08	−04 15	1 × 1	0.001	S 288
07 04	−10 25	20 × 20	0.081	IC 2177
07 05	−12 16	10 × 10	0.021	C 90
07 52	−26 28	8 × 7	0.007	NGC 2467
16 29	−26 37	60 × 40	0.546	IC 4606

Notes

This table is a selection from the *Catalogue of British Nebulae*, compiled by Beverly T. Lynds. The total number of nebulae listed is 1125 and may be found in *Astrophys. J. Suppl.*, 12, p.163(1965).

All nebulae listed here have a brightness of 1 (the brightest class).

In the last column, IC = Index Catalog; S = Sharpless HII Regions, 1959; C = Catalog of Diffuse Galactic Nebulae, Cederblad, 1959. Where an object has two designations, such as M16 (NGC 6611 and IC 4073), the NGC designation is used.

1923): "catalog" of 349 Dark Objects in the Sky in *A Photographic Atlas of Selected Regions in the Milky Way* (Carnegie Inst., Washington, DC, 1927). Note that, due to its far southerly declination, Barnard did not catalog what is perhaps the most famous dark nebula in the sky, the Coal Sack. It lies in the constellation of Crux at RA 12^h 53^m, Dec. −63°. It measures 400' × 300' in size.

The darkest dark nebulae

LDN	RA (J2000) h	m	Dec. (J2000) Deg (°)	Arcmin (′)	Area (degree²)
1506	04	20.0	+25	17	0.334
1535	04	35.5	+23	54	0.111
1517	04	55.2	+30	35	0.051
1544	05	04.1	+25	14	0.109
1627	05	46.6	+00	01	0.063
1622	05	54.6	+02	00	0.122
1757	16	31.9	−19	36	0.061
1709	16	33.0	−23	46	0.099
43	16	34.5	−15	50	0.070
260	16	47.7	−09	35	0.074
158	16	47.8	−14	05	0.056
204	16	47.8	−12	05	0.167
162	16	49.1	−14	15	0.124
65	17	13.0	−21	54	0.088
100	17	16.0	−20	53	0.075
219	17	39.5	−19	47	0.084
513	18	10.6	−01	33	0.127
429	18	16.7	−08	19	0.068
570	18	26.6	−00	28	0.066
557	18	38.6	−01	47	0.181
530	18	49.7	−04	47	0.124
581	19	07.4	−03	55	0.072
673	19	20.9	+11	16	0.199
704	19	26.8	+13	46	0.097
694	19	40.7	+10	57	0.109
1262	23	27.0	+74	17	0.066

Notes

This table is a selection from the Catalog of Dark Nebulae, compiled by Beverly T. Lynds. The total number of nebulae listed is 1791 and may be found in *Astrophys. J. Suppl.*, 7, p. 1 (1962).

All nebulae listed here have an opacity of 6 (the darkest class). The selected nebulae also have an area greater than 0.05 degree². The estimates were made by Lynds based on a comparison of the neighboring fields for the particular Palomar Observatory Sky Survey (POSS) photograph on which the cloud appeared. Thus the range in declination is from +90° to −33°. Both the red and the blue prints were used for this comparison. A cloud had to be visible on both photographs in order to be recorded.

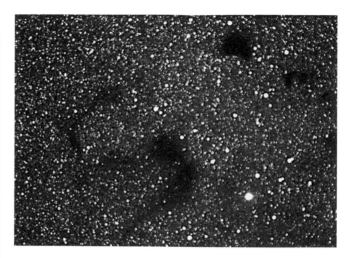

This is the Snake Nebula in Ophiuchus. Celestron Fastar 8 at f/1.95, PixCel 237 CCD. This is a 120-second image with the AP900 mount. Unguided at −3°C. (Image by Chris Anderson of Kentucky)

Observing project – The Horsehead Nebula

Once upon a time, the 33rd object in E. E. Barnard's catalog was considered one of the ultimate observing tests. Actually seeing B33 moved amateurs from "good" to "great." Well, times change and technology changes. Today, the Horsehead Nebula is still a test, but seeing it won't make you a legend. Here are some tips to help you locate this elusive object.

(1) Use the largest scope you can beg, borrow, or buy. The Horsehead Nebula has been seen in apertures as small as 125 mm, but not easily.

(2) Dark adapted eyes are crucial, especially when using medium-sized telescopes. (I have found that it is not that important if you're using a telescope with 750 mm aperture.) Some observers use a black cloth over their head to minimize stray light. Related to this, make certain you position ζ Ori (Alnitak) out of the field of view.

(3) Use a hydrogen beta filter. This filter is a superb help for locating the Horsehead. You can also use it on the California Nebula in Perseus and the Cocoon Nebula in Cygnus. That's pretty much it, however.

(4) A star chart will be handy, but an image (a copy of a photograph, etc.) of the area is better. Use the stars around the Horsehead to define its shape. For comparison purposes, pay particular attention to identifiable figures of stars. Volume 2 of *Burnham's Celestial Handbook* (Dover Publications, NY, 1978) has five photographs of B33. It has been suggested that the one taken by David Healy of Sierra Vista, Arizona (on page 1344), is the best for visual observation.

(5) Begin by searching for the nebulosity IC 434. If you can't see this steadily, you won't see the Horsehead, which is fairly low in contrast against this nebula.

(6) The Horsehead is not a miniscule object. It is larger than you think, especially at magnifications of 150× and up.

(7) The Horsehead is very dependent on sky conditions. If the seeing is bad, try for it on a different night.

The dark nebula Van den Bergh 142 in Cepheus. (12.5" Ritchey-Chretien, f/7.5 and a Finger Lakes Instrumentation IMG1024 camera. Two-frame Mosaic. Image by Robert Gendler, Avon, Connecticut)

Planetary nebulae

A planetary nebula is the most common end product of stellar evolution. Stars as massive or slightly more massive than the Sun eject mass in the red giant stage, near the end of the star's life. In this stage, helium is fusing to carbon at a higher temperature than that of hydrogen-to-helium fusion.

After a short period of time (well, astronomically speaking), the inner core collapses and the star pulsates, ejecting much of its outer envelope. This leaves a small star which is turning into a white dwarf. The ejected envelope becomes a spherical shell of cooler thin matter expanding into space at 10–40 km/s – a planetary nebula. The nebula has 10–20% the mass of the Sun.

The star (now called the central star of the nebula) is extremely hot, reaching temperatures of 200 000 K, emitting a large amount of the ultraviolet radiation. It is this radiation

Dark nebulae are interesting to observe. This one is number 86 in Edward Emerson Barnard's catalog, and has the designation B86. (Image by Steven Juchnowski, Balliang East in the State of Victoria, Australia. RGB: 6 x 300 s R, 6 x 300 s G, 6 x 450 s B through an ST-7E on a Celestron 11)

which ionizes the gas of the expanding planetary nebula and causes it to glow. Planetary nebulae have short lives, at least visually.

As the shell expands, the white dwarf can excite the gas, but only out to a certain distance. So after 10 000–50 000 years, the planetary nebula stops glowing. It's still expanding but we can no longer see it. Most astronomers agree that a planetary nebula is the fate of our Sun in approximately five billion years.

There are roughly 1000 planetary nebula known in our neighborhood of the Galaxy. A typical planetary nebula is less than one light-year across.

Great objects from Barnard's catalog

Barnard no.	Common name	RA h m	Dec Deg (°)	Size Arcmin (′)	Constel- lation
B33	Horsehead	05 41	−02.5	6×4	Ori
B42	Rho Ophiuchi	16 29	−24.3	12×12	Oph
B65/6/7	Pipe (stem)	17 21	−26.8	300×60	Oph
B72	Snake	17 24	−23.6	30×30	Oph
B78	Pipe (bowl)	17 33	−25.7	200×140	Oph
B86	Ink Spot	18 03	−27.8	5×3	Sgr
B87	Parrot's Head	18 04	−32.7	12×12	Sgr
B142/3	Barnard's E	19 41	+11.0	110×80	Aql
B348/9	Cygnus Rift	20 37	+42.2	240×240	Cyg

The Eastern Part of the Pipe Nebula. (SBIG ST7e NABG camera with the SBIG CLA Nikon lens adapter on a Vixen GP-DX EQ mount. 70 mm Nikon lens, 15-minute exposure. A set of 2" RGB filters were manually switched between shots. (Image by Chris Woodruff, of Valencia, California.)

The Horsehead Nebula in Orion. (Image by Mel and Betty Peterson and Adam Block/NOAO/AURA/NSF, using a 0.4 m Meade LX200 telescope)

History

On 12 July 1764, Charles Messier (1730–1817) discovered the first planetary nebula and made it number 27 on his list. We know it as the Dumbbell Nebula. It took 15 years before the next planetary was discovered. The object was the famous Ring Nebula, and it was found in January 1779, by Antoine Darquier (1718–1802), who compared it to a "fading planet."

It was none other than William Herschel who christened these objects planetary nebulae due to the fact that they resembled, in shape and color, Uranus, the planet he had discovered only a few years before. It was also Herschel who, in 1790, discovered what we now refer to as NGC 1514, in Taurus. Herschel noted a bright central star and made the connection between it and the bright nebulous material.

On the next page is the most widely used system for classifying planetary nebulae was devised in 1934 by the Russian astronomer Boris Alexandrovich Vorontsov-Velyaminov (1904–1994). More complex structures may be characterized by Vorontsov-Velyaminov combinations. For example, a planetary nebula showing both a ring and a disk would be classed 4+2. One with two rings would be 4+4.

Color of planetaries

In 1864, while studying a spectrum of NGC 6543, the English astronomer William Huggins (1824–1910) found that most of the light was being emitted as two wavelengths – 495.89 nm and 500.68 nm. To astronomers of the time,

The Medusa Nebula. 25 Jan 2001. (Photo by Shane Larson and Mike Murray of Bozeman, Montana, and Adam Block/NOAO/AURA/NSF, using a 0.4 m Meade LX200 telescope)

M97, the Owl Nebula. (Image by Adam Block/NOAO/AURA/NSF, using a 0.4 m Meade LX200 telescope)

The Vorontsov-Velyaminov classification scheme for planetary nebulae

1	Stellar image
2	Smooth disk
2a	Smooth disk, brighter towards center
2b	Smooth disk, uniform brightness
2c	Smooth disk, traces of a ring structure
3	Irregular disk
3a	Irregular disk, very irregular brightness distribution
3b	Irregular disk, traces of a ring structure
4	Ring structure
5	Irregular form, similar to a diffuse nebula
6	Anomalous form

this emission was a puzzle, caused, apparently, by an unknown substance or element. The substance was named nebulium, since it was only seen in the spectra of nebulae. It was identified in 1928 by the American physicist Ira S. Bowen (1898–1973) as being produced by "forbidden" (by the laws of chemistry) transitions between ionized states of already-identified elements. Specifically, 495.89 nm and 500.68 nm together form the light of doubly-ionized oxygen, OIII. These two spectral lines, along with the hydrogen beta line at 486.1 nm, give these nebulae their characteristic blue, green, bluish-green, or greenish-blue color.

What color does a particular planetary appear to you? This is a little tricky to answer. David Knisely of the Prairie Astronomy Club in Lincoln, Nebraska, sums it up well.

One problem is that the wavelength of the dominant OIII lines is close to the "border" between perceived green color and perceived blue color. Thus, even a very

NGC 5189. A very strange looking planetary nebula in Musca nicknamed the "Spiral Planetary". (The colour (LRGB) data of this image was taken with a C14 and ST-9E courtesy of Steve Crouch, the L data with a C11, ST-7E and AO-7 guider by Steven Juchnowski, Balliang East in the State of Victoria, Australia. The image is technically L(LRGB))

slight shift in a person's perceived color spectral response of the eye/brain could cause the color to appear either somewhat green or bluish. Another factor for a few planetaries is the amount of H-beta emission present. It usually isn't much in comparison to the booming OIII lines, but for a few planetary nebulae, it might be just enough to "shift" the color perception a bit towards the blue.

I would add that I believe that the altitude at which you observe a planetary nebula also affects its perceived color, though not to a large extent. Atmospheric reddening is real and any planetary viewed at a higher altitude will be slightly bluer than when it is viewed nearer the horizon.

The visually-elusive planetary nebula Jones-Emberson 1 in Lynx. (Celestron Fastar 8 at f/1.95, PixCel 237 CCD. This is a 600-second image consisting of five 120-second unguided exposures "Track and Accumulate" using the AP900 mount with Periodic Error correction engaged, 27.77°C. Image by Chris Anderson of Kentucky)

Planetary nebulae in clusters

Of the several hundred globular clusters in our galaxy, planetary nebulae have been discovered in only four of them. These planetaries are Pease 1 in M15 (Pegasus), IRAS 18333-2357 in M22 (Sagittarius), JaFu1 in Palomar 6 (Ophiuchus) and JaFu2 in NGC 6441 (Scorpius). (Note: JaFu refers to the two discoverers, George Jacoby and L. Kellar Fulton.) Observation of these planetaries, especially JaFu1 and JaFu2, are considered the ultimate planetary challenge by some amateurs. Comet discoverer and planetary nebula enthusiast Doug Snyder has dedicated some internet space to helping more amateurs see these planetaries. Find the specific page at http://www.blackskies.com/pn_gc_challenges.htm

Because planetary nebulae occur at the end of the lives of stars, only one has been found in an open cluster, which are generally young objects and which tend to break up before planetary nebulae can form. (If you are an advanced amateur, you've probably guessed "the planetary in M46." Actually, that appears to be a chance lineup.) The object is NGC 2818, which has been discovered to be a member of a rather old open cluster, NGC 2818A.

Observing planetary nebulae

The most numerous group of planetary nebulae are the so-called stellar planetaries. These nebulae have very small apparent sizes. More than half of all planetaries fall into this class. The majority of the rest have diameters less than one arcminute. It seems a general rule that the most interesting planetary nebulae have apparent diameters between 20″ and 40″.

Arild Moland, of Oslo, Norway, offers the following excellent tips for viewing small planetaries.

(1) To successfully identify planetary nebula too small to show a disk, use an OIII filter to "blink" compare the filtered and unfiltered field of view.

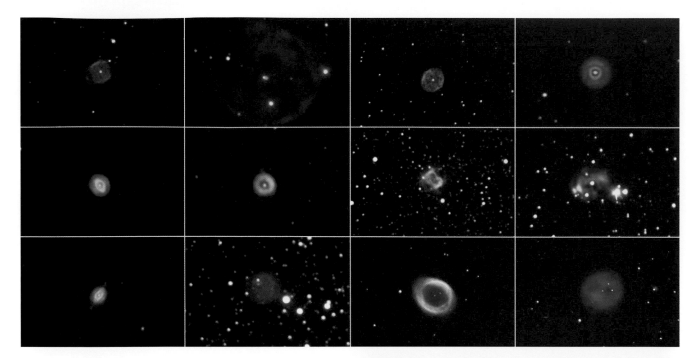

Planetary nebulae galore! All these images were made by Al Kelly of Danciger, Texas. (Visit Al's CCD Astronomy website at http://www.ghgcorp.com/akelly.)

Hold the filter between your eye and the eyepiece (sufficient eye relief is required!), and move the filter in and out of view. The "star" which retains its brightness or is brighter than the other stars when viewed through the filter, is the nebula.

(2) To successfully identify a planetary nebula too small to show a disc, use a diffraction grating. Screw the grating into the end of the eyepiece or hold it between the eye and eyepiece. The nebula — emitting light in discrete wavelengths only — will not display a continuous spectrum like the true stars in the field. Hence, the nebula will stand out from the crowd.

Kent Wallace, of Atascadero, California, contributes the following tip:

When using an OIII, UHC or H-beta filter to blink a field for a planetary nebula it is very important to have a dark cloth over your head to block any stray light from hitting the front surface of the filter. These filters are so reflective that starlight shining over your shoulder will significantly degrade the image of the object you are trying to observe. I use a 3 foot × 3 foot piece of dark cloth I got at the swap meet. It is one of the cheapest useful accessories that I own. Also it is useful for general observing by blocking stray light. I have heard that you gain a half magnitude of deeper observing by using a dark cloth. I don't know if this is true but using a dark cloth definitely helps me.

Those with medium to large telescopes may wish to try Eric

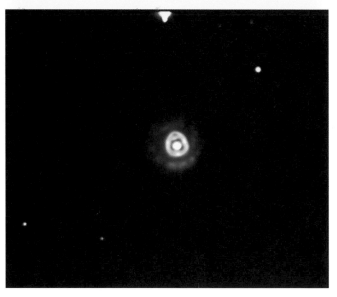

The Eskimo Nebula, NGC 2392. (Image by Adam Block/NOAO/AURA/NSF, using a 0.4 m Meade LX200 telescope)

Honeycutt's advanced lists of planetaries, available online at http://www.icplanetaries.com/advanced.html

One of the finest internet sites dealing with planetary nebulae is Doug Snyder's, online at http://www.blackskies.com

Finally, for fun, Kent Wallace has compiled a list of nicknames for planetary nebulae and possible proto-planetary nebulae. There is nothing "official" about this list. Some of these names (such as the Ring Nebula) are pretty universal while others (such as the Water Fountain Nebula) are not; some have more than one nickname. Now you, too, can observe this list and get to know these great planetaries by name, or make up your own names for them.

Planetary nebula observing project – the central star of M57

Locating a planetary nebula in one of the globular clusters is a real challenge. A tough challenge of a different sort is to see the central star of the Ring Nebula. The central star is easily revealed in larger apertures with high magnification, but with medium-sized telescopes it can be difficult to see. For lots of help accomplishing this feat, see the following internet page of the Huachuca Astronomy Club of Sierra Vista, Arizona: http://c3po.cochise.cc.az.us/astro/deepsky02.htm

A few additional notes, if I may. Better transparency and seeing happen at higher altitudes and, take my word for it, to see the central star you need great seeing. Wait until M57 is at least 60° high in the sky. Use a magnification in excess of 300×. What you are trying to do is emphasize the contrast between the star and the nebulosity within the ring. Make a note of the faintest field star you are able to see. If you cannot see the mag. 15.7 star just off the nebula's edge, you probably won't see the central star. Note the appearances of field stars. If their images are steady, the seeing is good. Believe it or not, at times you can predict the appearance of the central star simply by noting the seeing effects on field stars.

Observing Note: If your telescope has a mirror of 500 mm or larger, try to see a second, fainter star within the ring. Use high magnification.

Planetary nebulae and possible proto-PN nicknames
List compiled by Kent Wallace, Rev 1, 25 March 2002

Nickname	Catalog name	RA (2000) h	m	Dec. (2000) Deg (°)	Arcmin (′)
Ant Nebula	Mz 3	16	17.2	−51	59
Apple Core Nebula	NGC 6853 (M27)	19	59.6	+22	43
Baby Eskimo	IC 3568	12	33.1	+82	34
Barbell Nebula	NGC 650/651	01	42.3	+51	35
Blinking Planetary	NGC 6826	19	44.8	+50	32
Blue Flash Nebula	NGC 6905	20	22.4	+20	06
Blue Planetary	NGC 3918	11	50.3	−57	11
Blue Snowball	NGC 7662	23	25.9	+42	32
Boomerang Nebula	IRAS 12419-5414	12	44.8	−54	31
Bowtie Nebula	NGC 40	00	13.0	+72	31
Box Nebula	NGC 6309	17	14.1	−12	55
Bug Nebula	NGC 6302	17	13.7	−37	06
Butterfly Nebula	M 2-9	17	05.6	−10	09
Calabash Nebula	CRL 5237	07	42.3	−14	43
Campbell's Star	BD +303639	19	34.8	+30	31
Campbell's Hydrogen star	BD +303639	19	34.8	+30	31
Cat's Eye Nebula	NGC 6543	17	58.6	+66	38
CBS Eye	NGC 3242	10	24.8	−18	39
Cheerio Nebula	NGC 6337	17	22.3	−38	29
Cheese Burger Nebula	NGC 7026	21	06.3	+47	51
Cloverleaf Nebula	IRAS 19477 +2401	19	49.9	+24	09
Clown Face Nebula	NGC 2392	07	29.2	+20	55

Nickname	Catalog name	RA (2000) h	m	Dec (2000) Deg (°)	Arcmin (′)
Cork Nebula	NGC 650/651	01	42.3	+51	35
Cotton Candy	CRL 6815	17	18.3	−32	27
Crescent Nebula	NGC 6445	17	49.3	−20	01
Crystal Ball Nebula	NGC 1514	04	09.3	+30	47
Diamond Nebula	NGC 3242	10	24.8	−18	39
Double Bubble Nebula	NGC 2371/2372	07	25.6	+29	29
Double Headed Shot	NGC 6853 (M27)	19	59.6	+22	43
Dumbbell Nebula	NGC 6853 (M27)	19	59.6	+22	43
Etched Hourglass Nebula	MyCn 18	13	39.6	−67	23
Egg Nebula	CRL 2688	21	02.3	+36	42
Eight Burst Nebula	NGC 3132	10	07.0	−40	26
Emerald Nebula	NGC 6572	18	12.1	+06	51
Eskimo Nebula	NGC 2392	07	29.2	+20	55
Fetus Nebula	NGC 7008	21	00.5	+54	33
Footprint Nebula	M 1-92	19	36.3	+29	33
Frosty Leo Nebula	IRAS 09371+1212	09	39.9	+11	59
Ghost of Jupiter	NGC 3242	10	24.8	−18	39
Gomez's Hamburger	IRAS 18059-3211	18	09.2	−32	10
Green Rectangle	NGC 7027	21	07.0	+42	14

Planetary nebulae and possible proto-PN nicknames
List compiled by Kent Wallace, Rev 1, 25 March 2002 (*cont.*)

Nickname	Catalog name	RA (2000) h	m	Dec. (2000) Deg (°)	Arcmin (')	Nickname	Catalog name	RA (2000) h	m	Dec. (2000) Deg (°)	Arcmin (')
Headphones Nebula	JE 1	07	57.8	+53	25	Rotten Egg Nebula	CRL 5237	07	42.3	−14	43
Helix Nebula	NGC 7293	22	29.6	−20	50	Sakurai's Object	V 4334 Sgr	17	52.5	−17	41
Hourglass Nebula	MyCn 18	13	39.6	−67	23	Saturn Nebula	NGC 7009	21	04.2	−11	22
Hubble Double Bubble	Hb 5	17	47.9	−30	00	Seahorse Nebula	K 3-35	19	27.7	+21	30
Lemon Slice Nebula	IC 3568	12	33.1	+82	34	Silkworm Nebula	CRL 5385	17	47.2	−24	13
Little (or Mini) Dumbbell Nebula	NGC 650/ 651 (M76)	01	42.3	+51	35	Skull Nebula	NGC 246	00	47.1	−11	52
						Snowglobe Nebula	NGC 6781	19	18.5	+06	32
Little Gem Nebula	NGC 6818	19	44.0	−14	09	Spare Tyre Nebula	IC 5148/5150	21	59.6	−39	23
Little Ghost Nebula	NGC 6369	17	29.3	−23	46	Spindle Nebula	IRAS 17106-3046	17	13.9	−30	50
Magic Carpet Nebula	NGC 7027	21	07.0	+42	14	Spiral Planetary Nebula	NGC 5189	13	33.5	−65	68
Mask Nebula	MRSL 252	15	09.4	−55	34	Spirograph Nebula	IC 418	05	27.5	−12	42
Medusa Nebula	Abell 21	07	29.0	+13	15	Southern Crab Nebula	He 2-104	14	11.9	−51	26
Minkowski's Butterfly	M 2-9	17	05.6	−10	09	Southern Owl Nebula	K 1-22	11	26.7	−34	22
Minkowski's Footprint	M 1-92	19	36.3	+29	33	Southern Ring Nebula	NGC 3132	10	07.0	−40	26
Owl Nebula	NGC 3587 (M97)	11	14.8	+55	01	Stingray Nebula	He 3-1357	17	16.4	−59	29
						Sunflower Nebula	NGC 7293	22	29.6	−20	50
Oyster Nebula	NGC 1501	04	07.0	+60	55	Turtle Nebula	NGC 6210	16	44.5	+23	48
Peanut Nebula	CW Leo	09	47.9	+13	17	Walnut Nebula	IRAS 16594-4656	17	03.2	−47	00
Phantom Streak	NGC 6741	19	02.6	−00	27						
Raspberry Nebula	IC 418	05	27.5	−12	42	Water Fountain Nebula	IRAS 16342-3814	16	37.7	−38	20
Red Rectangle	CRL 915	06	20.0	−10	39						
Red Spider Nebula	NGC 6537	18	05.2	−19	51	Water Lily Nebula	IRAS 17245-3951	17	28.1	−39	54
Ring Nebula	NGC 6720 (M57)	18	53.6	+33	02	Westbrook Nebula	CRL 618	04	42.9	+36	07
						White Eyed Pea	IC 4593	16	11.7	+12	04

Robert Gendler (Avon, Connecticut) said the following. "The California Nebula in Perseus was imaged with a 300 mm Nikon lens attached to my Santa Barbara Instrument Group ST10 CCD camera. Many of my images are LRGB's, meaning that a luminance image, containing the detail and resolution data, was combined with the color data (RGB). In the case of the California Nebula, the luminance (L) consisted of both red-filtered and hydrogen alpha-filtered data because with emission nebulae all the detail is found at those wavelengths. Thus, the exposure is (HA+R) RGB = (90+70):10:10:20 minutes."

7.6 Star clusters

A star cluster is a group of stars loosely held together by gravity with a common center of mass. There are three types of star clusters, associations, open clusters and globular clusters. Associations contain on the order of tens of stars and in most ways resemble open clusters, except that they are poorer in star numbers. For this discussion, we will concentrate on the two other types of star clusters.

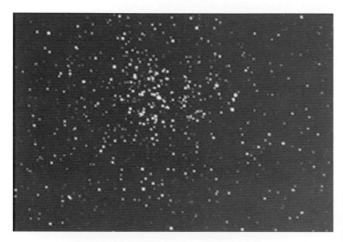

M37, an open cluster in Auriga. (Photo by Robert Kuberek of Valencia, California)

Open clusters

Located within the disk of our galaxy, the Milky Way, open clusters are also known as galactic clusters. They have been known since the earliest times. The Beehive (or Praesepe), the Pleiades and the Hyades have numerous historical references. Ptolemy mentioned Melotte 111 in Coma Berenices and M7 in Scorpius. Not until Galileo trained his telescope on the Beehive, however, was it discovered that these objects were made of individual stars.

Sparse open clusters contain less than a hundred stars while rich ones may have thousands. Open clusters form from nebulae, the beautiful clouds of dust and gas we can observe. All open clusters are relatively young objects, generally no more than a few hundred million years. The reason for this is that, given enough time, the stars in an open cluster disperse. This is due to gravitational interaction within the cluster. If we were able to follow a single open cluster from its formation, we would see stars dispersed throughout its entire galactic journey. Stars in open clusters are mainly hot and are rich in heavy elements (an astronomer's way of saying any elements other than hydrogen or helium). At the time of this writing, approximately 1700 open clusters are known.

The American astronomer Harlow Shapley (1885–1972)

originated a classification of open clusters. His system is quite simple, generally describing the richness and concentration of the cluster:

c – Very loose and irregular in shape
d – Loose and poor in stars
e – Intermediately rich in stars
f – Fairly rich in stars
g – Considerably rich in stars and concentrated

A more detailed classification system was introduced in 1930 by Robert Julius Trumpler (1886–1956) at the Lick Observatory. It has three parts. The first deals with the open cluster's concentration, the second with the range of magnitudes of the stars and the third with how rich the cluster appears.

NGC 1763/69, a cluster and nebulous region in Dorado. (LRGB: L= R+G+B, 600s R, 600s G, 900s B through an ST-7E on a Celestron 11. Image by Steven Juchnowski, Balliang East in the State of Victoria, Australia)

Part 1 – Concentration
  I   Detached; strong concentration toward center
  II Detached; weak concentration toward center
  III Detached; no concentration toward center
  IV Not well-detached from surrounding star field
Part 2 – Range of magnitudes
  1 Small range in brightness
  2 Moderate range in brightness
  3 Large range in brightness
Part 3 – Richness
  p Poor; less than 50 stars
  m Moderately rich; 50 to 100 stars
  r Rich; more than 100 stars

If the letter "n" follows the Trumpler class, it indicates that there is nebulosity associated with the cluster.

A combination object. NGC 281, an emission nebula and open cluster in Cassiopeia. (12.5" Ritchey-Chretien with a Finger Lakes Instrumentation IMG1024 CCD camera. LRGB of 40:10:10:10 minutes. Image by Robert Gendler, Avon, Connecticut)

A great reference for open clusters may be found on the internet at http://obswww.unige.ch/webda/navigation.html

(Scroll to the bottom and look for the box labeled "Archive Data Files." Click on "NGC.") Included on this site are a large number of scanned or plotted maps of open clusters. The extent of the clusters, their positions, and relative brightnesses may easily be found.

Observing open clusters

Open clusters are pretty. This makes them fun to look at by beginning amateur astronomers. But many advanced amateurs spend a great deal of time observing open clusters as well. As with the rest of amateur astronomy, you can take it as far as you choose.

Compared with other deep-sky objects, open clusters are relatively large. This means you will generally be using low-power eyepieces which provide wide fields. Some clusters are large enough and bright enough that binoculars will provide more satisfying views. I have observed hundreds of clusters through my 7 × 50 and 15 × 70 binoculars. My

The Beehive, Praesepe or M44 in Cancer, a cluster which can be easily seen with the naked eye from a reasonably dark location. (Photo by Robert Kuberek of Valencia, California)

M11, a rich open cluster in Scutum. (31 minutes on hypered Technical Pan 2415 film, Celestron 14 @f/7, 22 Oct 1989, Naco, Arizona. Image by David Healy, Sierra Vista, Arizona)

friend Bob Kuberek in Los Angeles has Miyauchi 20 × 100 mm binoculars with two sets of eyepieces. The first set provides a magnification of 20 and a huge 2.5° field of view. The "high-power" set renders 37× and a 1.8 field. This may be the ultimate open cluster instrument.

You can find interesting objects within the boundaries of open clusters. Take M46 for example. This rich open cluster in the constellation Puppis has a bright planetary nebula within it which a medium-sized telescope will easily reveal. Other clusters have planetaries, diffuse nebulae, even smaller clusters within or near the boundaries of the main cluster. In addition, double stars and interesting asterisms abound.

Take your time when observing an open cluster. Examine the field of view closely. Try to discern the members of the cluster as opposed to actual field stars. This is easy in many cases, but can be a real trick within the area of the Milky Way. Also, large-aperture telescopes can sometimes hinder the identification of cluster stars, making so many background stars visible that confusion ensues.

Observing Challenge: Try to see the nebulosity around the star Merope in the Pleiades.

Various apertures and open clusters

Jeff Medkeff of Sierra Vista, Arizona, has the following to say.

I've noticed that at least several of the open clusters in

the Herschel 400 list are real dogs with a 450 mm or 500 mm telescope, but the same clusters are often gems in something that is in the 125 mm to 200 mm range (or even in a big finder).

I suppose the reason has something to do with cluttering the background with faint trash stars in a bigger telescope – i.e., you get the best view with the aperture that includes the most cluster stars and the fewest background stars. It also seemed to me as though some of those clusters looked a lot better with extensive real estate around them – field sizes that the big scopes just can't supply. I suppose that has to do with maximizing the apparent density of a looser cluster, but I'm not sure. Anyway it is another reason why not all observers prefer larger apertures.

Globular clusters

The other main cluster type is the globular cluster. These objects are spherical in shape with the greatest concentration of stars toward their centers. Globular clusters have many more stars than open clusters, with a range of about ten thousand stars to about a million stars. Globular clusters are old. Most have ages well in excess of ten billion years with stars poor in heavy elements.

What really sets globulars apart, however, is their location. Open clusters lie within the spiral arms (disk) of our Milky Way. Globular clusters surround the Milky Way in a spherical distribution. This structure is often called the halo, but that term also can describe the outer regions of the central bulge of a galaxy. Studying the distribution of globular clusters around the

50 bright open clusters

Object	Constellation	Magnitude	Object	Constellation	Magnitude
M45	Tau	1.6	M37	Aur	6.2
M7	Sco	3.3	NGC 1545	Per	6.2
M44	Cnc	3.9	M50	Mon	6.3
NGC 869/884	Per	4.4	NGC 6940	Vul	6.3
NGC 2244	Mon	4.8	NGC 457	Cas	6.4
NGC 2362	CMa	4.8	NGC 7243	Lac	6.4
M41	CMa	5.0	M21	Sgr	6.5
M47	Pup	5.0	M36	Aur	6.5
M39	Cyg	5.3	M93	Pup	6.5
NGC 2244	Mon	5.3	NGC 129	Cas	6.5
NGC 6633	Oph	5.3	NGC 654	Cas	6.5
M6	Sco	5.5	NGC 752	And	6.5
M35	Gem	5.6	NGC 663	Cas	6.5
NGC 7686	Cas	5.6	NGC 1528	Per	6.5
M34	Per	5.8	M16	Ser	6.6
M23	Sgr	5.9	M46	Pup	6.6
M48	Hya	6.0	NGC 1027	Cas	6.7
NGC 1647	Tau	6.0	NGC 2343	Mon	6.7
NGC 1746	Tau	6.0	NGC 2423	Pup	6.7
NGC 1981	Ori	6.0	NGC 7209	Lac	6.7
NGC 2264	Mon	6.0	NGC 7789	Cas	6.7
NGC 2301	Mon	6.0	M11	Sct	6.8
M67	Cnc	6.1	NGC 7036	Cyg	6.8
NGC 7160	Cep	6.1	M103	Cas	6.9
M25	Sgr	6.2	M38	Aur	7.0

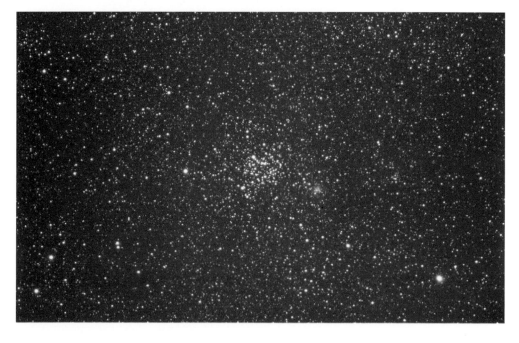

M35 in Gemini. (Kodak Elite Chrome 200, pushed 1 stop. Nikon F2 with a 105 mm Nikkor lens. Photo by Ulrich Beinert of Kronberg, Germany)

M22, the fabulous globular cluster in Sagittarius. This was the first globular ever discovered. (Image by Adam Block/NOAO/AURA/NSF, using a 0.4 m Meade LX200 telescope)

Scorpius-Sagittarius area led Harlow Shapley to correctly conclude that the center of our galaxy must lie in that direction. Most globulars move in orbits that have high eccentricities.

M22, a beautiful object in Sagittarius, was the first globular cluster to be discovered, in 1665. Twelve years later, the great Omega Centauri was observed by Halley. Today, about 200 globular clusters are known to surround our galaxy. Globulars have been observed around many other galaxies as well. As an example, the giant elliptical galaxy M87 in Virgo contains over a thousand globular clusters.

There exists a classification scheme for globular clusters devised by Harlow Shapley and Helen Sawyer (1905–1993) based on their studies of large numbers of clusters during the twentieth century. In the Shapley-Sawyer classification system, the Roman numerals from I to XII indicate star cluster concentration, I being a globular having the highest star concentration and XII having the lowest.

Observing globular clusters

In the entire deep-sky menagerie, there is no group that excites new observers more than globular clusters. They are considered by even seasoned amateur astronomers to be the most rewarding objects to observe. It's easy to see why. Many are bright enough to be seen even from urban settings. But when observed from a dark site, these objects explode with detail. The use of higher magnifications brings whole new levels to these clusters. Faint fuzz resolves itself into individual sparkling points of light which form intricate patterns. Careful observers with large enough telescopes may even be rewarded with views of faint planetary nebulae entangled within the starry vistas.

When observing globular clusters, begin by concentrating

M92, the "overlooked" globular in Hercules. (Image by Adam Block/NOAO/AURA/NSF, using a 0.4 m Meade LX200 telescope)

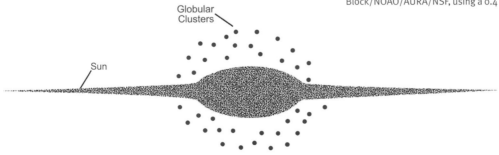

The spherical distribution of globular clusters around the center of the Milky Way. (Illustration by Holley Y. Bakich)

The 25 brightest globular clusters

Object	Other designation	Visual magnitude
NGC 5139	Omega Centauri	3.9
NGC 104	47 Tuc	4.0
NGC 6656	M22	5.2
NGC 6397		5.3
NGC 6752		5.3
NGC 6121	M4	5.4
NGC 5904	M5	5.7
NGC 6205	M13	5.8
NGC 6218	M12	6.1
NGC 2808		6.2
NGC 6809	M55	6.3
NGC 6541		6.3
NGC 5272	M3	6.3
NGC 7078	M15	6.3
NGC 6266	M62	6.4
NGC 6341	M92	6.5
NGC 6254	M10	6.6
NGC 7089	M2	6.6
NGC 362		6.8
NGC 6723		6.8
NGC 6388		6.8
NGC 6273	M19	6.8
NGC 7099	M30	6.9
NGC 3201		6.9
NGC 6626	M28	6.9

M12, a globular cluster in Ophiuchus. (Image by Ed Grafton, Houston, Texas, using a Celestron 14 and an ST5c CCD camera)

reds of stars which are apart from the central condensation. For at least a couple of globulars (Omega Centauri and 47 Tucanae), carefully counting stars will lead you into the thousands.

Brian Skiff of Lowell Observatory states:

The point at which a globular cluster is "well-resolved" is when your telescope has a limiting magnitude at or below the level of the horizontal branch. The reason is simply that the number of stars in any given magnitude interval takes a sudden leap at the magnitude of the horizontal branch.

Let's explore what this means. A plot of a globular cluster's stars on a Hertzsprung-Russell diagram is illustrated in the next figure.

Generally, the vertical axis is given in terms of absolute magnitude, but since a globular's stars are all at roughly the same distance, we can use apparent brightness equally as well. Note that if your telescope's limiting magnitude is 13, you will see a "nice" number of stars. At 14, things start to

on the constellations Ophiuchus, Scorpius, and Sagittarius. Nearly 70 globular clusters may be found in these three constellations. That's right, more than a third of all known Milky Way globulars are located in an area which comprises only 5.6% of the sky.

There's a lot more to observing globular clusters than just stating that they're round. Carefully note the shape of each. Some are slightly elliptical. Some seem to have "arms" which extend beyond the general concentration of stars. You can even apply the three delimiters of the Trumpler classification scheme for open clusters to globulars:

(1) What is the concentration of the cluster?
(2) What is the range of brightnesses of the stars?
(3) How rich is the cluster?

Observing Note: Many globular clusters, especially those on Messier's list, are visible naked-eye. How many can you see?

As far as resolution is concerned, the larger your telescope the more resolved will be the globular. With medium-sized telescopes, you may be able to count dozens or even hund-

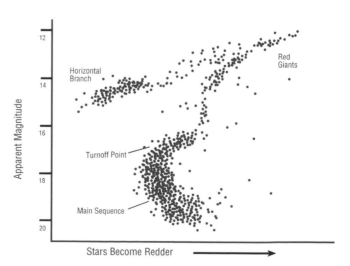

Hertzsprung–Russell diagram of a globular cluster. (Illustration by Holley Y. Bakich)

47 Tucanae (NGC 104), considered by some the greatest of the globular clusters, lies at declination 72° S, very near the Small Magellanic Cloud. (RG: 600s R, 600s G, 900s B through an ST-7E on a Celestron 11. Image by Steven Juchnowski, Balliang East in the State of Victoria, Australia)

get interesting. However, if the limiting magnitude is 15, all of the stars of the horizontal branch are observable. At this point, the globular cluster is "well-resolved."

As stated earlier, many other galaxies possess systems of globular clusters. Observing extra-galactic globulars is a fine project for a large amateur telescope. An excellent resource devoted to this purpose may be found on the "Adventures in Deep Space" website run by Jim Shields. Find the specific page on the internet at http://www.angelfire.com/id/jsredshift/gcextra.htm

7.7 Galaxies

NGC 4565. (Image by Adam Block/NOAO/AURA/NSF, using a 0.4 m Meade LX200 telescope)

One might wonder how objects composed of up to a trillion or more individual suns could be so difficult to observe. Of course, the answer is distance. Galaxies are so far away that, except for a small number of exceptions, they all appear small and faint. Advanced observers regard observing faint galaxies as a challenge. I know one amateur who progressed from a 200 mm (8-inch) telescope to a 300 mm (12-inch), then to a 400 mm (16-inch) and, as I write this, now uses a 600 mm (24-inch). With each new telescope the galaxies he observed became fainter and fainter, always just at the very limit of what the telescope would reveal.

This type of compulsive, competitive observing may not be for you. That's ok. There are lots of galaxies out there, and some are relatively bright. Brian Skiff of Flagstaff, Arizona, has compiled a list of the brightest galaxies. Among these are the showpieces of the sky. Brian's list is reproduced in Appendix E.

In addition to the magnitude of a galaxy, you must also take into account its surface brightness. This is the overall brightness of the galaxy divided by its area. Surface brightness is a measure of brightness per unit area, and is usually expressed in terms of magnitude per square arc-minute or per square arcsecond.

Jeff Medkeff of Sierra Vista, Arizona, has developed a simple rule of thumb to help him determine how difficult a galaxy will be to see. He multiplies the integrated magnitude by the surface brightness. This gives him a number (no units, just a number). The higher that number, the tougher the galaxy will be to observe. Jeff stresses that this is not a hard and fast rule but that it does seem to provide a good indicator of difficulty.

In my experience, there are a number of factors that will determine how well your galaxy observing session goes. Six of these are now briefly discussed here.

(1) Size does matter

You can be the greatest observer this planet has ever seen but if you're trying to observe galaxies with a 100 mm (4 inch) telescope your observing log is

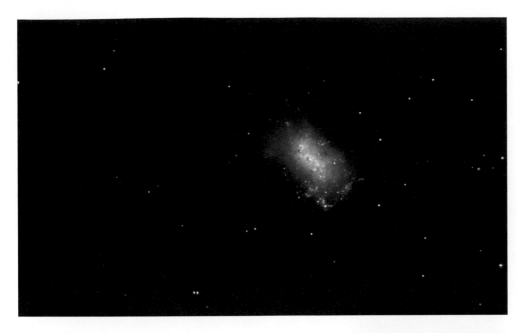

NGC 4449. (Image by Adam Block/NOAO/AURA/NSF, using a 0.4 m Meade LX200 telescope)

going to be filled with reports of rough shapes and central condensations and descriptors like "hinted at" and "small and faint." There's no way around it. If you want to observe galaxies, and I mean really get something out of the time you put in at the eyepiece, you are going to have to use a large telescope.

(2) Be patient

Rome was not built in a day. Likewise, a galaxy will not give up its secrets in a minute. Begin by noting the overall shape of the galaxy. Is it round, oval, rectangular, triangular? (At this stage, no description is a bad one.) Next, look at the way the brightness is distributed. Is there a central condensation? Are there brighter areas elsewhere? Note the stars in the field of view. How does the galaxy, or its brighter parts, compare with the field stars?

(3) Crank up the power

Use whatever power eyepiece you must to locate the galaxy you want to observe. Then start to increase the magnification. In my opinion, one of the biggest mistakes that new observers make is believing that a lower power will make the galaxy easier to see. In fact, using higher power will increase the contrast between the galaxy and the sky background, making the galaxy easier to see. A lengthy discussion as to why this is true is given in Roger N. Clark's superb book, *Visual Astronomy of the Deep Sky* (Cambridge University Press, 1990).

(4) Keep at it

Experience at the eyepiece is never more important than when viewing galaxies. What, to me, looked like an amorphous blob when I first began observing, now reveals a wealth of (albeit faint) details.

Observing Tip: When observing galaxies, especially in rich areas such as Coma Berenices and Virgo, check the field of view carefully. Your telescope may be able to see other galaxies not on your star map.

NGC 1300. (Image by Adam Block/NOAO/AURA/NSF, using a 0.4 m Meade LX200 telescope)

(5) Filters and galaxies

Galaxies are composed of many different types of objects. Because of this, their spectra are essentially continuous. Using a filter of any kind on such an object simply removes some of the light the galaxy is emitting. This makes a faint object even more difficult to observe. In a few cases, light pollution filters may aid the observer; however, they are essentially useless for galaxies if you have a small telescope. For the fortunate with large scopes, a lot of deep-sky observers use a light blue 82A filter to observe galaxies because of what it does to suppress auroral glow (the natural glow of the upper atmosphere – it does not imply there is an auroral outburst).

Nebraska amateur and eyepiece filter expert David Knisely has this to say:

At low to moderate powers on some galaxies, broad-band filters are of at least a slight benefit, although not to the same degree as with emission nebulae. I have

NGC 891 is bright enough to be glimpsed naked-eye, but this is a better view. (Image by Adam Block/NOAO/AURA/NSF, using a 0.4 m Meade LX200 telescope)

noted slight improvements in the views for the larger or more diffuse objects like M33, M101, M81, NGC 253, NGC 2403, etc. For others, increased power seems to be somewhat more effective in bringing out the detail (it dilutes the faint sky background and increases the image scale to make seeing faint detail a bit easier). The spiral arms in some galaxies do contain HII regions, but these do not contribute nearly as much to the visible light from the arms as do the brighter star clouds in the arms. The continuous emission from the brighter spiral arm tracing star groups is continuous enough to not be greatly enhanced by the filters. The emission from the brighter bluish stars in the galaxy is not "peaked" enough to allow these filters to enhance them the way they do emission line nebulae.

Jeff Medkeff adds:

and yet, blue sensitive emulsions do perk up the arms in photographs. So perhaps a blue filter would help. I've used both light and dark blue filters on brighter spiral galaxies for some years with varying success in boosting the contrast of the arms. The results seem somewhat hit or miss, but there is a definite benefit in the case of some galaxies which is not quite as dramatic as using a narrowband nebular filter on nebulae, but a good deal more dramatic than using an LPR on galaxies. What really needs to happen is for

The Sombrero Galaxy, M104. (Image by Don Stotz, Mike Ford, and Adam Block/NOAO/AURA/NSF, using a 0.4 m Meade LX200 telescope)

some observer with a large telescope – say, ahem, a 22-incher under dark skies – to undertake to correlate the best filter to a large sample of targets. I steal all of my large-scope time from friends, so I'm not set up to do this, but I assure you it will not be a waste of time.

(6) The sky will decide

Whether or not your telescope will provide superb views on any given night is a factor totally out of your control. Just as with planets, double stars, or anything else, the seeing is what sets the limit on how much detail is available.

Some amateurs believe that, since galaxies are extended objects, seeing does not affect them as much as, say, double stars. While it is true that galaxies as a whole are relatively large, the details you are trying to see in them (spiral arm structure, stellar condensations, etc.) are not, and these details are at the mercy of the night's seeing.

Classification of galaxies

Today, we are indebted to the great American astronomer Edwin Hubble (1889-1953) for developing a simple classification scheme for galaxies. Hubble first mentioned this in a paper he wrote in 1922. Four years later he expanded it and added some illustrations. Finally, in 1936, Hubble provided a slightly expanded explanation of the classification scheme in his book *The Realm of the Nebulae* (Yale University Press, New Haven, CT). It was in this book that the famous "tuning fork" diagram first appeared.

Hubble described several different main types of galaxies. His classification scheme has three types. These are ellipticals, spirals, and barred spirals. An argument for a fourth Hubble type (off the tuning fork) could be made. The fourth class contained "irregular galaxies," essentially every galaxy not on the fork.

Spirals like our own galaxy fell into several classes depending on their shape and the relative size of the bulge: ordinary spirals were labeled either Sa, Sb and Sc, while those which had developed a bar in the interior region of the spiral arms were SBa through SBc. Spiral galaxies were characterized by the presence of gas in their disks which indicated that star formation remained active, hence the younger population of stars. Spirals are usually found in an area of low

The Hubble tuning fork diagram. Although now out of date, it is the most widely recognized classification scheme for galaxies. (Illustration by Holley Y. Bakich)

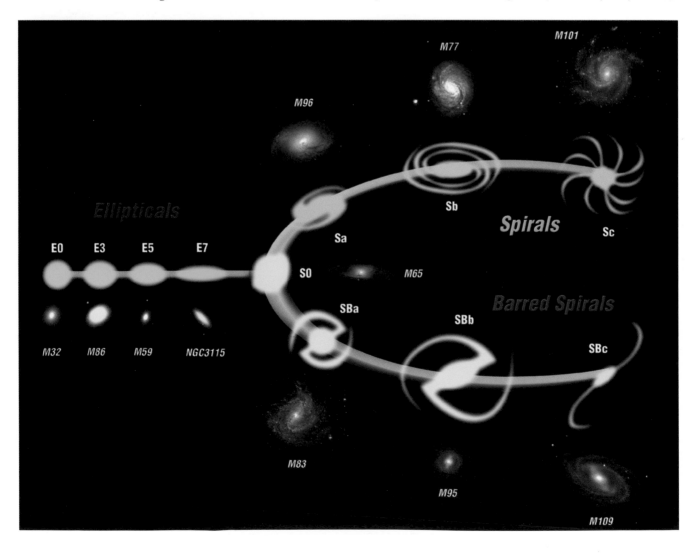

M87, ruler of the Virgo Cluster. 25 Jan 2001. (Photo by Shane Larson and Mike Murray of Bozeman, Montana, and Adam Block/NOAO/AURA/NSF, using a 0.4 m Meade LX200 telescope)

Galaxy classification schemes compared

Original Hubble classification

Type	E	SO	SA SB	SB	Irregular	Peculiar
Frequency (%)	23.4	21.0	24.4	26.3	3.4	1.5

de Vaucouleurs classification

Type	E	Sa, SBa	Sb, SBb	Sc, SBc	I
Frequency (%)	17	19	25	36	2.5

galaxy density where their delicate shape can avoid disruption by tidal forces from neighboring galaxies.

Ellipticals were placed in the categories E0, E1, through E7 depending on their degree of ellipticity. They have a uniform brightness and are similar to the bulge in a spiral galaxy, but with no disk. The stars in these galaxies are old and there is no gas present. Ellipticals are usually found in areas of high galaxy density, or at the center of clusters.

The tuning fork updated

When I asked Brian Skiff of Lowell Observatory for an example of one of the types of galaxies represented on the tuning fork diagram, he provided some insight as to why that system is no longer used by astronomers:

> Unfortunately, the Hubble system is based mostly on luminous, symmetrical galaxies, which are very much in the minority in nearby space. This is the reason the "Tuning Fork" isn't good enough. Hubble hadn't looked at enough galaxies!

In 1959, the French astronomer Gerard de Vaucouleurs published an update of Hubble's classification scheme. This is a much more detailed analysis of types of galaxies and includes four classes of galaxies. The Magellanic class (of which the Large Magellanic Cloud is the prototype) was later separated out and made its own class, separate from the irregulars. The percentage of types of galaxies was also revised. The latest data indicate that about one-third of all galaxies have bars, about one-third have none, and about one-third are mixed cases.

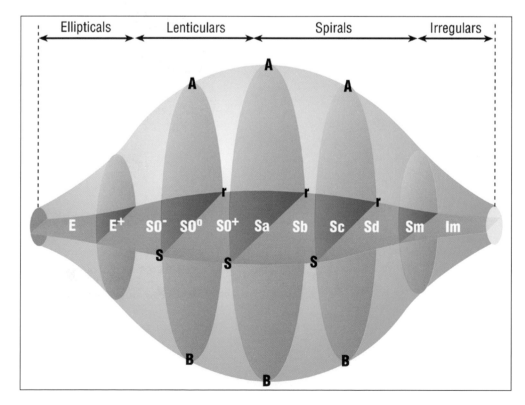

The galaxy classification scheme of Gerard de Vaucouleurs. At the top are the ordinary families of spirals (labeled A); toward the bottom are the barred spiral families (labeled B); on the near side are the S-shaped varieties (labeled S), on the far side the ringed varieties (labeled r). On the far right, Im represents the Magellanic-type irregulars. (Illustration by Holley Y. Bakich)

NGC 4725. (Image by Adam Block/NOAO/AURA/NSF, using a 0.4 m Meade LX200 telescope)

Brian continued:

The de Vacouleurs system essentially takes the fork and spins it along the handle, so that, instead of two tines, there is a circle of transitional cases. Sweep one way around from the "ordinary" to the barred galaxies and you go through the transitional cases with inner rings. In the other direction are the transitional cases without the inner rings/lenses.

Observing galaxies

As you may have gleaned from earlier chapters, I enjoy observing many different kinds of objects. Right near the top of my list, however, are spiral galaxies. Unfortunately, my 100mm refractor does not allow me to examine many of the details visible in larger telescopes. In practical terms, the lower limit for a telescope which could reveal more than just the minimum detail on galaxies is 200 mm. And, as with many of the objects described in this book, there is no upper limit. Exceptions to this rule are the larger galaxies such as M31.

Visual observation of detailed spiral structure, like that seen in images, requires a large telescope. My personal preference is to use 500mm telescopes and above for such work. With smaller scopes, I often see what many term "mottling," which may indicate the presence of spiral arms but does not constitute a true observation of them.

For success in observing detail in medium-sized telescopes (apart from M31) I like the Messier galaxies M33 (Tri), M51 (CVn), M64 (Com), M81 (UMa), M83 (Hya), M101 (UMa), M106 (UMa), and M108 (UMa). I am very much indebted to Meade Instrument Corporation for the loan of a 300mm LX200 GPS during the writing of this book. This telescope is at the upper end of "medium-sized," and from my dark-sky site, it provides wonderful views of

these objects. M101 has the lowest surface brightness of any of these, but from a true dark site a patient observer at a quality telescope will discover nice detail. For observers with large telescopes, the Messier galaxies M61, M91, and M95, as well as NGC 1395 are barred spirals. With adequate magnification, both the bar and the extent of the spiral arms are easy to detect.

Observing Tip: With a medium-sized telescope, a very dark sky is essential for an observer. Look for

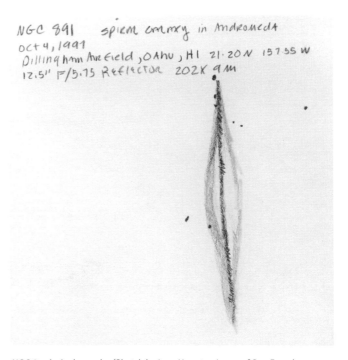

NGC 891 in Andromeda. (Sketch by Jane Houston Jones of San Francisco, California)

The Large and Small Magellanic Clouds. (Image by Steven Juchnowski, Balliang East in the State of Victoria, Australia)

the central condensation, subtle brightness variations, overall shape and the extent of the spiral arms.

Irregular galaxies are the smallest class of galaxies but, in my opinion, are far more interesting to observe than ellipticals. Most irregular galaxies are quite faint, but there are exceptions. The king of irregular galaxies for northern amateur astronomers is M82 in Ursa Major. In the southern hemisphere, well, it's pretty obvious. The Large and Small Magellanic Clouds are, by virtue of their distance, the greatest galaxies – of any type – period. Take as long as you need to study these objects. It will be time well spent. NGC 55 (Scl), NGC 625 (Phe), NGC 4449 (CVn), NGC 5128 (Cen), and NGC 6822 (Sgr) are other, relatively bright, irregular galaxies.

Clusters of galaxies

Do you have a large telescope and plenty of time? Observing clusters of galaxies may be for you. (It's amazing that, in a few short chapters, we have moved from discussing observing star clusters to observing galaxy clusters.) Clusters of

A cluster of galaxies in Hercules. The brightest member is NGC 6166. (Taken the morning of 14 May 2002, 0.4 m Meade LX200. This fabulous LRGB CCD image has exposures of 105 minutes luminance, 20 minutes each of R, G, and B. Image by Jeff Hapeman of Madison, Wisconsin, and Adam Block/NOAO/AURA/NSF)

The galaxies of the Local Group

Name	RA (2000)			Dec.		Visual magnitude	Dimensions Arcmin (')	Type	Distance kpc
	h	m	s	Deg (°)	Arcmin (')				
WLM	0	01	58	−15	27.7	10.6	12×4	IB(s)m V	950
IC 10	0	20	20	+59	18.0	11	6.3×5.1	IB(s)m IV	660
Cetus	0	26	11	−11	02.5	14		dE	780
NGC 147	0	33	12	+48	30.5	9.5	13×8	dE5	660
And III	0	35	29	+36	30.5	15	4.5×3.0	dE	760
NGC 185	0	38	58	+48	20.2	9.2	12×10	dE3	660
NGC 205	0	40	22	+41	41.1	8.1	22×11	SAo-	760
NGC 221	0	42	42	+40	51.9	8.1	8.7×6.5	cE2	760
NGC 224	0	42	44	+41	16.1	3.4	190×60	SA(s)b I-II	760
And I	0	45	40	+38	02.2	13.6	2.5×2.5	E3	810
SMC	0	52	44	−72	49.7	2.3	320×185	IB(s)m V	60
Sculptor	1	00	09	−33	42.5	9	40×31	dE	90
Pisces	1	03	54	+21	53.0	18	2×2	Im VI?	810
IC 1613	1	04	46	+02	07.1	9.2	16×15	IB(s)m V	720
And V	1	10	17	+47	37.7	15		dE	810
And II	1	16	30	+33	25.2	13	3.6×2.5	dE	680
NGC 598	1	33	51	+30	39.6	5.7	71×42	SA(s)cd II-III	790
Phoenix	1	51	07	−44	26.7	12.5	4.9×4.1	IAm V	400
Fornax	2	39	59	−34	27.0	9	17×13	dEo	140
LMC 5	2	3	34	−69	45.4	0.6	645×550	IB(s)m IV	50
Carina 6	4	1	37	−50	58.0	16?	23×16	E3	100
Leo A	9	59	26	+30	44.8	12	5.1×3.1	IBm V	690
Leo I	10	08	27	+12	18.5	10	9.8×7.4	dE	250
Sextans	10	13	03	−01	36.9	12		dE	90
Leo II	11	13	29	+22	09.3	12.0	12×11	dEo	210
UMi	15	09	11	+67	12.9	11	30×19	dE	60
Draco	17	20	12	+57	54.9	10	36×25	dE	80
Milky Way	17	45	40	−29	00.5			SB:(rs?)bc: II?	
Sgr I	18	55	03	−30	28.7			dE?	
SagDIG	19	29	59	−17	40.7	15	2.9×2.1	IB(s)m: V	30
NGC 6822	19	44	56	−14	48.1	9	16×14	IB(s)m: V	500
Aquarius	20	46	52	−12	50.9	13.9	2.2×1.1	IB(s)m: V	950
Tucana	22	41	50	−64	25.2	15	2.9×1.2	dE4	870
And VII	23	26	31	+50	41.5	13	2.5×2.0	dE?	690
Pegasus	23	28	35	+14	44.6	13	5.0×2.7	Im V	760
And VI	23	51	46	+24	35.0	14	4.0×2.0	dE?	780

galaxies may be made up of a handful of galaxies up to a thousand. As with star clusters, clusters of galaxies are held together by mutual gravitational attraction.

   Note: A galactic cluster and a cluster of galaxies are totally different. The first is a group of stars (also known as an open cluster). The second is a group of galaxies.

The most famous cluster of galaxies, I suppose, is the one which contains our Milky Way. Known as the Local Group, it comprises 36 members. The following list is courtesy of Brian Skiff, Lowell Observatory, Flagstaff, Arizona, from an article by Sidney van den Bergh which appeared in the May 2000 issue of *Publications of the Astronomical Society of the Pacific*. Observing every member of the Local Group is a real challenge.

An interacting pair of galaxies. NGC 7253. (Image by Harihar Padmanabh and Adam Block/NOAO/AURA/NSF, using a 0.4 m Meade LX200 telescope)

Two other well-known and often observed clusters of galaxies are the Virgo Cluster and the Fornax Cluster. The Virgo Cluster lies nearly 50 million light years from Earth. This is a bright cluster. A number of Messier objects are members, including M49, M60, M84, M85, M86, M87, M88, M89, M90, M98, M99, and M100. In addition, well over a hundred NGC galaxies belong to the Virgo Cluster.

Many observers think that the Fornax Cluster is the finest for amateur astronomers. Eighteen galaxies make up the cluster. The brightest is NGC 1316 at visual magnitude 8.8 with a surface brightness of 12.7. This galaxy is also a radio source called Fornax A. NGC 1399 (visual magnitude 9.8, surface brightness 12.3) is the next easiest to see. NGC 1365

is third (visual magnitude 9.5, surface brightness 13.7); this is a barred spiral seen face on, with open spiral arms. In a wide-field, high-power eyepiece, up to ten members of this cluster of galaxies can be viewed.

The most famous name related to clusters of galaxies is that of the American astronomer George O. Abell (1927–1983). In 1958, he published a list of 2712 northern clusters of galaxies (to declination −27°) which he had identified from the Palomar Observatory Sky Survey (POSS) plates. A list of southern clusters of galaxies (numbers 2713–4076) was begun by Abell and Harold Corwin in 1975. This was completed (after Abell's death) by Corwin and Ronald Olowin in 1987. These were based on plates taken by the

The wonderful group of galaxies known as Stephan's Quintet in Pegasus. (Image by Adam Block/NOAO/AURA/NSF, using a 0.4 m Meade LX200 telescope)

United Kingdom's 1.2m Schmidt Telescope at Siding Spring, Australia. A supplementary catalog of southern objects contains 1174 additional galaxy clusters considered either not rich enough or too distant for inclusion in the main catalog. They are designated by the suffix "s", as in Abell 696s.

For another very challenging list of galaxy clusters, try the Hickson Catalog. Paul Hickson (at this writing a professor in

The Whale. NGC 4631. 25 Jan 2001. (Photo by Shane Larson and Mike Murray of Bozeman, Montana, and Adam Block/NOAO/AURA/NSF, using a 0.4 m Meade LX200 telescope)

the Department of Physics and Astronomy at the University of British Columbia) has authored the *Atlas of Compact Groups of Galaxies* (Gordon & Breach Science Publishers, London, 1994). This book contains 100 compact galaxy groups, including

M63. (Image by Adam Block/NOAO/AURA/NSF, using a 0.4 m Meade LX200 telescope)

Observing project – finding M81 naked-eye

Several months ago (at this writing), a discussion ensued on the amastro e-mail list concerning the furthest object that could be observed with the naked-eye. From our dark-sky observing site near El Paso, my wife, who has exceptional eyes, was able to pick out NGC 253 (and that when it was only 20 degrees up!). I thought this might have been the furthest naked-eye sighting, but was corrected by Brian Skiff of Lowell Observatory. He later followed up with a wonderful description of how to observe M81 (slightly more distant than NGC 253) naked-eye. Here it is.

M81. (Image by Adam Block/NOAO/AURA/NSF, using a 0.4 m Meade LX200 telescope)

As mentioned briefly in a previous post, I was able to see M81 naked-eye last night from Lowell's Anderson Mesa dark-sky site. In case others would like to try this, here are some details. For reference, have a look at *Uranometria* chart 23, or the relevant chart from the new edition of *Sky Atlas 2000*.

The two galaxies lie in a string of faint stars that starts with 24 UMa (also the variable DK UMa) at the west end, and arcs eastward and a bit south past M81 and NGC 3077 to a mag. 6 star appearing on chart 24 (brightest of a triangle). M81 in fact is one element of

the string. To have a hope of seeing M81, and to be sure to avoid confusion with a star, you need first to identify several of these stars. The brighter star about 1.5 degrees west of M81 is HD 83489, at V mag. 5.7. Next identify the fainter star at the east end of the string (the one on U2000 chart 24), which is HD 89343 = EN UMa, a very small-amplitude delta-Scuti star, at V mag. 6.0. Now, between HD 83489 and 89343 I consistently saw (glimpsed is a better term, however) at least three star-like objects. First, about 1.5 degrees east of M81 (and just west of HD 89343), is HD 87703, which is V mag. 7.1. If this star isn't pretty readily visible (at say the 20–30 percent detection level), then you probably won't be able get the galaxy. Much more difficult (at the 510 percent detection threshold) is the group of stars shown on the U2000 chart near NGC 3077. These are HD 86458 (V = 8.0), HD 86574 (V = 8.2), and HD 86677 (V = 7.9). I saw these three (or perhaps just the closer pair 86458/86677) as a single object. The combined brightness of the three stars is V = 6.8, but the extended nature of the trio means the surface brightness will be lower to the naked eye than a single star of that magnitude, making it more difficult to spot.

The third object, repeatedly spotted in the same (correct) place, is M81! Again, this is a threshold object, which I detected only 5–10 percent of the time with optimally-averted vision together with the other faint stars just mentioned. On occasion I also seemed to pick up another star, HD 85828 (V = 7.7), which is about 40′ south of the galaxy. Both NGC 253 and NGC 5128 (Cen A) are somewhat closer than M81, which is probably the most distant discrete object visible to the unaided eye, at something like 3.6 megaparsecs (11.8 million light-years). Having tried this observation seriously (and unsuccessfully) twice before, I found the key this time was getting all the field stars sorted out. Because there are several stars of similar brightness nearby, you really must be able to identify each of these in order to securely locate the galaxy.

some famous ones such as Stephan's Quintet. Don't purchase this book if you're just starting out. This is a list for serious amateur astronomers with large telescopes.

For completeness, I will mention the fact that there is yet another listing of clusters of galaxies. This one may take you a while to complete, however, and you'll need a very large telescope. The Zwicky Catalog of Galaxy Clusters contains 9134 separate clusters with some as faint as twentieth magnitude!

Superclusters of galaxies

The largest separate structures in the universe, and, some would say, the structures which define the shape of our universe, are superclusters of galaxies. The Local Group,

which contains the Milky Way, belongs to a supercluster of galaxies called the Local Supercluster (creative, eh?). The Local Group is at one end and the Virgo Cluster is near its center. The major axis of the Local Supercluster is about 125–150 million light years. Other, nearby, superclusters of galaxies include the Hydr–Centaurus supercluster, which lies at a distance of about 150 million light years and the Perseus supercluster, which is roughly 220 million light years away.

Galaxy catalogs

Many catalogs of galaxies exist. Some of the better ones are the Morphological Catalog of Galaxies (MCG) which contains 30 642 objects, the Zwicky Catalog of Galaxies

NGC 6946. (Image by Adam Block/NOAO/AURA/NSF, using a 0.4 m Meade
LX200 telescope)

(ZC) with 19 367 objects, the Uppsala General Catalog
(UGC) galaxies with 12 921 objects in total, and the
Southern Galaxy Catalogue (SGC), compiled by the
European Southern Observatory with 5476 objects. All of
these catalogs have positions in right ascension and
declination for the included galaxies good to a few
arcseconds or better.

The best reference source for galaxies – bar none – is the
NASA/IPAC Extragalactic Database (NED). If you need
information about a galaxy, try NED first. At last count it
included data for 4.3 million galaxies. The positions of most
of these galaxies are given with arcsecond accuracy.
Thankfully, this wonderful resource can be found on the
internet at http://nedwww.ipac.caltech.edu/

Appendix A

The constellations

Constellation	Pronunciation	Best seen	Constellation	Pronunciation	Best seen
Andromeda	an draw' meh duh	9 Oct	Gemini	gem' in eye	5 Jan
Antlia	ant' lee ah	24 Feb	Grus	groose	28 Aug
Apus	ape'us	21 May	Hercules	her' cur leez	13 Jan
Aquarius	uh qwayr'ee us	25 Aug	Horologium	hor uh low' gee um	10 Nov
Aquila	ak'will uh	16 Jul	Hydra	hi' druh	15 Mar
Ara	air' uh	10 Jun	Hydrus	hi' druss	26 Oct
Aries	air' eeze	30 Oct	Indus	in' dus	12 Aug
Auriga	or ey' guh	21 Dec	Lacerta	luh sir' tuh	28 Aug
Boötes	bow owe' teez	2 May	Leo	lee' owe	1 Mar
Caelum	see' lum	1 Dec	Leo Minor	lee' owe my' nor	23 Feb
Camelopardalis	kam uh low par' dah liss	23 Dec	Lepus	lee' pus	14 Dec
Cancer	kan' sir	30 Jan	Libra	lye' bruh	9 May
Canes Venatici	kay' neez ven ah tee' see	7 Apr	Lupus	loo' pus	9 May
Canis Major	kay' niss may' jor	2 Jan	Lynx	links	19 Jan
Canis Minor	kay' niss my' nor	14 Jan	Lyra	lie' ruh	4 Jul
Capricornus	kap rih kor' nus	8 Aug	Mensa	men' suh	14 Dec
Carina	kuh ree' nuh	31 Jan	Microscopium	my krow scop' ee um	4 Jul
Cassiopeia	kass ee oh pee' oh	9 Oct	Monoceros	mon oss' sir us	5 Jan
Centaurius	sen tor' us	30 May	Musca	mus' kuh	30 Mar
Cepheus	see' fee us	29 Sep	Norma	nor' muh	19 May
Cetus	see' tus	15 Oct	Octans	ok' tans	——
Chamaeleon	kuh meel' ee un	1 Mar	Orphiuchus	off ee oo' kus	11 Jun
Circinus	sir sin' us	30 Apr	Orion	or eye' on	13 Dec
Columba	kol um' buh	18 Dec	Pavo	pah' voe	15 Jul
Coma Berenices	koe' muh bear uh nye' seez	2 Apr	Pegasus	peg' ah sus	1 Sep
Corona Australis	kor oh' nuh os tral' iss	30 Jun	Perseus	pur' see us	7 Nov
Corona Borealis	kor oh' nuh boar ee al' iss	19 May	Phoenix	fee' niks	4 Oct
Corvus	kor' vus	28 Mar	Pictor	pik' tor	16 Dec
Crater	kray' ter	12 Mar	Pisces	pie' seez	27 Sep
Crux	kruks	28 Mar	Piscis Austrinus	pie' siss os try' nus	25 Aug
Cygnus	sig' nus	30 Jul	Puppis	pup' iss	8 Jan
Delphinus	dell fee' nus	31 Jul	Pyxis	pik' siss	4 Feb
Dorado	dor ah' doe	17 Dec	Reticulum	reh tik' yoo lum	19 Nov
Draco	dray' koe	24 May	Sagitta	suh gee' tuh	16 Jul
Equuleus	ek woo oo' lee us	8 Aug	Sagittarius	sa ji tare' ee us	7 Jul
Eridanus	air uh day' nus	10 Nov	Scorpius	skor' pee us	3 Jun
Fornax	for' nax	2 Nov	Sculptor	skup' tor	26 Sep

The constellations (*cont.*)

Constellation	Pronunciation	Best seen
Scutum	skoo' tum	1 Jul
Serpens	sir' pens	6 Jun
Sextans	sex' tans	22 Feb
Taurus	tor' us	30 Nov
Telescopium	tel es koe' pee um	10 Jul
Triangulum	try ang' yoo lum	23 Oct
Triangulum Australe	try ang' yoo lum os trail'	23 May
Tucana	too kan' uh	17 Sep
Ursa Major	er' suh may' jor	11 Mar
Ursa Minor	er' suh my' nor	13 May
Vela	vay' luh	13 Feb
Virgo	ver' go	11 Apr
Volans	ver' lans	18 Jan
Vulpecula	vul pek' yoo lah	25 Jul

Note

The column headed "Best seen" gives the midnight culmination dates for the central points of the constellations. This is when a constellation is opposite the Sun, in terms of right ascension. Octans is circumpolar and has no date given.

Appendix B

The 30 brightest stars

		Magnitude
α CMa	Sirius	−1.46
α Car	Canopus	−0.72
α Cen	Rigil Kentaurus	−0.27
α Boo	Arcturus	−0.04
α Lyr	Vega	0.03
α Aur	Capella	0.08
β Ori	Rigel	0.12
α CMi	Procyon	0.38
α Eri	Achernar	0.46
α Ori	Betelgeuse	0.50
β Cen	Hadar	0.61
α Aql	Altair	0.77
α Cru	Acrux	0.79
α Tau	Aldebaran	0.85
α Vir	Spica	0.98
β Gem	Pollux	1.14
α PsA	Fomalhaut	1.16
α Sco	Antares	1.22
β Cru	Mimosa	1.25
α Cyg	Deneb	1.25
α Leo	Regulus	1.35
ε CMa	Adhara	1.50
α Gem	Castor	1.58
γ Cru	Gacrux	1.63
λ Sco	Shaula	1.63
γ Ori	Bellatrix	1.64
β Tau	Elnath	1.65
β Car	Miaplacidus	1.68
ε Ori	Alnilam	1.70
α Gru	Alnair	1.74

Appendix C

The Messier marathon
(the order is typical of a mid-northern latitude site)

M number	NGC number	Constellation	Type	Approximate magnitude	M number	NGC number	Constellation	Type	Approximate magnitude
M77	NGC 1068	Cet	Glx	8.9	M109	NGC 3992	UMa	Glx	9.8
M74	NGC 628	Psc	Glx	8.5	M40	NGC (Win4)	UMa	DS	9.0, 9.6
M33	NGC 598	Tri	Glx	5.7	M106	NGC 4258	CVn	Glx	8.3
M31	NGC 224	And	Glx	3.4	M94	NGC 4736	CVn	Glx	8.2
M32	NGC 221	And	Glx	8.2	M63	NGC 5055	CVn	Glx	8.6
M110	NGC 205	And	Glx	8.0	M51	NGC 5194	CVn	Glx	8.4
M52	NGC 7654	Cas	OC	6.9	M101	NGC 5457	UMa	Glx	7.9
M103	NGC 581	Cas	OC	7.4	M102	NGC 5866	Dra	Glx	10.0
M76	NGC 650	Per	PN	10.1	M53	NGC 5024	Com	GC	7.7
M34	NGC 1039	Per	OC	5.2	M64	NGC 4826	Com	Glx	8.5
M45	—	Tau	OC	1.5	M3	NGC 5272	CVn	GC	5.9
M79	NGC 1904	Lep	GC	7.7	M98	NGC 4192	Com	Glx	10.1
M42	NGC 1976	Ori	N	3.7	M100	NGC 4321	Com	Glx	9.3
M43	NGC 1982	Ori	N	6.8	M85	NGC 4382	Com	Glx	9.1
M78	NGC 2068	Ori	N	8.0	M84	NGC 4374	Vir	Glx	9.1
M1	NGC 1952	Tau	SNR	8.0	M86	NGC 4406	Vir	Glx	8.9
M35	NGC 2168	Gem	OC	5.1	M87	NGC 4486	Vir	Glx	8.6
M37	NGC 2099	Aur	OC	5.6	M89	NGC 4552	Vir	Glx	9.7
M36	NGC 1960	Aur	OC	6.0	M90	NGC 4569	Vir	Glx	9.5
M41	NGC 2287	CMa	OC	4.5	M88	NGC 4501	Com	Glx	9.6
M93	NGC 2447	Pup	OC	6.2	M91	NGC 4548	Com	Glx	10.1
M47	NGC 2422	Pup	OC	5.7	M58	NGC 4579	Vir	Glx	9.6
M46	NGC 2437	Pup	OC	6.1	M59	NGC 4621	Vir	Glx	9.6
M50	NGC 2323	Mon	OC	5.9	M60	NGC 4649	Vir	Glx	8.8
M48	NGC 2548	Hya	OC	5.8	M49	NGC 4472	Vir	Glx	8.4
M44	NGC 2632	Cnc	OC	3.1	M61	NGC 4303	Vir	Glx	9.6
M67	NGC 2682	Cnc	OC	6.0	M104	NGC 4594	Vir	Glx	8.0
M95	NGC 3351	Leo	Glx	9.7	M68	NGC 4590	Hya	GC	7.6
M96	NGC 3368	Leo	Glx	9.2	M83	NGC 5236	Hya	Glx	7.5
M105	NGC 3379	Leo	Glx	9.3	M5	NGC 5904	Ser	GC	5.7
M65	NGC 3623	Leo	Glx	8.8	M13	NGC 6205	Her	GC	5.3
M66	NGC 3627	Leo	Glx	9.0	M92	NGC 6341	Her	GC	6.5
M81	NGC 3031	UMa	Glx	6.9	M57	NGC 6720	Lyr	PN	8.8
M82	NGC 3034	UMa	Glx	8.4	M56	NGC 6779	Lyr	GC	8.4
M97	NGC 3587	UMa	PN	9.9	M29	NGC 6913	Cyg	OC	6.6
M108	NGC 3556	UMa	Glx	10.0	M39	NGC 7092	Cyg	OC	4.6

The Messier marathon (*cont.*)

M number	NGC number	Constellation	Type	Approximate magnitude
M27	NGC 6853	Vul	PN	7.3
M71	NGC 6838	Sge	GC	8.0
M107	NGC 6171	Oph	GC	7.8
M12	NGC 6218	Oph	GC	6.8
M10	NGC 6254	Oph	GC	6.6
M14	NGC 6402	Oph	GC	7.6
M9	NGC 6333	Oph	GC	7.8
M4	NGC 6121	Sco	GC	5.4
M80	NGC 6093	Sco	GC	7.3
M19	NGC 6273	Oph	GC	6.8
M62	NGC 6266	Oph	GC	6.7
M6	NGC 6405	Sco	OC	4.2
M7	NGC 6475	Sco	OC	2.8
M11	NGC 6705	Sct	OC	5.3
M26	NGC 6694	Sct	OC	8.0
M16	NGC 6611	Ser	N	6.0
M17	NGC 6618	Sgr	N	6.0
M18	NGC 6613	Sgr	OC	6.9
M24	NGC 6603	Sgr	SC	2.5
M25	IC 4725	Sgr	OC	4.6
M23	NGC 6494	Sgr	OC	5.5
M21	NGC 6531	Sgr	OC	5.9
M20	NGC 6514	Sgr	N	6.3
M8	NGC 6523	Sgr	N	3.0
M28	NGC 6626	Sgr	GC	6.9
M22	NGC 6656	Sgr	GC	5.2
M69	NGC 6637	Sgr	GC	7.4
M70	NGC 6681	Sgr	GC	7.8
M54	NGC 6715	Sgr	GC	7.2
M55	NGC 6809	Sgr	GC	6.3
M75	NGC 6864	Sgr	GC	8.6
M15	NGC 7078	Peg	GC	6.0
M2	NGC 7089	Aqr	GC	6.3
M72	NGC 6981	Aqr	GC	9.2
M73	NGC 6994	Aqr	OC	8.9
M30	NGC 7099	Cap	GC	6.9

Key

DS = double star	GC = globular cluster
Glx = galaxy	N = nebula
OC = open cluster	PN = planetary nebula
SC = star cloud	SN = supernova remnant

Appendix D

The Caldwell catalogue

C number	NGC/IC number	Constellation	Type	RA		Dec.	
				h	m	Deg (°)	Arcmin (′)
1	NGC 188	Cep	OC	00	44.4	+85	20
2	NGC 40	Cep	PN	00	13.0	+72	32
3	NGC 4236	Dra	SG	12	16.7	+69	28
4	NGC 7023	Cep	BN	21	01.8	+68	12
5	IC 342	Cam	SG	03	46.8	+68	06
6	NGC 6543	Dra	PN	17	58.6	+66	38
7	NGC 2403	Cam	SG	07	36.9	+65	36
8	NGC 559	Cas	OC	01	29.5	+63	18
9	Sh2-155	Cep	BN	22	56.8	+62	37
10	NGC 663	Cas	OC	01	46.0	+61	15
11	NGC 7635	Cas	BN	23	20.7	+61	12
12	NGC 6946	Cep	SG	20	34.8	+60	09
13	NGC 457	Cas	OC	01	19.1	+58	20
14	NGC 869/884	Per	OC	02	20.0	+57	08
15	NGC 6826	Cyg	PN	19	44.8	+50	31
16	NGC 7243	Lac	OC	22	15.3	+49	53
17	NGC 147	Cas	EG	00	33.2	+48	30
18	NGC 185	Cas	EG	00	39.0	+48	20
19	IC 5146	Cyg	BN	21	53.5	+47	16
20	NGC 7000	Cyg	BN	20	58.8	+44	20
21	NGC 4449	CVn	IG	12	28.2	+44	06
22	NGC 7662	And	PN	23	25.9	+42	33
23	NGC 891	And	SG	02	22.6	+42	21
24	NGC 1275	Per	IG	03	19.8	+41	31
25	NGC 2419	Lyn	GC	07	38.1	+38	53
26	NGC 4244	CVn	SG	12	17.5	+37	49
27	NGC 6888	Cyg	BN	20	12	+38	21
28	NGC 752	And	OC	01	57.8	+37	41
29	NGC 5005	CVn	SG	13	10.9	+37	03
30	NGC 7331	Peg	SG	22	37.1	+34	25
31	IC 405	Aur	BN	05	16.2	+34	16
32	NGC 4631	CVn	SG	12	42.1	+32	32
33	NGC 6992/5	Cyg	SN	20	56.4	+31	43
34	NGC 6960	Cyg	SN	20	45.7	+30	43
35	NGC 4889	Com	EG	13	00.1	+27	59
36	NGC 4559	Com	SG	12	36.0	+27	58

The Caldwell catalogue (*cont.*)

C number	NGC/IC number	Constellation	Type	RA		Dec.	
				h	m	Deg (°)	Arcmin (′)
37	NGC 6885	Vul	OC	20	12.0	+26	29
38	NGC 4565	Com	SG	12	36.3	+25	59
39	NGC 2392	Gem	PN	07	29.2	+20	55
40	NGC 3626	Leo	SG	11	20.1	+18	21
41	——	Tau	OC	04	27.0	+16	00
42	NGC 7006	Del	GC	21	01.5	+16	11
43	NGC 7814	Peg	SG	00	03.3	+16	09
44	NGC 7479	Peg	SG	23	04.9	+12	19
45	NGC 5248	Boo	SG	13	37.5	+08	53
46	NGC 2261	Mon	BN	06	39.2	+08	44
47	NGC 6934	Del	GC	20	34.2	+07	24
48	NGC 2775	Cnc	SG	09	10.3	+07	02
49	NGC 2237-9	Mon	BN	06	32.3	+05	03
50	NGC 2244	Mon	OC	06	32.4	+04	52
51	IC 1613	Cet	IG	01	04.8	+02	07
52	NGC 4697	Vir	EG	12	48.6	−05	48
53	NGC 3115	Sex	EG	10	05.2	−07	43
54	NGC 2506	Mon	OC	08	00.2	−10	47
55	NGC 7009	Aqr	PN	21	04.2	−11	22
56	NGC 246	Cet	PN	00	47.0	−11	53
57	NGC 6822	Sgr	IG	19	44.9	−14	48
58	NGC 2360	CMa	OC	07	17.8	−15	37
59	NGC 3242	Hya	PN	10	24.8	−18	38
60	NGC 4038	CrV	SG	12	01.9	−18	52
61	NGC 4039	CrV	SG	12	01.9	−18	53
62	NGC 247	Cet	SG	00	47.1	−20	46
63	NGC 7293	Aqr	PN	22	29.6	−20	48
64	NGC 2362	CMa	OC	07	18.8	−24	57
65	NGC 253	Scl	SG	00	47.6	−25	17
66	NGC 5694	Hya	GC	14	39.6	−26	32
67	NGC 1097	For	SG	02	46.3	−30	17
68	NGC 6729	CrA	BN	19	01.9	−36	57
69	NGC 6302	Sco	PN	17	13.7	−37	06
70	NGC 300	Scl	SG	00	54.9	−37	41
71	NGC 2477	Pup	OC	07	52.3	−38	33
72	NGC 55	Scl	SG	00	14.9	−39	11
73	NGC 1851	Col	GC	05	14.1	−40	03
74	NGC 3132	Vel	PN	10	07.7	−40	26
75	NGC 6124	Sco	OC	16	25.6	−40	40
76	NGC 6231	Sco	OC	16	54.0	−41	48
77	NGC 5128	Cen	EG	13	25.5	−43	01
78	NGC 6541	CrA	GC	18	08.0	−43	42

The Caldwell catalogue (*cont.*)

C number	NGC/IC number	Constellation	Type	RA		Dec.	
				h	m	Deg (°)	Arcmin (′)
79	NGC 3201	Vel	GC	10	17.6	−46	25
80	NGC 5139	Cen	GC	13	26.8	−47	29
81	NGC 6352	Ara	GC	17	25.5	−48	25
82	NGC 6193	Ara	OC	16	41.3	−48	46
83	NGC 4945	Cen	SG	13	05.4	−49	28
84	NGC 5286	Cen	GC	13	46.4	−51	22
85	IC 2391	Vel	OC	08	40.2	−53	04
86	NGC 6397	Ara	GC	17	40.7	−53	40
87	NGC 1261	Hor	GC	03	12.3	−55	13
88	NGC 5823	Cir	OC	15	05.7	−55	36
89	NGC 6087	Nor	OC	16	18.9	−57	54
90	NGC 2867	Car	PN	09	21.4	−58	19
91	NGC 3532	Car	OC	11	06.4	−58	40
92	NGC 3372	Car	BN	10	43.8	−59	52
93	NGC 6752	Pav	GC	19	10.9	−59	59
94	NGC 4755	Cru	OC	12	53.6	−60	20
95	NGC 6025	TrA	PC	16	03.7	−60	30
96	NGC 2516	Car	OC	07	58.3	−60	52
97	NGC 3766	Cen	OC	11	36.1	−61	37
98	NGC 4609	Cru	OC	12	42.3	−62	58
99	—	Cru	DN	12	53	−63	
100	IC 2944	Cen	OC	11	36.6	−63	02
101	NGC 6744	Pav	SG	19	09.8	−63	51
102	IC 2602	Car	OC	10	43.2	−64	24
103	NGC 2070	Dor	BN	05	38.7	−69	06
104	NGC 362	Tuc	GC	01	03.2	−70	51
105	NGC 4833	Mus	GC	12	59.6	−70	53
106	NGC 104	Tuc	GC	00	24.1	−72	05
107	NGC 6101	Aps	GC	16	25.8	−72	12
108	NGC 4372	Mus	GC	12	25.8	−72	40
109	NGC 3195	Cha	PN	10	09.5	−80	52

Key

BN = bright nebula

EG = elliptical galaxy

IG = irregular galaxy

PN = planetary nebula

SN = supernova remnant

DN = dark nebula

GC = globular cluster

OC = open cluster

SG = spiral galaxy

Appendix E

The brightest galaxies
(compiled by Brian Skiff)

RA			Dec.			Name	Blue magnitude
h	m	s	(°)	(′)	(″)		
00	01	57.0	−15	27	01	WLM	11.03
00	15	08.6	−39	13	13	NGC 55	8.42
00	33	11.7	+48	30	28	NGC 147	10.47
00	34	46.6	−08	23	48	NGC 157	11.00
00	38	58.1	+48	20	18	NGC 185	10.10
00	40	22.6	+41	41	11	NGC 205	8.92
00	42	42.0	+40	51	55	NGC 221	9.03
00	42	44.5	+41	16	08	NGC 224	4.36
00	47	08.8	−20	45	38	NGC 247	9.67
00	47	33.2	−25	17	18	NGC 253	8.04
00	52	38.2	−72	48	01	SMC	2.70
00	54	53.9	−37	40	57	NGC 300	8.72
01	00	09.5	−33	42	33	Sculptor	9.50
01	04	48.5	+02	07	10	IC 1613	9.88
01	33	51.0	+30	39	37	NGC 598	6.27
01	34	17.6	−29	24	58	NGC 613	10.73
01	36	41.8	+15	47	00	NGC 628	9.95
01	59	20.4	+19	00	22	NGC 772	11.09
02	22	33.2	+42	20	48	NGC 891	10.81
02	23	04.7	−21	14	00	NGC 908	10.83
02	27	16.9	+33	34	41	NGC 925	10.69
02	39	59.5	−34	26	57	Fornax	8.40
02	40	24.2	+39	03	46	NGC 1023	10.35
02	42	40.3	−00	00	48	NGC 1068	9.61
02	46	19.0	−30	16	21	NGC 1097	10.23
03	09	45.4	−20	34	52	NGC 1232	10.52
03	17	17.7	−41	06	28	NGC 1291	9.39
03	18	15.5	−66	29	51	NGC 1313	9.20
03	22	41.7	−37	12	28	NGC 1316	9.42
03	33	36.7	−36	08	17	NGC 1365	10.32
03	36	27.0	−34	58	33	NGC 1380	10.87
03	38	29.1	−35	26	58	NGC 1399	10.55
03	38	29.7	−23	01	40	NGC 1395	10.55
03	38	51.8	−35	35	36	NGC 1404	10.97
03	38	51.9	−26	20	11	NGC 1398	10.57

The brightest galaxies (*cont.*)

RA			Dec.			Name	Blue magnitude
h	m	s	(°)	(′)	(″)		
03	40	12.5	−18	34	52	NGC 1407	10.70
03	42	01.4	−47	13	17	NGC 1433	10.70
03	46	49.7	+68	05	45	IC 342	9.10
04	12	05.6	−32	52	28	NGC 1532	10.65
04	15	45.1	−55	35	31	NGC 1549	10.72
04	16	10.4	−55	46	51	NGC 1553	10.28
04	17	37.4	−62	47	04	NGC 1559	11.00
04	20	00.4	−54	56	18	NGC 1566	10.33
04	45	42.2	−59	14	57	NGC 1672	10.28
05	05	15.2	−37	58	47	NGC 1792	10.87
05	07	42.9	−37	30	51	NGC 1808	10.74
05	23	34.7	−69	45	22	LMC	0.91
06	44	48.6	−27	38	20	NGC 2280	10.90
07	27	04.4	+80	10	41	NGC 2336	11.05
07	36	54.5	+65	35	58	NGC 2403	8.93
08	19	06.0	+70	42	51	Holmberg II	11.10
08	52	41.1	+33	25	03	NGC 2683	10.64
08	53	33.1	+51	18	53	NGC 2681	11.09
08	55	38.7	+78	13	28	NGC 2655	10.96
09	10	20.6	+07	02	19	NGC 2775	11.03
09	11	37.7	+60	02	22	NGC 2768	10.84
09	17	53.1	−22	21	20	NGC 2835	11.01
09	22	01.8	+50	58	31	NGC 2841	10.09
09	32	09.8	+21	30	02	NGC 2903	9.68
09	45	39.5	−31	11	28	NGC 2997	10.06
09	47	15.6	+67	54	50	NGC 2976	10.82
09	55	33.5	+69	04	00	NGC 3031	7.89
09	55	54.0	+69	40	57	NGC 3034	9.30
10	03	06.8	−26	09	32	NGC 3109	10.39
10	03	21.1	+68	44	02	NGC 3077	10.61
10	05	14.2	−07	43	07	NGC 3115	9.87
10	08	27.4	+12	18	27	Leo I	10.7
10	13	03	−01	37		Sextans	11
10	14	14.5	+03	28	08	NGC 3169	11.08
10	18	17.4	+41	25	26	NGC 3184	10.36
10	19	55.0	+45	33	09	NGC 3198	10.87
10	23	31.6	+19	51	48	NGC 3227	11.10
10	28	22.5	+68	24	59	IC 2574	10.80
10	43	30.9	+24	55	25	NGC 3344	10.45
10	43	58.1	+11	42	15	NGC 3351	10.53
10	46	37.8	+63	13	22	NGC 3359	11.03
10	46	45.3	+11	49	16	NGC 3368	10.11

The brightest galaxies (*cont.*)

RA			Dec.			Name	Blue magnitude
h	m	s	(°)	(')	(")		
10	47	50.0	+12	34	57	NGC 3379	10.24
10	48	17.3	+12	37	49	NGC 3384	10.85
11	00	23.7	+28	58	33	NGC 3486	11.05
11	05	49.0	−00	02	15	NGC 3521	9.83
11	11	31.9	+55	40	15	NGC 3556	10.69
11	13	17.0	−26	45	20	NGC 3585	10.88
11	16	54.2	+18	03	12	NGC 3607	10.82
11	18	16.9	−32	48	49	NGC 3621	10.28
11	18	55.4	+13	05	35	NGC 3623	10.25
11	20	15.1	+12	59	29	NGC 3627	9.65
11	20	16.4	+13	35	22	NGC 3628	10.28
11	21	02.8	+53	10	17	NGC 3631	11.01
11	26	07.9	+43	35	06	NGC 3675	11.0
11	33	21.0	+47	01	39	NGC 3726	10.91
11	51	02.2	−28	48	23	NGC 3923	10.80
11	52	49.9	+44	07	26	NGC 3938	10.90
11	53	49.6	+52	19	39	NGC 3953	10.84
11	57	36.3	+53	22	31	NGC 3992	10.60
12	01	52.8	−18	51	54	NGC 4038	10.91
12	01	53.8	−18	53	06	NGC 4039	11.10
12	03	09.7	+44	31	55	NGC 4051	10.83
12	08	07.3	+65	10	22	NGC 4125	10.65
12	13	48.3	+14	53	43	NGC 4192	10.95
12	15	39.6	+36	19	39	NGC 4214	10.24
12	15	53.2	+13	08	58	NGC 4216	10.99
12	16	43.5	+69	27	56	NGC 4236	10.05
12	17	30.1	+37	48	27	NGC 4244	10.88
12	18	49.5	+14	25	07	NGC 4254	10.44
12	18	58.0	+47	18	16	NGC 4258	9.10
12	20	07.3	+29	16	47	NGC 4278	11.09
12	21	54.8	+04	28	20	NGC 4303	10.18
12	22	55.3	+15	49	23	NGC 4321	10.05
12	24	28.0	+07	19	06	NGC 4365	10.52
12	25	03.8	+12	53	15	NGC 4374	10.09
12	25	24.8	+18	11	27	NGC 4382	10.00
12	25	50.0	+33	32	46	NGC 4395	10.64
12	26	11.9	+12	56	49	NGC 4406	9.83
12	26	27.6	+31	13	29	NGC 4414	10.96
12	27	26.5	+11	06	29	NGC 4429	11.02
12	27	45.6	+13	00	36	NGC 4438	11.02
12	28	11.5	+44	05	40	NGC 4449	9.99
12	28	29.5	+17	05	05	NGC 4450	10.90

The brightest galaxies (*cont.*)

RA			Dec.			Name	Blue magnitude
h	m	s	(°)	(')	(")		
12	29	46.6	+07	59	58	NGC 4472	9.37
12	30	36.8	+41	38	23	NGC 4490	10.22
12	30	49.8	+12	23	24	NGC 4486	9.59
12	31	24.4	+25	46	25	NGC 4494	10.71
12	31	59.7	+14	25	17	NGC 4501	10.36
12	32	45.7	+00	06	44	NGC 4517	11.10
12	34	03.0	+07	42	01	NGC 4526	10.66
12	34	20.4	+08	11	53	NGC 4535	10.59
12	35	26.4	+14	29	49	NGC 4548	10.96
12	35	40.0	+12	33	25	NGC 4552	10.73
12	35	57.9	+27	57	36	NGC 4559	10.46
12	36	20.7	+25	59	05	NGC 4565	10.42
12	36	50.2	+13	09	48	NGC 4569	10.26
12	37	44.3	+11	49	11	NGC 4579	10.48
12	39	59.5	−11	37	22	NGC 4594	8.98
12	40	00.4	+61	36	33	NGC 4605	10.89
12	42	02.6	+11	38	49	NGC 4621	10.57
12	42	07.8	+32	32	28	NGC 4631	9.75
12	42	49.9	+02	41	17	NGC 4636	10.43
12	43	40.4	+11	32	58	NGC 4649	9.81
12	43	56.7	+13	07	33	NGC 4654	11.10
12	45	06.3	+03	03	26	NGC 4665	10.50
12	48	36.0	−05	48	02	NGC 4697	10.14
12	49	02.4	−08	39	52	NGC 4699	10.41
12	50	27.0	+25	30	01	NGC 4725	10.11
12	50	53.7	+41	07	10	NGC 4736	8.99
12	52	22.9	−01	11	57	NGC 4753	10.85
12	56	44.4	+21	41	05	NGC 4826	9.36
13	05	26.3	−49	28	15	NGC 4945	9.30
13	08	38.4	−49	30	17	NGC 4976	11.04
13	10	56.3	+37	03	29	NGC 5005	10.61
13	13	28.1	+36	35	38	NGC 5033	10.75
13	15	49.4	+42	02	06	NGC 5055	9.31
13	18	55.4	−21	02	21	NGC 5068	10.70
13	21	57.9	−36	37	47	NGC 5102	10.35
13	25	29.1	−43	01	00	NGC 5128	7.84
13	29	53.4	+47	11	48	NGC 5194	8.96
13	29	58.8	+47	16	21	NGC 5195	10.45
13	37	00.4	−29	52	04	NGC 5236	8.20
13	37	32.0	+08	53	08	NGC 5248	10.97
13	38	03.7	−17	52	56	NGC 5247	10.50

The brightest galaxies (*cont.*)

RA			Dec.			Name	Blue magnitude
h	m	s	(°)	(′)	(″)		
13	39	56.0	−31	38	41	NGC 5253	10.87
13	56	07.2	+05	15	19	NGC 5363	11.05
14	03	12.6	+54	20	55	NGC 5457	8.31
14	32	41.5	−44	10	24	NGC 5643	10.74
15	06	29.5	+01	36	25	NGC 5846	11.05
15	06	30.3	+55	45	46	NGC 5866	10.74
15	09	11.3	+67	12	51	Ursa Minor	11.5
15	15	53.9	+56	19	46	NGC 5907	11.12
16	52	468	−59	12	59	NGC 6221	10.66
17	16	59.4	−62	49	11	NGC 6300	10.98
17	20	18.7	+57	54	48	Draco	11
17	49	27.8	+70	08	41	NGC 6503	10.91
19	09	45.5	−63	51	22	NGC 6744	9.14
19	44	58.0	−14	48	11	NGC 6822	9.31
20	34	52.1	+69	09	15	NGC 6946	9.61
22	02	41.9	−51	17	44	IC 5152	11.06
22	07	52.3	+31	21	35	NGC 7217	11.02
22	09	17.0	−47	09	58	NGC 7213	11.01
22	37	05.3	+34	25	10	NGC 7331	10.35
22	57	10.6	−36	27	45	IC 1459	10.97
22	57	18.1	−41	04	09	NGC 7424	10.96
23	34	27.4	−36	06	05	IC 5332	11.09
23	57	49.6	−32	35	24	NGC 7793	9.63

Appendix F

The planets: angular size

	Maximum (equatorial)	Minimum (equatorial)
Mercury	10″	4.9″
Venus	64″	10″
Mars	25.16″	3.5″
Jupiter	50.11″	30.467″
Saturn	20.75″	18.44″
Uranus	3.96″	3.60″
Neptune	2.52″	2.49″
Pluto	0.11″	0.065″

Notes

The diameter of Saturn's rings is 225% of its equatorial diameter. Pluto reached perihelion 4 Sep 1989 and closest approach to Earth 7 May 1990. At that moment, its disk diameter was at a maximum. Its size is now shrinking and will continue to decrease each year – ever so slightly – until 2115, when Pluto's apparent size will again begin to grow.

Appendix G

The planets: brilliancy at opposition

	Maximum visual magnitude	Minimum visual magnitude
Mars	−2.9	−1.0
Jupiter	−2.9	−2.0
Saturn	−0.3	+0.9
Uranus	+5.65	+6.06
Neptune	+7.66	+7.70
Pluto	+13.6	+15.95

Appendix H

The planets: distance from the Sun

	Maximum	Mean	Minimum
Mercury	69 815 900 km 0.4667 AU	57 910 000 km 0.3871 AU	46 003 500 km 0.3075 AU
Venus	108 934 900 km 0.7282 AU	108 200 000 km 0.7233 AU	107 463 300 km 0.7184 AU
Earth	152 104 980 km 1.016 759 AU	149 597 870 km 1.0000 AU	147 085 800 km 0.983 208 AU
Mars	249 251 000 km 1.666 14 AU	227 940 000 km 1.5237 AU	206 615 600 km 1.3811 4 AU
Jupiter	816 056 400 km 5.455 AU	778 330 000 km 5.2028 AU	740 659 100 km 4.591 AU
Saturn	1 506 750 000 km 10.072 AU	1 429 400 000 km 9.5388 AU	1 347 877 000 km 9.010 AU
Uranus	3 007 665 000 km 20.105 AU	2 870 990 000 km 19.1914 AU	2 734 799 000 km 18.281 AU
Neptune	4 534 406 000 km 30.324 AU	4 504 300 000 km 30.0611 AU	4 458 765 000 km 29.805 AU
Pluto	7 524 587 000 km 50.299 AU	5 913 520 000 km 39.5294 AU	4 435 128 000 km 29.647 AU

Appendix I

The planets: orbital inclination and eccentricity

	Inclination	Eccentricity
Mercury	7.004°	0.2056
Venus	3.394°	0.0068
Earth	0.000°	0.0167
Mars	1.850°	0.0934
Jupiter	1.308°	0.0483
Saturn	2.488°	0.0560
Uranus	0.774°	0.0461
Neptune	1.774°	0.0097
Pluto	17.148°	0.2482

Appendix J

The planets: mass

	Mass (kg)	Ratio (Earth = 1)
Mercury	3.303×10^{23}	0.055
Venus	4.869×10^{24}	0.815
Earth	5.976×10^{24}	1.000
Mars	6.421×10^{23}	0.107
Jupiter	1.900×10^{27}	317.94
Saturn	5.688×10^{26}	95.18
Uranus	8.686×10^{25}	14.54
Neptune	1.024×10^{26}	17.14
Pluto	1.29×10^{22}	0.002

Appendix K

The planets : orbital period and velocity

	Sidereal period	Velocity (km/h)	Synodic period
Mercury	$0^y 87^d 23.3^h$	172 368	115.88^d
Venus	$0^y 224^d 16.8^h$	126 072	583.92^d
Earth	$1^y 0^d 0^h$	107 224	—
Mars	$1^y 320^d 18.2^h$	86 868	779.94^d
Jupiter	$11^y 315^d 1.1^h$	47 052	398.88^d
Saturn	$29^y 167^d 6.7^h$	34 812	378.09^d
Uranus	$84^y 3^d 15.66^h$	24 516	369.66^d
Neptune	$164^y 288^d 13^h$	19 620	367.49^d
Pluto	$248^y 197^d 5.5^h$	17 064	366.73^d

Appendix L

The planets: rotational period and velocity

	Period	Velocity (km/h)
Mercury	$59^d15^h30.5^m$	10.891
Venus	$243^d0^h26.9^m$	6.520
Earth	$0^d23^h56.1^m$	1650.8
Mars	$1^d0^h37.4^m$	866.9
Jupiter	$0^d9^h55.5^m$	45 259.5
Saturn	$0^d10^h14^m$	37 004.9
Uranus	$0^d17^h14^m$	8 971.5
Neptune	$0^d16^h6.6^m$	9 667.1
Pluto	$6^d9^h17.6^m$	47.6

Appendix M

The planets: size

	Equatorial diameter (km)	Ratio (Earth = 1)	Polar diameter
Mercury	4 879.4	0.383	4 879.4
Venus	12 104	0.949	12 104
Earth	12 576.28	1.000	12 533
Mars	6 794.4	0.533	6 794.4
Jupiter	142 984	11.209	133 717
Saturn	120 536	9.449	107 566
Uranus	51 118	4.007	49 584
Neptune	49 572	3.880	48 283
Pluto	2 320	0.182	2 320

Appendix N

The planets: volume

Mercury	$6.084 \times 10^{10}\,km^3$ 5.8% that of Earth
Venus	$9.2843 \times 10^{11}\,km^3$ 85.7% that of Earth
Earth	$1.042 \times 10^{12}\,km^3$
Mars	$1.643 \times 10^{11}\,km^3$ 15.8% that of Earth
Jupiter	$1.377 \times 10^{15}\,km^3$ 1321 times that of Earth
Saturn	$8.183 \times 010^{14}\,km^3$ 785 times that of Earth
Uranus	$6.995 \times 010^{13}\,km^3$ 67.1 times that of Earth
Neptune	$6.379 \times 10^{13}\,km^3$ 61.2 times that of Earth
Pluto	$6.545 \times 10^{9}\,km^3$ 0.628% that of Earth

Appendix O

Satellite distances from the planets

		Distance from planet (km)
Earth	Moon	3.844×10^5
Mars	Phobos	9.377×10^3
	Deimos	2.3436×10^4
Jupiter	Amalthea	1.813×10^5
	Io	4.216×10^5
	Europa	6.709×10^5
	Ganymede	1.07×10^6
	Himalia	1.148×10^7
Saturn	Prometheus	1.3935×10^5
	Hyperion	1.4811×10^5
	Mimas	1.8552×10^5
	Enceladus	2.3802×10^5
	Tethys	2.9466×10^5
	Dione	3.774×10^5
	Rhea	5.2704×10^5
	Titan	1.22185×10^6
	Iapetus	3.5613×10^6
Uranus	Miranda	1.298×10^5
	Ariel	1.912×10^5
	Umbriel	2.66×10^5
	Titania	4.358×10^5
	Oberon	5.826×10^5
Neptune	Proteus	1.17647×10^5
	Triton	3.5476×10^5
	Nereid (average)	5.5134×10^6
Pluto	Charon	1.9405×10^4

Appendix P

Satellite size and orbital period

		Equatorial diameter (km)	Orbital period
Earth	Moon	3476	$27^d07^h43.7^m$
Mars	Phobos	26 × 18	$07^h39.2^m$
	Deimos	16 × 10	$01^d06^h17.9^m$
Jupiter	Amalthea	262 × 146 × 134	$11^h57.4^m$
	Io	3630	$01^d18^h27.6^m$
	Europa	3120	$03^d13^h14.6^m$
	Ganymede	5268	$07^d03^h42.6^m$
	Callisto	4806	$16^d16^h32.2^m$
	Himalia	180	$250^d13^h35.3^m$
Saturn	Prometheus	1528	$14^h42.7^m$
	Hyperion	370 × 280 × 226	$21^d06^h38.3^m$
	Mimas	397.6	$22^h37.1^m$
	Enceladus	498.2	$01^d08^h53.1^m$
	Tethys	1059.8	$01^d21^h18.4^m$
	Dione	1120	$02^d17^h41.2^m$
	Rhea	1530	$04^d12^h25.2^m$
	Titan	5150	$21^d06^h38.3^m$
	Iapetus	1436	$79^d07^h55.5^m$
Uranus	Miranda	480 × 468 × 465	$01^d09^h54.7^m$
	Ariel	1162 × 1156 × 1156	$02^d12^h28.8^m$
	Umbriel	1169.4	$04^d03^h27.4^m$
	Titania	1577.8	$08^d16^h56.6^m$
	Oberon	1522.8	$13^d11^h06.7^m$
Neptune	Proteus	436 × 416 × 402	$01^d02^h56.1^m$
	Titon	2705.2	$05^d21^h02.7^m$ (r)
	Nereid	340	$360^d03^h16.11^m$
Pluto	Charon	1172	$06^d09^h17.3^m$

Appendix Q

Useful magnification ranges for visual observing in astronomical telescopes (David Knisely, Lincoln, Nebraska)

- LOW POWER (3.7× to 9.9× per inch of aperture/6.9 mm to 2.6 mm exit pupil): Useful for finding objects and for observing ones of large angular size like open clusters, large faint nebulae, or some larger galaxies. For lunar work, it is generally somewhat on the low side, but can show the thin Crescent Moon with background starfields well. This is also the range where Nebula filters tend to perform the best.

- MEDIUM POWER (10× to 17× per inch of aperture/2.5 mm to 1.4 mm exit pupil): Useful for observing details in many deep-sky objects such as galaxies, some diffuse nebulae, smaller open clusters, and moderate to large planetary nebulae, as well as getting at least partial resolution on many brighter globular star clusters. Also useful in detecting very small galaxies which may be invisible at low powers and for revealing star-like galactic nuclei. Very useful for wide area views of the Moon, or for showing the Moon systems and occasionally, some of the larger features of the planets.

- HIGH POWER (18× to 29.9× per inch of aperture/1.4 mm to 0.8 mm exit pupil): A very useful power range for observing the planets and for high-power lunar viewing (this is the range where the full theoretical resolving power of the telescope is becoming visible). Also useful for getting better star resolution in tight globular clusters or for viewing detail in the smaller planetary nebulae, as well as resolving tight double stars. This power range is sometimes compromised in apertures larger than 5 inches by seeing effects (i.e. disturbances in the Earth's atmos-phere which can blur fine detail).

- VERY HIGH POWER (30× to 41.9× per inch of aperture/0.8 mm to 0.6 mm exit pupil): Useful for high-resolution study of planetary detail, and resolving double stars near or just above the theoretical resolution limit of the instrument. Also useful for resolving the cores of some very tight globular clusters or for detecting the finer detail and faint central stars in the smaller planetary nebulae. Quite useful for telescope collimation tests or rough star-testing. This power range is not as frequently used with larger apertures as somewhat lower powers are due to seeing disturbances. Eye defects like motes and floaters also begin to become more noticeable and annoying in the upper half of this range.

- EXTREME POWER (42× to 75× per inch of aperture/0.6 mm to 0.3 mm exit pupil): Mainly used for resolution of double stars at the resolution limit of the instrument, or for detecting elongation of unresolved doubles. Powers up to 60× per inch are sometimes usable in rather small instruments for making gross planetary detail easier for beginners to see (i.e. Jupiter's main belts or the Cassini Division in Saturn's rings). This power range is not often used in apertures above 6 inches due to seeing limitations, and requires good optical quality in the instrument. Even when conditions are good, lunar and planetary views using this power range can sometimes seem less pleasing overall than at somewhat lower powers due to the lower light intensity and increasing interference from eye defects like floaters. However, this range can be somewhat useful for certain "specific" targets or details which require extreme scale. Examples include (for large apertures) seeing Encke's Division in Saturn's rings, the central star in M57, detail in some brighter planetary nebulae, or for resolving a few small specific lunar details. Powers from 75× to 90× per inch are occasionally used for very close double star elongation or for optical testing, but otherwise, powers beyond 75× per inch can be nearly useless.

- EMPTY MAGNIFICATION (100× per inch of aperture and above). Basically useless powers. Used mainly as a ploy by unscrupulous telescope retailers or manufacturers to sell small over-powered telescopes to beginners.

Appendix R

Approximate telescopic limiting magnitude

mm	Magnitude
75	13.1
100	13.7
125	14.2
150	14.6
200	15.2
250	15.7
300	16.1
350	16.5
400	16.7
450	17.0
500	17.2
600	17.6
750	18.1
900	18.5

Nebulosity surrounding Gamma Cygni. Robert Gendler (Avon, Connecticut) said the following about his image: "The mosaic was taken with my 300 mm Nikon lens attached to my ST10 CCD camera. It's made up of two frames. Each frame is an LRGB where the luminance (L) is a 60 minute "red" filtered exposure and the RGB was 10:10:10 minutes."

Glossary

Abbe number a number indicating the dispersion of an optical substance; the Abbe number is large when the glass is low dispersion.

aberration a defect in the image formed by an optical system. Such defects include astigmatism, chromatic aberration, coma, curvature of field, distortion, and spherical aberration.

absolute magnitude the apparent magnitude of a celestial object if the object were at a distance of 10 parsecs; the true brightness of a celestial object; luminosity.

absolute magnitude of a comet the brightness of a comet when it is at 1 AU from both the Earth and Sun. As this virtually never happens, this quantity is calculated from the comet's light curve. Unfortunately, this quantity is far from absolute. It can be different pre- and post-perihelion. It can also change from apparition to apparition (for periodic comets).

absorption lines black lines corresponding to missing discrete colored lines in continuous spectra due to light from behind being absorbed by atoms (or molecules). Study of these Fraunhofer (1787–1826) lines is helpful in identifying non-radiating atoms in space, though 15 000 lines have been observed in the Sun.

absorption nebula a region of dust and gas which has no stars to illuminate it from within; these objects are completely dark and obscure any light from stars or emission nebulae which may lie behind them; typical temperatures lie in the range of 10 to 20 K.

accretion disk a disk-like structure of dust and gas which forms around the massive object at the centre of a region undergoing gravitational collapse. Accretion disks are, primarily, found around protostars and black holes.

achromat *see* achromatic lens.

achromatic doublet *see* achromatic lens.

achromatic lens a two-element lens that is corrected to bring two wavelengths (usually of red and blue light) to a common focal point to reduce the defect found in single lenses known as chromatic aberration. The term achromatic literally means "without color," but such lenses do exhibit slight chromatic aberration.

active galactic nucleus a galactic nucleus in which some kind of extraordinary activity (perhaps a black hole) is liberating vast amounts of energy in the form of electromagnetic radiation.

active galaxy a galaxy in which some kind of unusual, perhaps violent, phenomenon is releasing energy, usually from the nucleus.

active optics a type of corrective system applied to mirrors of large reflectors to correct for distortions in the mirror itself, or, as in the case of newer multi-mirrored telescopes, the system of mirrors.

adaptive optics a type of corrective system applied to mirrors of large reflectors through a system of fast reacting electronic supports which deform the mirror to compensate for the twinkling of stars.

afocal system in a telescope, this is achieved when the object and the secondary image are both (theoretically) infinitely distant; in astrophotography, this is a telescope focused at infinity combined with a camera with the lens on it also focused on infinity.

airglow a faint glow in the Earth's upper atmosphere caused by light emitted during the recombination of atoms and molecules following collisions with high-energy particles and photons, mainly from the Sun. Termed airglow generally, but nightglow when seen at night.

Airy disk the image of a point source of light formed by a perfect lens as a minute pattern of concentric and progressively fainter rings of light surrounding a central dot; the size of the Airy disk sets a limit to the resolving power of a telescope.

albedo the reflectance of a planet, satellite, or other non-luminous object; the ratio of the total amount of light reflected in all directions. A perfect reflector would have an albedo of 1.0; a black surface which absorbs all light would have an albedo of 0.0. Albedo may be divided into two types: Bond albedo and geometric albedo.

altazimuth (alt-az) mounting a type of telescope mount where one axis points to the zenith and allows rotation along the horizon and the other allows changes in altitude, or distance above the horizon; used for most small telescopes; a Dobsonian mount is an alt-az system; also used for larger telescopes with the addition of computer-controlled drives for both axes.

altitude the angular distance of a point or celestial object above or below the horizon. It is measured along the vertical circle through the body from 0° (on the horizon) to 90° (at the zenith). Negative values correspond to objects which lie below the horizon.

aluminizing the process by which a thin coating of aluminum is vacuum-deposited on an unfinished mirror blank.

anastigmatic system one which is free from astigmatism; this is accomplished by the use of a compound lens in which each lens corrects for the astigmatic effects of the other lens.

angular diameter the apparent size of a celestial object, measured in degrees, minutes, and seconds, as seen from the Earth; for example, the average angular size of the Sun, as seen from Earth, is 0.53°.

angular distance the observed distance between two celestial bodies expressed in degrees, minutes, and/or seconds of arc.

ansae the portion of Saturn's rings visible on either side of the planet; Latin for "handles"; also, protrusions of some planetary nebulae.

antapex see solar antapex.

anti-tail (or anomalous tail) an apparently sunward-pointing comet's tail. In reality, the tail only appears to be pointing toward the Sun. The anti-tail is produced by large ("heavy") dust particles left along the comet's orbit instead of being pushed away from the Sun by light pressure. When a comet is close to the Sun, and if the Earth–comet–Sun geometry is correct, the dust in the comet's orbit will appear to point toward the Sun.

apastron the maximum distance between the components of a binary star system.

aperture the diameter of the main lens or mirror in an optical telescope or the antenna size in a radio telescope.

apex see solar apex.

aphelion the position of an object in solar orbit when it is furthest from the Sun; the instant in a given orbit of a planet (or other body) when it is furthest from the Sun.

apochromatic lens an optical design, usually incorporating three elements, which is virtually free of residual color.

apogee the position of the Moon or an object in Earth orbit when it is furthest from the Earth; the instant in a given orbit of the Moon (or other object) when it is furthest from the Earth.

aplanatic system an optical system designed and manufactured to minimize spherical aberration and coma.

apparent field of view (afov) the angular size of the light cone able to enter through the eyepiece; the range is from 25°–80° or so.

apparent magnitude the brightness of a celestial object as seen from Earth, irrespective of its true brightness.

apparition — the period of time during which a celestial body may be observed.

appulse — the close approach of one celestial object to another without the occurrence of an occultation or an eclipse.

arcminute — *see* minute of arc.

arcsecond — *see* second of arc.

ascending node — *see* nodes.

ashen light — a faint luminosity on the nightside of Venus, first reported in 1643, similar in appearance to "earthshine" on the Moon but generally not as bright.

aspect — the position of the Moon or planets relative to the Sun, as seen from Earth. Conjunctions, oppositions and quadratures are examples of aspects. Also known as configuration.

aspheric lens — a lens that has at least one surface not spherically symmetric.

association — a large, loose grouping of young stars of similar spectral type; a much looser group than an open cluster.

asterism — an unofficial, recognizable grouping of visible stars. The stars within an asterism may belong to one constellation (e.g., the "Big Dipper," within the constellation Ursa Major) or several constellations (e.g., the "Summer Triangle," composed of stars from the constellations Lyra, Cygnus, and Aquila).

asteroid — small bodies composed of rock and metal which orbit the Sun. Most (95%) lie in a belt between the orbits of the planets Mars and Jupiter. Also known as minor planet.

asteroid belt — an area between the orbits of Mars and Jupiter where most of the asteroids are found; also referred to as the main belt.

asteroid designation — when an asteroid is discovered, it is given a provisional designation by the Minor Planet Center (MPC). When it has been observed enough to determine a good orbit (usually a period of two to five oppositions) the MPC will assign it a number. In addition, the discoverer has the right to give the asteroid its official name.

astigmatism — a defect in an optical system which causes light to not be focused to a point; this aberration is most noticeable when off-axis light enters the system.

astrometric binary — a double star where only one of the components is visible; the system is determined to be binary due to the gravitational influence the unseen member exerts on the visible star.

astrometry — the measurement of the positions of celestial objects.

astronomical horizon — the great circle on the celestial sphere whose every point lies 90° from the zenith.

astronomical twilight — the condition of solar illumination at a point on the Earth's surface before sunrise or after sunset when the Sun's zenith distance is 108° (that is, when the Sun is 18° below the horizon).

astronomical unit (AU) — a unit of distance which is approximately (within about 3/100 000 000) the average distance from the Earth to the Sun; this distance is approximately 149 597 870 km.

astronomy — the study (both observational and theoretical) of the physical and chemical properties of celestial objects which inhabit the universe, or of the universe as a whole.

astrophotography — the imaging, onto photographic emulsion, of celestial objects; separate from astroimaging, which generally refers to images acquired by CCD cameras.

atmospheric refraction — *see* refraction, atmospheric

AU — *see* astronomical unit.

aurora — a glow in a planet's ionosphere caused by the interaction between the planet's magnetic field and charged particles from the Sun (the solar wind).

autoguider — a CCD camera attached to a small telescope (usually the guide scope) which, with an associated computer drive system, keeps a star centered during an exposure, thus compensating for the rotation of the Earth.

autumn (or autumnal) equinox — *see* September equinox.

axis — the imaginary line around which a rotating body (such as a planet or satellite) turns.

azimuth — the angular distance to an object measured eastwards along the horizon from the north to the intersection of the object's vertical circle; varies from 0° to 360°. Thus, an object due east would have an azimuth of 90°, and an object due west would have an azimuth of 270°.

Baily's Beads — during a total solar eclipse, the effect seen just before and just after totality when only a few points of sunlight are visible at the edge of the Moon, caused by the irregularity of the lunar surface.

Barlow lens — a negative lens which increases the effective focal length of an objective lens or mirror; rated by magnification factor 2×, 3×, etc.

Barnard objects — another name for dark nebulae which obscure the light from background stars. This name is used to honor an American astronomer who discovered them in about 1900.

barred spiral galaxy — a type of spiral galaxy where the arms do not originate at the center but rather from the ends of a bar-like structure which protrudes from the nucleus.

Bayer letter — a letter of the Greek alphabet assigned to visible stars within constellations; first use was by Johannes Bayer in his 1603 work, *Uranometria*.

Bertele eyepiece — a multi-element (usually four) eyepiece that provides a very wide apparent field of view.

bias frame — a zero time exposure of a CCD chip; it is the image of the readout noise from a CCD camera.

Big Bang — a theory which states that the universe came into being in an instantaneous event around 15 billion years ago; everything was created in that initial event; subsequently, the universe has expanded and the contents have evolved into the stars and galaxies of today.

Big Crunch — the theorized demise of a closed universe; if the universe contains enough mass, the expansion will reverse, and all matter will be drawn together by gravity.

binary star — a double star; a pair of stars in space; in most cases, the stars are revolving around a common center of mass, but occasionally there are stars which lie in the same direction (line of sight) which appear to be bound together – these are known as optical binaries.

binning — related to a CCD chip, the combining of multiple pixel charges in both the horizontal and vertical direction into a single larger charge. Binning 1 × 1 means that the individual pixel is used as is. Binning 2 × 2 means that an area of four adjacent pixels have been combined into one larger pixel. In 2 × 2 binning, the sensitivity to light has been increased by four times (the four pixel contributions), but the resolution of the image has been cut in half.

binoculars — a hand-held optical instrument composed of two telescopes and a focusing device which uses prisms to bring the images from each eye to a common focus.

bipolar nebula — a nebula with two lobes; often described as "butterfly" or "hourglass" shaped.

black drop — an extension of the dark disk of either Mercury or Venus during a transit at second and third contacts; a meniscus between the planet and the solar limb.

black dwarf — the cold remnant of a dead, low-mass star, formed by the cooling of a white dwarf.

black holes — stellar core remnants so dense that not even light can escape from them, formed in supernova explosions of the most massive stars.

black ligament — *see* black drop.

blazars — a type of active galaxy caused by jets of gas being expelled from the active galactic nucleus, at speeds close to the speed of light, almost directly into the line of sight from Earth. The name is a concatenation of BL Lacertae object and quasar.

blink comparitor — a device which alternates between two views (images) of the same star field in a search for moving objects. This may be done mechanically, as with two slides, or digitally via computer.

BL Lacertae — the first discovered blazar.

bloomed lens — lens with a coating that reduces reflections from its surface and in which interference between the coating and the lens produces some false color.

blueshift — a shift in the frequency of a photon (light) toward higher energy and shorter wavelength; produced by relative motion of the emitter toward the observer, as in the Doppler effect.

bolide — an exploding meteor.

brightness — *see* magnitude.

brilliancy at opposition — the apparent magnitude of a celestial body when at opposition, 180° from the Sun, as seen from the Earth.

brown dwarfs — objects formed by gravitational collapse, which are less than 0.08 solar masses, the lower limit to ignite the nuclear fusion of hydrogen into helium at its core, and which radiate energy in the infrared.

canals — imaginary features on Mars supposedly observed by Percival Lowell and several other astronomers around the end of the nineteenth and beginning of the twentieth centuries.

cardinal points — the four main directions, north, south, east and west, generally specified to lie on the horizon.

Cassegrain telescope — a telescope design in which light collected by the primary mirror (a paraboloid) reflects light to a secondary mirror (a hyperboloid) and that light is brought to a focus behind the primary mirror, through a hole in the center of the primary.

Cassini Division — a relatively empty area of Saturn's ring system, 4200 km wide, located between the A and B rings.

catadioptric telescope — one which uses a spherical primary mirror with the addition of a correcting plate; a compound telescope, employing elements of both refracting and reflecting telescopes.

CCD — acronym for charge coupled device; an electronic detector which collects light with a much higher quantum efficiency than photographic emulsions; technically, the CCD is just the light-sensitive chip, but the entire assembly is usually called a CCD camera.

cD galaxy — a huge elliptical galaxy found at the centre of a cluster of galaxies; the cD designation stands for "cluster dominating."

celestial equator — the intersection of the equatorial plane of the Earth with the celestial sphere; the projection of the Earth's equator onto the sky.

celestial longitude — the angular distance from the vernal equinox to the longitude circle of any celestial object, measured eastward along the ecliptic from 0°–360°.

celestial sphere — the apparent background of the stars, assumed to be of infinite extent in all directions; the sky.

central meridian — an arbitrary reference line on a planet (not Earth) which intersects both poles.

Cepheid variable — any of a class of pulsating stars whose very regular light variations are related directly to their luminosities by the period–luminosity relationship. Cepheids are named after the prototype star, Delta Cephei.

Ceres
asteroid number 1, discovered on 1 Jan 1801; the largest known asteroid, having a diameter of 930 km.

chromatic aberration
an aberration whereby light of different wavelengths is focused at different distances from the objective.

chromosphere
the region of the atmosphere of a star (such as the Sun) between the star's photosphere and its corona.

circumpolar star
a star whose apparent daily path through the sky lies completely above an observer's horizon. At the equator no star is circumpolar. At the North or South Pole all stars are circumpolar. For an observer at any other latitude a star whose declination is greater than 90° minus the observer's latitude will be circumpolar.

civil twilight
the condition of solar illumination at a point on the Earth's surface before sunrise or after sunset when the Sun's zenith distance is 96° (that is, when the Sun is 6° below the horizon).

cloud features
visible features in the atmosphere of a planet or satellite; may be temporary or permanent.

cluster
physically related group of stars held together by gravity; see globular cluster, open cluster.

cluster of galaxies
a physically related group of galaxies held together by gravity. NOTE: not the same as a galactic cluster.

coaltitude
90 minus the altitude, expressed in degrees; also known as zenith distance.

coating, anti-reflection
a thin dielectric or metallic film (applied as one or many layers) applied to lenses that reduces reflections and increases the transmission of the lens.

cold camera
a camera used for astrophotography in which the film is cooled (usually by chilled water, dry ice, or liquid nitrogen), while the image is being taken; the cooling of the film is done to prevent reciprocity failure. NOTE: CCD chips are often cooled but this is to dissipate the heat they produce, as well as to increase the thermal efficiency.

collimate
to adjust the line of sight or lens axis of an optical instrument so that it is in its proper position relative to other parts of the instrument.

collimator
an optical device for artificially creating a target at infinite distance used in testing and adjusting certain optical instruments, usually consisting of a converging lens and a target, and a system or arrangement of crosshairs, placed at the principal focus of the lens.

color index
the difference in brightness of a star, as measured at two different wavelengths, usually blue and visual (giving a number for the quantity B−V).

coma
(1) an optical defect associated mainly with reflecting telescopes which affects off-axis images and is more pronounced near the edges of the field of view; the images have a comet-shaped appearance, hence the name; (2) the gaseous, usually spherical area that surrounds and hides a comet's true nucleus; composed of gas and dust, the coma is evaporated off the comet's nucleus.

combined magnitude
a single magnitude expressing the brightness of both stars in a binary system.

comet
one of many relatively small solar system objects which orbit the Sun, composed of frozen gases and dust.

concave
hollowed or curving inward; in astronomy, this generally refers to the shape of a mirror or lens element.

conjunction
the alignment of two celestial objects such that the difference in their longitude, as seen from Earth, is 0°. Two objects may also be in conjunction in right ascension. When one of the objects is the Sun, "conjunction" denotes when the other object is in line with the Sun, and therefore usually invisible.

constellation
one of 88 arbitrary configurations of stars; the (officially recognized) area of the celestial sphere containing one of these configurations.

convex bowed or curving outward; in astronomy, this generally refers to the shape of a mirror or lens element.

Copernican system a model of our solar system introduced by Nicolas Copernicus. It was published in 1543, in the book *De Revolutionibus Orbium Coelestium* (*On the Revolutions of the Celestial Spheres*). It placed the Sun in the center of the solar system, rather than the Earth as was previously thought.

corona the thin shell of illuminated gas that extends out some distance from the surface of the Sun or a star.

coronal mass ejection (CME) gigantic eruptions of gas from the Sun's corona, more often seen during times of solar maximum.

corrector an aspheric lens used to correct the aberrations which would be caused by using only a spherical mirror in a catadioptric telescope (e.g., Schmidt, Schmidt–Cassegrain, Maksutov).

cosmology the study of why the universe is like it is, how it came into being, how it will evolve and how it will end.

counterglow *see* gegenschein.

craters roughly circular depressions on the surface of many objects in the solar system; most craters were created through meteoritic impact, with the remainder being volcanic or caused by surface collapse.

Crepe ring a faint ring of Saturn between the brightest (B) ring and the ball of the planet; also called the C ring.

crescent a phase of the Moon or other celestial body where the percentage of visible surface illumination is greater than 0% but less than 25% (or 50% of the side currently being observed).

crown lens glass with an Abbe number higher than about 55.

culmination the passage of a celestial body across an observer's meridian; upper culmination (also called transit) is the crossing nearer to the observer's zenith; lower culmination is the crossing further from the observer's zenith.

curvature of field an optical system defect present when the sharpest focus is formed along a curved surface rather than a flat plane.

dark adaptation when the eye becomes adapted to luminances of less than about 0.034 candlepower/m².

dark frame a raw image which is taken with no light incident on a CCD chip; this is to determine the internal noise of the CCD camera; generally, the length of a dark frame exposure is equal to the exposure of the image, or scaled to it.

dark nebula a region of space in which the material is visible because it blocks out the light from background stars and emission nebulae.

Dawes limit approximately the smallest separation of equally-bright stars that can be achieved with a given telescope; the formula given is 118 divided by the aperture in millimeters or 4.54 divided by the aperture in inches.

day generally defined as one rotation of a planet on its axis. A sidereal day is measured with respect to the stars, while a solar day is measured with respect to the Sun. Many other types of days are defined.

dayglow the resonant absorption and emission of sunlight at a particular wavelength in the atmosphere of a planet or satellite.

December solstice that instant when the Sun achieves minimum declination; the point on the ecliptic where the Sun's declination is at a minimum, having celestial coordinates of RA = 18^h, Dec. = $-23.5°$, approximately. NOTE: this point is the "winter" solstice only in the northern hemisphere of Earth.

declination an Earth-centered angle measured perpendicularly from the celestial equator to a point on the celestial sphere. Declination is positive if the object or point is north of the celestial equator and negative if the object or point is south of the celestial equator.

degree of arc	1/360 of a circle; one degree of arc contains 60 minutes of arc; designated; rather than; the sign for seconds of arc thus 85°18′08″ = 85 degrees, 18 minutes, 8 seconds.
degree of condensation (DC)	an indicator of how much the surface brightness of the coma of a comet increases toward the center of the coma; DC = 0 indicates totally diffuse and DC = 9 means stellar.
descending node	*see* nodes.
detection level	the point at which a star or other celestial object becomes observable; sometimes given as a percentage of the time the object is observable, whether because of unstable conditions or due to the true faintness of the object.
dew cap	an attached extension to the optical tube of a telescope (generally a refractor or catadioptric) which prevents condensation from forming on the lens or corrector plate.
diagonal	*see* diagonal mirror.
diagonal mirror	a flat mirror used in a Newtonian telescope to reflect the image formed by the primary mirror into the eyepiece; when used in this way, the diagonal mirror is often called the secondary.
diameter	the length from the surface of a celestial object, through its center, to the surface on the other side; diameter equals radius times two.
diamond ring	the effect just prior to second contact, or just after third contact of a total solar eclipse when a small portion of the solar photosphere plus the corona produce an effect similar to a ring with a brilliant diamond.
diaphragm	a baffle (generally more than one is employed) in the light path used for enhancing image contrast by reducing the incidence of unwanted scattered light; may also refer to an aperture mask to effectively increase the focal ratio of the telescope.

dichotomy	the moment when a planet (Mercury or Venus) appears exactly half illuminated.
diffraction	the effect which causes a star to appear as a small disk (called the Airy disk) in a telescope. This is surrounded by bright rings called diffraction rings. With perfect optics, 84% of the light from the star will be in the bright center, another 7% in the first ring, 3% in the second ring, etc. Poor optics, or a telescope with a central obstruction, will spread the light out into the diffraction rings.
diffraction grating	a first-surface mirror (reflection grating) or optically flat glass (transmission grating) with a large number of parallel, closely spaced grooves, causing light to be separated into its component wavelengths.
diffraction-limited	a claim made by a number of telescope manufacturers, technically referring to a telescope that meets the 1/50th wave root-mean-square wavefront criterion, corresponding to a Strehl ratio of 98.4%.
direct motion	the apparent (west to east) motion of a planet or other celestial object on the celestial sphere, as seen from Earth.
disk	the visible surface of any heavenly body projected against the sky.
Dobsonian mount	a simple altazimuth mount invented by John Dobson.
Doppler effect	a physical process which alters the wavelength of electromagnetic radiation (or sound) because the source of emission and the observer are in relative motion. Motion toward the source produces a shift toward shorter wavelengths (a blue shift); motion away produces a shift toward longer wavelengths (a red shift).
double star	any pair of stars that appear close to one another on the celestial sphere. Double stars may be physically linked by gravity to one another, in which case are known as binary stars, or simply chance alignments.
doublet	a lens containing two elements.

dust lane — a dark band (composed of gas and dust) which blocks out the light from stars in a galaxy.

dust tail — one of two types of comet tails, composed of dust and shining by reflected sunlight; dust tails are more curved than ion tails.

dwarf galaxy — a very small galaxy or one with very low surface brightness; such objects contain at most a few million stars.

dwarf star — a main sequence star too small to be classified as a giant star or a supergiant star; the Sun is classified a dwarf star.

eccentricity — a measurement (from 0 to 1) which is the amount that the orbit of any solar system object is not circular. An object in a circular orbit would have an eccentricity of 0. Mathematically, this is defined as the distance between the focal points of an ellipse divided by twice the length of the major axis.

eclipse — the obscuration of light from a celestial body as it passes through the shadow of another body; such obscuration may be total or partial.

eclipse season — the period of time when the Sun is near alignment with a lunar node, during which eclipses may take place. For solar eclipses, this time window of $37\frac{1}{2}$ days occurs every 173 days.

eclipse year — the length of time it takes for a lunar node to return to its original alignment with respect to the Sun (about 346.6 days).

eclipsing binary — a binary star system in which chance alignment of the orbits mean that, from Earth, the stars periodically pass in front of one another causing eclipses, resulting in a variability of the light from the system.

ecliptic — the great circle described by the Sun's annual path on the celestial sphere; the mean plane of the Earth's orbit around the Sun.

ED glass — an abbreviation for extra low dispersion, a glass type that does not disperse light into its component colors as easily as regular glass, and therefore shows less chromatic aberration.

effective focal length — the focal length of a compound lens; the focal length of the system is related to the focal lengths of the component lenses and the distance between them.

element — an individual lens in a multi-element design.

ellipse — a type of conic section with an eccentricity less than one (a circle is an ellipse with an eccentricity of zero).

elliptical galaxy — a galaxy with an ellipsoidal shape and no spiral arms.

elongation — the apparent angle subtended by the Sun and a planet or by a planet and one of its satellites as seen from Earth; measured from 0° to 180° east or west of the Sun and from 0° east or west of the planet.

emersion — the reappearance of a celestial body after an occultation.

emission nebulae — a cloud of interstellar gas which is glowing because it absorbs – then re-emits – ultraviolet radiation from hot stars embedded within it.

Encke Division — a division in the Saturnian ring system which separates the A ring and the F ring; the Encke Division is approximately 325 km in width. NOTE: it is incorrect to call this feature the Encke Gap.

ephemeris — a listing of predicted positions of celestial objects on certain dates and/or at certain times; a published work that contains positions of the Sun, Moon and planets over a certain time, usually one year (plural = ephemeredes).

epoch — a date which is used for determining a standard reference time for star catalogs; the current standard epoch is J2000.0; in the past, standard epoch designations have remained for 50 years but recently a trend toward a 25-year standard epoch has been seen.

equation of time — the difference in time between a sundial and a clock, with no consideration given to Daylight Saving Time; technically, this is the correction applied to apparent solar time to obtain mean solar time.

equator the great circle on the surface of a rotating celestial body that lies in the plane that passes through the center of the body and is perpendicular to its axis of rotation.

equatorial coordinate system a system in which the vernal equinox and the Earth's equator form the reference circles for measurements in right ascension and declination.

equatorial diameter the diameter of a celestial object measured through its rotating plane (equator); as in the case of Jupiter, this can be different from the polar diameter.

equatorial mounting a mounting for a telescope, designed so that the two axes are aligned one to the polar axis and one to the equator of the Earth.

equinox either of two points on the ecliptic, lying at right ascension 0 hours (March equinox) and 12 hours (September equinox).

Erfle eyepiece an eyepiece design utilizing two doublets and a singlet in 2–1–2 configuration or three doublets.

eruptive variable any star which increases its brightness unpredictably. Examples of eruptive variables are novae and supernovae as well as red dwarf stars which can suffer unpredictable and violent stellar flares.

evening star a term often used to describe the appearance of a bright planet (Mercury, Venus, Mars, Jupiter, Saturn) in the western evening sky.

exit pupil the diameter of the light cone exiting from the eyepiece; technically, exit pupil equals the objective diameter in millimeters divided by the magnification.

extinction a reduction in the amount of radiation received from a celestial object due to absorption and/or scattering by an intervening medium; in space this is due to interstellar dust grains; closer to home, this is due to the Earth's atmosphere.

eye relief the distance from the vertex of the eye lens to the location of the exit pupil. This quantity can be made to vary for a given eyepiece by, e.g., the introduction of a Barlow lens, but this will be more pronounced with long focal length eyepieces, and in general is difficult to detect casually.

eyepiece the lens (or combination of lenses) at the eye end of an optical instrument through which the image is viewed.

facula a bright region of the Sun's photosphere associated with sunspots; seen most easily when near the Sun's limb (plural = faculae).

field curvature a lens aberration that causes a flat surface to be imaged onto a curved surface rather than a plane.

field galaxy a galaxy which is not a member of an observed cluster of galaxies.

field of view the area visible through the lens of an optical instrument; measured in terms of angular diameter (e.g., 20 minutes of arc).

field star a star which is not a member of an observed celestial object, but which appears in the field of view.

filament a cloud of gas in the Sun's chromosphere that is visible as a dark feature against the brighter disk.

filar micrometer a device used to measure the separation and position angle of a double star.

filter a material that absorbs certain wavelengths of light (visible and other) and transmits others.

finder a low (or unity) power telescope attached to a larger telescope whose optical axis is aligned to the larger telescope; finder scopes usually have larger fields of view than the telescopes to which they are attached.

fireball a meteor bright enough to cast a shadow; the magnitude limit (visual) for labeling a meteor a fireball is generally given as −3.

first contact	during an eclipse, the moment that the shadow of the eclipsing body first makes contact with the body being eclipsed; the beginning of the eclipse; during a transit, the moment that the disk of the planet first makes contact with the Sun.
first point of Aries	a rather outdated term for the vernal equinox; today, due to precession, the vernal equinox lies within the constellation Pisces; the point at which right ascension and declination are both equal to zero.
First Quarter	the lunar phase, lying approximately a week after New Moon and a week prior to Full Moon, at which one-quarter of the entire Moon (or one-half of the side facing the Earth) is visible to us.
Flamsteed number	a designation for stars (brighter than about seventh magnitude) arising from John Flamsteed's catalog, entitled *Historia Coelestis Britannica*, published in 1725 by H. Meere, London (an unauthorized version appeared in 1712); approximately 3000 stars are so numbered.
flare	a bright, short-lived area of the Sun's chromosphere, best viewed through an hydrogen alpha filter.
flare stars	red dwarf stars which undergo very rapid and unpredictable increases in brightness, possibly due to flares just above their photosphere.
fluorite	a type of ED glass which has low refractive index and low dispersion; chemical notation CaF_2.
focal length	the distance from the objective (lens or mirror) to the focal point in an optical system.
focal plane	the plane through the focus and perpendicular to the optical axis of the system in which the image will be formed.
focal point	the point which lies at the focal length of a lens or mirror; also called the focus.
focal ratio	the ratio between the focal length and the diameter of a mirror or lens; e.g., a 100 mm diameter lens with a 1500 mm focal length would be f/15.
focus	*see* focal point.
fork mounting	a type of equatorial telescope mounting (very popular with SCTs) which supports the tube on either end of its minor axis.
fourth contact	during an eclipse, the moment that the shadow of the eclipsing body breaks contact with the body being eclipsed; the end of the eclipse.
frequency	a measure of the number of wavelengths that pass a specific point in space within a specific time. The unit of frequency is the hertz (Hz) which is 1 per second.
full	a phase of the Moon or other celestial body where the percentage of the total visible surface illumination is 50% (or 100% of the side currently being observed).
galactic center	the central region of a galaxy; in the Milky Way, the center lies in the direction of the constellation Sagittarius.
galactic cluster	*see* open cluster.
galactic disk	the plane in which the spiral arms of spiral or barred spiral galaxies exist; lenticular galaxies also possess disks but no spiral arms.
galactic equator	the great circle which is defined by the galactic plane, the central plane of the Milky Way galaxy; the galactic plane and the celestial equator are inclined to one another at an angle of approximately 63°.
galactic halo	a spherical region around our own and other spiral galaxies which consists of dim stars, brown dwarfs, and globular clusters.
galactic nucleus	the central bulge of (mostly) older Population II stars that is found in the center of all spiral galaxies and which surrounds the galactic center.
galactic pole	one of two points on the celestial sphere lying either 90° north or 90° south of any point on the galactic equator; the two

poles form the axis around which the Milky Way galaxy rotates.

galaxy a collection of up to thousands of billions of stars, dust, and gas, held together by gravity.

Galilean satellites the four large satellites of Jupiter (Io, Europa, Ganymede, Callisto) discovered in 1610 by Galileo.

Galilean telescope the first type of astronomical telescope, developed by Galileo – a refractor with a single objective lens and a simple eyepiece lens.

gas giant a type of planet which is composed of vast quantities of gas. In our solar system, there are four gas giant planets: Jupiter, Saturn, Uranus, and Neptune.

gegenschein (counterglow) a very faint glow of light visible at the position of the ecliptic 180° from the Sun. It is believed to be caused by sunlight reflected from tiny interplanetary particles. Most easily detected in the constellations of Pisces (in September) and western Virgo (in March) as those two areas of the ecliptic are furthest from the bright Milky Way.

geometric albedo how bright a solar system object is relative to a sphere of equal size made of perfectly reflecting diffuse white material.

German mounting a popular type of equatorial telescope mounting which places the declination axis at one end of the polar axis, forming a 'T' shape.

giant a type of star much larger than the Sun, and also more massive, but not proportionately so, having a very thin outer atmosphere.

gibbous a phase of the Moon or other celestial body where the percentage of visible surface illumination is greater than 25% but less than 50% (or greater than 50% but less than 100% of the side currently being observed).

globular cluster a spherical collection of old (Population II) stars found within galactic halos of spiral galaxies and, more commonly, around elliptical galaxies.

granulation a mottled appearance of the photosphere of the Sun. This is caused by the convection of gas cells which are rising and falling.

gravity the attractive force of all bodies possessing mass.

Great Red Spot (GRS) an oval-shaped storm in the atmosphere of Jupiter, located 22° south of Jupiter's equator.

greatest brilliancy the points in the orbits of Venus and Mercury when they appear brightest as seen from Earth.

greatest elongation the maximum angular distance of Mercury or Venus from the Sun, as seen from Earth.

Greenwich Mean Time (GMT) the time at Greenwich, England, which is used as the basis for standard time throughout the world.

Gregorian calendar the calendar, implemented in 1582 by Pope Gregory XIII, now in use throughout most of the world.

Gregorian telescope a reflecting telescope invented by James Gregory, using a paraboloid primary mirror and an ellipsoidal secondary mirror; the secondary reflects light through a hole in the primary mirror.

guide telescope any telescope mounted on a (usually) larger telescope which is used to maintain the position of a celestial object for imaging through the larger scope; small corrections which are needed are manually applied.

HI region a cloud of neutral hydrogen (i.e., atoms) in interstellar space that emits photons of wavelength 21 cm.

HII region a cloud of hot ionized hydrogen in interstellar space which usually forms a bright nebula around young hot stars (or clusters).

head the nucleus and coma of a comet; sometimes refers to the main drive assembly of an equatorial telescope, upon which the optical tube assembly is mounted.

Hertzsprung–Russell (H–R) diagram	a graph which has the absolute magnitudes (or luminosities) of stars plotted against their spectral classification (or effective temperatures).	Union (IAU)	astronomers worldwide. On the internet at http://www.iau.org
horizon	where the celestial sphere intersects the Earth at every point; where the sky meets the Earth.	interplanetary medium	the material contained in the solar system in the space between the planets. One of the main components of this space is the solar wind.
hour angle	the angle measured along the celestial meridian westward to the hour circle intersecting any celestial body; measured in hours, minutes, and seconds from 0 to 24 hours.	interstellar absorption	a process by which visible light is dimmed by the interstellar medium.
		interstellar medium	diffuse material found between the stars within most types of galaxies.
hour circle	any of an infinite number of circles on the celestial sphere which intersect both the north and south celestial poles.	ion tail	also called the plasma tail, one of two types of comet tails, composed of ionized molecules; ion tails are generally straight, bluer than dust tails, and can reach lengths of tens of millions of kilometers.
Huygens eyepiece	an eyepiece consisting of two lenses, each plano-convex, with a field stop between them; a simple type of eyepiece with a great deal of spherical aberration, except when used in long focal length refractors.	irregular galaxy	one of the main galaxy classifications; a galaxy whose shape is neither elliptical or spiral, but random and unordered; irregular galaxies are generally young and are smaller than either ellipticals or spirals.
image tube	an electronic device used to enhance the signal (light or other radiation) coming from a (usually faint) celestial object.	Jovian	of or pertaining to Jupiter.
immersion	the disappearance of a celestial body during an occultation.	Jovian planets	the four large outer planets; Jupiter, Saturn, Uranus, Neptune; named due to the similarities in size and composition with Jupiter.
inclination	the angle between the orbital plane of a planet and the plane of the ecliptic.	Julian date	a designation pertaining to the number of days since 1 Jan 4713 BC; Julian days start at noon, UT; e.g., at noon UT on 1 Jan 2000, the Julian day number was JD2 451 605; at midnight, UT on 2 Jan 2000, the Julian day number was JD2 451 605.5.
inferior conjunction	the position of an inferior planet (Mercury or Venus) when it is in conjunction with the Sun and between the Sun and Earth.		
inferior planet	any planet whose orbit around the Sun is closer than that of the Earth; Mercury or Venus.		
inner planets	planets closer to the Sun than the asteroid belt; Mercury, Venus, Earth, and Mars.	June solstice	that instant when the Sun achieves maximum declination; the point on the ecliptic where the Sun's declination is at a maximum, having celestial coordinates of $RA = 6^h$. Dec. $= +23.5°$, approximately. NOTE: this point is the "summer" solstice only in the northern hemisphere of Earth.
integrated magnitude	the brightness an extended celestial object (comet, nebula, galaxy, etc.) would have if all its light originated from one point.		
intergalactic medium	diffuse matter found in the space between galaxies.	Keeler Gap	a division 35 km in width, lying in the outer A ring of the Saturnian ring system.
International Astronomical	the main decision-making body for astronomy-related matters, composed of	Kellner eyepiece	an eyepiece with a single field lens and a doublet eye group invented in 1849. This

eyepiece is orthoscopic in its traditional configuration. Has about 0.5× its focal length in eye relief.

Konig one of a class of eyepieces in which one design is a simplification of the Erfle and has a two-element field group and a single-element eye lens.

Kuiper Belt a flared disk of icy bodies (KBOs) which has its inner boundary just beyond the orbit of Pluto.

lanthanum a type of exotic glass, developed using lanthanum oxide, which, when added to glass gives it a higher refractive index and Abbe number than crown or flint.

Large Magellanic Cloud an irregular galaxy and a satellite of the Milky Way; the closest galaxy to the Milky Way, lying 180 000 light years away.

Last Quarter the lunar phase, lying approximately a week after Full Moon and a week prior to New Moon, at which one-quarter of the entire Moon (or one-half of the side facing the Earth) is visible to us.

lateral color an aberration which is off-axis chromatic aberration; it results in a star imaged at the edge of the field being smeared out into a rainbow or showing color fringing.

latitude angular distance north or south from the Earth's equator, with values 0°–90°; celestial latitude is the angular distance of a celestial object from the ecliptic.

lens a piece of transparent material (such as glass, quartz, plastic) that has two opposite regular surfaces either both curved or one curved and the other plane that is used either singly or combined in an optical instrument for forming an image by focusing rays of light; the name for a combination of lenses may also be lens, though, more correctly, compound lens.

lenticular galaxy an intermediate form of galaxy, between elliptical galaxies and spiral galaxies. They possess flattened forms and galactic discs but have no spiral arms.

libration a small angular change in the face that a synchronously rotating satellite presents towards the center of its orbit; the phenomenon whereby more than 50% of the Moon's surface is visible to a terrestrial observer.

light adaptation when the eye becomes adapted to luminances of more than about 3.4 candlepower/m².

light curve a graph showing the change in brightness of a celestial object plotted against time.

light gathering power the measure of a telescope's ability to gather light, solely dependent on the clear aperture of the telescope.

light year the distance the light, moving at approximately 300 000 km/s, travels in one year.

limb the outer edge of the apparent disk of a celestial body.

limb darkening a phenomenon whereby the edge of the solar disk appears darker than the center, due to the light rays from the edge having to move through more of the solar atmosphere to reach us than light rays near the center of the disk.

limiting magnitude the faintest magnitude which can be seen either visually, telescopically, or with a photographic or electronic detector.

Local Group a group of galaxies containing a few dozen galaxies (of which the Milky Way is a member) with a radius of approximately three and a quarter million light years.

local sidereal time the right ascension of a celestial object (or, simply, a point) on the meridian at a particular location.

Local Supercluster a very large grouping of galaxies containing the Local Group, centered on the Virgo cluster of galaxies with a radius of approximately 50 million light years.

long-period comet a comet with an orbital period greater than 200 years.

long-period variable a giant or supergiant star that undergoes a (usually) periodic variation in brightness over a time greater than about

	80 days; omicron Ceti (Mira) is the proto-typical long-period variable star, having a period of approximately 332 days.
lower culmination	*see* culmination.
luminosity	the total energy radiated into space every second by a celestial object such as a star.
luminosity classes	a grouping of star types based upon whether they are supergiants (luminosity class I), bright giants (II), giants (III), subgiants (IV), or main sequence stars (V).
lunar	of or pertaining to the Moon.
lunar eclipse	an eclipse at Full Moon when the Moon passes either totally or partially through the shadow of the Earth.
lunation	the interval from one New Moon to the next.
Magellanic clouds	two satellite galaxies of the Milky Way, both irregular.
magnitude	a measure of the amount of light flux (or other radiation) received from a luminous celestial object; *see also* absolute magnitude and apparent magnitude.
main belt	*see* asteroid belt.
main sequence	a region on the Hertzsprung–Russell diagram in which stable, middle-aged, hydrogen-burning stars are found.
major axis	the greatest distance across an ellipse; the distance from edge to edge through the center and both foci. For a circle, this distance would equal the diameter.
major planet	*see* planet.
Maksutov telescope	a telescope which uses a meniscus lens as a corrector plate and has a spherical mirror as the primary.
Maksutov–Newtonian telescope	a telescope comprising a meniscus corrector, concave primary, and flat secondary, assembled into a Newtonian configuration.

March equinox	that instant when the Sun, moving northerly, crosses the equatorial plane of the Earth; one of two points where the ecliptic and celestial equator meet, having celestial coordinates of RA = 0^h. Dec. = $0°$. NOTE: this point is the "spring" equinox only in the northern hemisphere of Earth.)
mare	Latin word for "sea"; the term is still applied to the basalt–filled impact basins common on the face of the Moon; any large, relatively smooth area on a planet or satellite.
Martian	of or pertaining to Mars.
Maxwell Gap	a division, 270 km wide, lying between the B and C rings of the Saturnian ring system.
mean solar time	an arrangement which allows clocks to move at a regular and constant speed, rather than following apparent solar time, which is not constant; one mean solar day is defined at 24 hours 3 minutes and 56.555 seconds of mean sidereal time.
mean Sun	a fictitious Sun which moves at a constant speed; created so as to institute mean solar time.
meridian	the great circle passing through the observer's zenith and the celestial poles.
meridian passage	the passage of a celestial object across one's meridian; also known as "meridian transit"; also known as "culmination."
Messier, Charles	a French astronomer and comet hunter whose most famous work was a list of objects published to help avoid confusion with possible comets.
Messier object	one of the objects on Charles Messier's list.
meteor	a streak of light in the night sky caused by a meteoroid entering the Earth's upper atmosphere and burning due to friction with the atmosphere; sometimes called a "shooting star" or "falling star."
meteor shower	the annual result of the passage of the Earth through a meteoroid stream.

meteor storm	a meteor shower in which the average is greater than 1000 meteors per hour; this rate must be maintained for 20 minutes or more.
meteorite	any meteoroid which (after becoming a meteor) lands on the surface of the Earth.
meteoroid	small bodies made of rock, metal or a combination of both which orbit the Sun; most are extremely small, with masses between 1/1000 and 1/1000000 of a gram.
Milky Way	the name of our own galaxy.
minor axis	the distance from the edge of an ellipse, through the center, to the other edge, perpendicular to the line connecting the foci.
minor planet	*see* asteroid.
minute of arc	1/60 of one degree of arc. There are 60 seconds of arc in one minute of arc; designated ′; thus $45°06′14″ = 45$ degrees, 6 minutes, 14 seconds.
mirror	an optical component, generally covered on its figured end with a very thin coat of aluminum, designed to reflect light.
mirror cell	the device which enables the primary mirror to be mounted within a telescope.
molecular cloud	a cold (approximately 10 K), interstellar region of gas molecules.
monochromatic	light of a single wavelength; an example is the wavelength of hydrogen alpha light at 656.3 nm.
moon	a naturally-occurring, relatively large body which is in orbit around a planet.
morning star	a term often used to describe the appearance of a bright planet (Mercury, Venus, Mars, Jupiter, Saturn) in the eastern morning sky.
mounting	the support structure of a telescope; composed of everything except the optical tube assembly; may include head, tripod, pier, etc.

multi-coated	referring to the air-to-glass surfaces of lenses when they have received more than one layer of coatings.
multiple star	two or more stars in a close system with a common center of gravity; maximum number according to theory is six; when only two stars are involved, this is called a double star.
mutual phenomena (of satellites)	the situation that occurs when one satellite – usually of Jupiter – occults or is eclipsed by a second satellite.
nadir	the point on the celestial sphere that lies directly beneath the observer, on the observer's meridian; the point opposite the zenith.
Nagler eyepiece	a type of eyepiece design introduced in 1982, which has a very wide apparent field of view, good contrast and eye relief, and sharp images.
Nasmyth focus	one of two focal points along the altitude axis of an altazimuth-mounted telescope, primarily used in conjunction with large imaging equipment.
nautical twilight	the condition of solar illumination at a point on the Earth's surface before sunrise or after sunset when the Sun's zenith distance is 102° (that is, when the Sun is 12 degrees below the horizon).
Near-Earth Object	an asteroid with a statistical chance of colliding with the Earth.
nebula	Latin word for "cloud;" a cloud of interstellar dust and/or gas.
negative shadow	the extension of the umbra of an annular eclipse that delineates the path from which observers may see the ring of Sun of the annular eclipse.
neutron star	the central remains of a star which has undergone a supernova explosion and whose electrons have been fused with protons to produce neutrons; a neutron star has a typical radius of 10 km but a density of up to $10^{18} kg/m^3$.
New General Catalogue (NGC)	a catalog of non-stellar celestial objects that was compiled by J. L. E. Dreyer of

Armagh Observatory and published in 1888. The original listed 7840 objects and was followed by a supplement which now includes 5386 not on the original. Objects on the original list are identified by the letter NGC followed by the catalog number. Those objects on the supplement are denoted by the letters IC followed by the catalog number.

new — a phase of the Moon or other celestial body where the percentage visible surface illumination is 0%.

Newtonian telescope — the first practical reflecting optical telescope, invented by Isaac Newton, consisting of a concave, paraboloidal primary mirror and a flat secondary mirror that diverts light out the side of the tube.

NGC number — a number given to any object (galaxy, star cluster or nebula) found in J. L. E. Dreyer's *A New General Catalogue of Nebulae and Clusters of Stars, being the Catalogue of the late Sir John F.W. Herschel, Bart., revised, corrected, and enlarged*. This work was published as *Memoirs of the Royal Astronomical Society*, vol. xlix, part 1, London, 1888. Two supplements, also *Memoirs of the Royal Astronomical Society*, were published by Dreyer. These were entitled *Index Catalog of Nebulae found in the Years 1888 to 1894, etc.* (London, 1895) and *Second Index Catalogue of Nebulae and Clusters, etc.* (London, 1908). These three catalogs are now published as one work, known as the *Revised New General Catalog of Nonstellar Astronomical Objects*.

nightglow — *see* airglow.

nodes — the two points at which the orbital plane of a celestial body intersect a reference plane, such as the ecliptic or celestial equator. If the body is seen to move across the reference plane from south to north, the node is referred to as an ascending node. If the body is seen to move from north to south, the node is a descending node.

north polar distance — an angular measurement of a celestial object from the north celestial pole, equal to 90° − the object's declination.

north polar — a group of stars near the north celestial

sequence — pole which once formed the standardization basis for the magnitude system.

nova — a star which suddenly brightens by up to ten magnitudes and then gradually declines over a period of months; novae are close binary star systems in which one of the members is a white dwarf star; matter is transferred to the white dwarf, forming an accretion disk, gradually spiraling onto its surface where it builds up until the temperature and pressure become sufficient to ignite nuclear fusion causing catastrophic nuclear detonation.

nucleus — the icy body which is a comet; when close to the Sun the nucleus is surrounded by a cloud of dust and gas (the coma).

nutation — a periodic, irregular motion of the Earth caused by the gravitational attraction of the Sun and Moon, along with their varying distances and relative directions. This motion is superimposed upon the motion of precession.

objective — the primary lens of a refractor telescope.

oblateness — the measurement of the non-sphericity of a planet or other celestial object. Mathematically, this is the difference between the equatorial and polar diameters of a planet divided by the planet's equatorial diameter.

obliquity — the angle between the equatorial plane of a planet or other celestial object and its orbital plane; the angle between a planet's axis of rotation and the pole of its orbit.

observatory — a building or enclosure constructed for the purpose of housing instruments for astronomical observation.

observing log — a record (short or long term) of what an observer sees.

occultation — the obscuration (total or partial) of any celestial object by another of larger apparent diameter.

occulting bar — a bar placed in the focal plane of a telescope eyepiece to cover part of the field of view, usually to cover a bright object in order to permit observation of a nearby faint object.

occulting disk a device often used to study the solar corona by blocking out the light from the Sun's photosphere.

off-axis used most frequently among amateur astronomers to denote something that is not at the center of the field of view.

Oort Cloud a spherical region theorized to surround the solar system, containing a vast number of comets; thought to exist between the distances of 30 000 to 100 000 AU.

open cluster mostly young systems of stars usually containing between a few hundred and a few thousand stars.

opposition the position of two celestial objects when their longitude (as seen from Earth) differs by 180°. When one of the objects is the Sun, opposition means the other object is opposite the Sun in the sky, therefore visible all night long.

opposition effect the moment when the phase angle of a superior planet is 0°. This occurs when the Earth is in transit across the face of the Sun, as seen from the planet.

optical axis an imaginary line in an optical system which passes through the center of all optical elements and upon which is the focal point.

optical binary a pair of stars which happen to lie close to one another on the celestial sphere because of a chance alignment. They are not physically associated with one another and probably exist at vastly different distances. Optical binaries are also known as visual binaries.

orbit the path of a celestial object through space as influenced by the gravity of some primary body or bodies.

orbit–orbit resonances a phenomenon in which two or more satellites interact gravitationally such that their motions follow certain repetitive patterns.

orbital elements six parameters which specify the position and motion of a celestial body in its orbit and that can be established by observation. The six elements are eccentricity, semi-major axis, inclination, longitude of the ascending node, argument of perihelion, and epoch (sometimes referred to as time of perihelion passage).

orthoscopic eyepiece a lens that is free of spherical aberration, and magnifies an image uniformly throughout the field (i.e., the lens satisfies the tangent condition in which the ratio of the tangent of a' to the tangent of a is a constant for every ray on the lens).

outburst (cometary) an unexpected increase in brightness over a short period of time due to the release of dust and gas into the coma from the nucleus.

outer planets planets further from the Sun than the asteroid belt; Jupiter, Saturn, Uranus, Neptune, and Pluto.

Palomar Sky Survey one of the great photographic star atlases ever made, completed by the 48″ Palomar Schmidt camera, covering the sky to a declination of −33°.

Panoptic an eyepiece introduced by Tele-Vue which is a modified Erfle design, having wide field, very sharp images and low eye relief in some models.

parabola one of the conic sections, an open curve.

paraboloid a three dimensional parabola; the curve which forms the surface of the mirror of a reflecting telescope, bringing all light to a single focus.

parallax *see* stellar parallax.

paraxial near the optical axis of an optical system.

parfocal a property of two or more eyepieces, the focal planes of which are the same distance from the barrel top.

parsec the distance at which a star would have an annual parallax of one arcsecond; at this distance the semi-major axis of the Earth's orbit (one astronomical unit) subtends an angle of one second of arc. One parsec =

3.0857×10^{13} km $= 206\,265$ AU $= 3.2616$ light years.

path of totality the path (up to 200 miles wide) that the Moon's shadow traces on the Earth during a total solar eclipse.

Paul–Baker telescope a telescope design using a paraboloid primary mirror, an ellipsoid secondary mirror and a spherical tertiary mirror to produce a very wide field of view.

peculiar galaxy any of about a dozen galaxy types not easily categorized by the old Hubble sequence.

penumbra the less dark outer region of a shadow cast by a solar system object illuminated by the Sun.

penumbral eclipse a lunar eclipse during which the Moon never enters the darker, inner part of the Earth's shadow, but remains in the outer, lighter penumbra.

periastron the minimum distance between the components of a binary star system.

perigee the position of the Moon or an object in Earth orbit when it is closest to the Earth; the instant in a given orbit of the Moon (or other object) when it is closest to the Earth.

perihelion the position of an object in solar orbit when it is closest to the Sun; the instant in a given orbit of a planet (or other body) when it is closest to the Sun.

period with regard to a celestial object, the time interval between two successive, similar events.

periodic comet these comets are indicated by a "P/" before the names. For example, P/Halley is Halley's Comet or more properly known as periodic Comet Halley. Recently, the International Astronomical Union has started numbering periodic comets that have been seen at more than one apparition. Thus, Halley's Comet is 1P/Halley and P/de Vico is now known as 122P/de Vico.

perturbation the disturbing effect, when small, on the orbit of a planet or satellite as predicted by theory, produced by (1) another planet or satellite or (2) a group of planets and/or satellites.

phase the percentage of illumination of the Moon or other solar system object at a particular time during its orbit.

phase angle the angle (from 0°–180°) formed by the Moon–Earth–Sun or by a planet–Earth–Sun.

photodetector any of various devices for detecting and measuring the intensity of radiant energy through photoelectric action.

photographic magnitude the magnitude of a star or other celestial body as measured by a detector with a maximum spectral response in the blue region, centered on a wavelength of 425 nm.

photometer an instrument for measuring luminous intensity, luminous flux, illumination, or brightness in which the radiation is converted to an electrical signal which can be precisely measured.

photometry a branch of science that deals with measurement of the intensity of light by use of a photometer.

photon a massless elementary particle which carries energy, momentum and angular momentum created in reactions which involve electromagnetism, and which travels at the speed of light.

photopic vision color-sensitive vision where the principal receptors are the cones of the retina; requires higher levels of light than scotopic vision.

photosphere the surface of a star, including the Sun, which is the layer emitting visible light, about 500 kilometers deep.

pixel any of the small discrete elements that together constitute an image (on a VDU, for example); any of the detecting elements of a charge-coupled device used as an optical sensor.

plage   bright areas in the chromosphere of the Sun which are at a higher temperature than the surrounding areas; also known as bright flocculi.

planet   one of nine solar system objects which orbit the Sun and shine by reflected light; a similar object in orbit around another star. In order of distance from the Sun, the planets of our solar system are Mercury, Venus, Earth, Mars, Jupiter, Saturn, Uranus, Neptune, and Pluto.

planetary nebula   the outer, gaseous layers of a red giant star which have been gently blown off into space, and which glow because the gas is excited by radiation from the central, collapsing star.

plasma tail   *see* ion tail.

Plossl   an eyepiece design consisting of two groups of two elements each, in which all air–glass surfaces are convex and the crown elements face one another.

polar diameter   the distance between the poles of a celestial object, measured through the center of the object.

poles   two points on a planet, satellite or other celestial object which lie 90° above or below a given great circle, generally the object's equator.

Population I   stars which are luminous, hot and young, concentrated in the disks and arms of spiral galaxies, and which tend to have heavy elements formed by earlier generation stars; the Sun is an example.

Population II   stars which are found in globular clusters and the nucleus of galaxies. They are older, less luminous, cooler and contain fewer heavy elements than Population I stars.

position angle (PA)   (1) in a double star system, the angle, measured from north through east, of the line joining the primary with the companion star; (2) the direction on the sky (in degrees from north) toward which a comet's tail is pointing.

precession   the sweeping out of a cone by the spin axis of a rotating body when acted upon by a torque perpendicular to its spin axis.

primary   in a system of two or more orbiting celestial bodies, the one nearest to the center of mass and the one around which the others seem to revolve.

primary mirror   the main light-gathering mirror of a reflecting telescope.

prime focus   the focal point of the primary mirror; a type of through-the-telescope astrophotography where there are no intervening optics between the primary mirror and the film.

prism diagonal   a diagonal assembly where the light is diverted 90° by the use of a prism, rather than a mirror.

prograde motion   orbiting in the same direction as the prevailing direction of motion; the opposite of retrograde.

prominence   a large-scale, gaseous formation above the surface of the Sun (or, theoretically, a star), usually occurring over regions of solar activity such as sunspot groups.

proper motion   the apparent angular motion per year of a star or other celestial object in a direction perpendicular to the line of sight.

provisional designation   *see* asteroid designation.

pulsar   a rotating neutron star, visible because of two beams of electromagnetic radiation which emanate from the magnetic poles and which happen to sweep across our line of sight.

Purkinje effect   the phenomenon encountered during the viewing of two stars of equal magnitude – one red and the other blue – where the blue star looks brighter; if viewed for an extended period, however, the red star will appear to increase in brightness.

quadrant   any of the four quarters into which something is divided by two real or imaginary lines that intersect each other at right angles.

quadrature the position of the Sun and another celestial object when their longitude (as seen from Earth) differs by 90°.

quasar (quasi-stellar radio source) compact extragalactic objects at extreme distance which are highly luminous, thought to be active galactic nuclei.

radial velocity the velocity of a star or other (non-solar system) celestial object along the line of sight of the observer.

radiant the apparent origin of a meteor shower, when the paths of all shower meteors are traced backward in the sky.

radio galaxy a type of active galaxy (always an elliptical) which is an intense source of radio waves.

radius the length from the center of a celestial object to its surface; the radius equals the diameter divided by two.

radius of curvature the semidiameter of a circle whose circumference matches the curvature of the lens or mirror in question.

Ramsden eyepiece a simple eyepiece consisting of two plano-convex lenses with the plane side of the lower lens nearer the objective.

reciprocity failure the breakdown of the normal relationship between aperture and shutter speed related to the density of an exposed image on film.

recurrent nova a nova which has been observed to erupt more than once.

red dwarf a small, cool star with mass near the lower limit of 0.08 solar masses.

red giant any star which is burning helium in its core and has a spectral classification of K or M.

red supergiant an evolved star which has an even larger radius than a red giant.

reddening an effect whereby a star's color is reddened due to interstellar clouds which more effectively block blue light.

reflecting telescope a telescope which uses mirrors to focus the light.

reflection the production of an image by a mirror.

reflection nebula an interstellar cloud which scatters starlight into our line of sight. It will often appear to be blue because blue light, having a smaller wavelength, scatters more efficiently than red light.

refracting telescope a design of telescope which uses lenses to focus the light from distant celestial objects.

refraction the deflection from a straight path undergone by an energy wave in passing from one medium into another.

refraction, atmospheric the bending of light passing obliquely through a body's atmosphere; in the case of the Earth, the result is that celestial bodies appear to be displaced towards the zenith, with the amount of displacement increasing with the object's zenith distance.

refractive index the ratio of the velocity of propagation of an electromagnetic wave in a vacuum, to its velocity in a medium; how much a particular substance bends light.

regression the movement of points in an orbit in the direction opposite from the motion of the orbiting body. For example, the Moon travels from west to east, but its nodes are regressing from east to west.

residual the difference between the observed and the predicted (theoretical) position of a planet.

resolution the amount of detail visible in an image.

resolving power usually defined as the ability to separate two stars of equal brightness; *see* Dawes limit.

reticule a set of lines or circles, either illuminated or not, which forms a pattern in an eyepiece's field of view.

retrograde orbiting opposite to the prevailing direction of motion; the opposite of prograde.

retrograde motion the false (east to west) apparent motion of a planet or other celestial object on the celestial sphere, as seen from Earth. This occurs when a superior planet is near opposition. The Earth, moving faster, overtakes the planet and the planet appears to move backwards. A similar phenomenon occurs when one automobile passes another – even though both vehicles are traveling in the same direction, the slower automobile seems to be going the other way.

revolution — the orbital motion of a planet or other celestial object around the Sun, or of a satellite around a planet.

rhodopsin — the photopigment of the rods in the eye's retina.

rich-field telescope (RFT) — a telescope with a large field of view.

right ascension (RA) — a geocentric spherical coordinate that is an angle measured eastwards along the celestial equator from the vernal equinox to the intersection of the hour circle passing through the body; usually expressed in hours, minutes, and seconds from 0 hours to 24 hours, where one hour of right ascension equals 15°.

ring galaxy — a rare type of galaxy thought to be caused by a compact galaxy passing through the center of a normal spiral galaxy.

rising — the appearance of a celestial object above the horizon due to the rotation of a planet or satellite.

RKE — an eyepiece derived from the Kellner design, except that the field group is of two elements instead of the eye lens, and the addition of low dispersion glass; exhibits a moderately curved field and moderate distortion, and less off-axis astigmatism than the traditional Kellner.

rotation — the spinning motion of a planet or other celestial object around an axis.

rotational period — the period of time which a planet or satellite takes to spin once, measured at its equator; this is thus the equatorial period of rotation.

Saros — the eclipse cycle with a period of 223 synodic months, or 6 585.32 days (18 years and about 11 days).

satellite — a body that revolves around a larger body.

Schmidt–Cassegrain telescope (SCT) — a telescope design comprising a spherical primary mirror, a full-aperture corrector plate, and a negative secondary in a Cassegrain configuration.

Schroeter effect — the fact that the dichotomy of Venus does not coincide with the time of greatest elongation; first described by the German amateur astronomer Johann Hieronymus Schroeter (1745–1816). This effect has also been noted for the Moon and Mercury.

scintillation — rapid changes in the brightness of a celestial body caused by our atmosphere; twinkling.

scotopic vision — vision where the principal receptors are rods; not color-sensitive but operates at low light levels.

SCT — acronym for Schmidt–Cassegrain telescope.

second contact — during a total eclipse, the moment of total obscuration of the Sun or Moon; the instant totality begins; during a transit, the moment the disk of the planet is totally within the disk of the Sun.

second of arc — 1/60 of one minute of arc. Thus, there are 3600 seconds of arc in one degree of arc; designated ″; thus 22°31′46″ = 22 degrees, 31 minutes and 46 seconds.

secondary — in a system of two or more orbiting celestial bodies, any body not the closest to the center of mass and any which revolve around the primary.

secondary spectrum — the residual chromatic aberration manifested in the image produced by an achromatic lens (and other multi-element lenses).

seeing — a generalized measure of the overall steadiness of the Earth's atmosphere at a particular location; often expressed in seconds of arc.

semi-apo — a refractor that approaches apochromatic performance.

semi-diameter — the apparent radius of a generally spherical celestial body.

semimajor axis — one-half the greatest distance across an ellipse; the distance from the center to the edge through one of the foci. For a circle, this distance would equal the radius.

semi-minor axis — the distance from the center of an ellipse to the edge perpendicular to the line connecting the foci.

semi-regular variable — a variable whose light curve exhibits random increases in intensity superimposed over a recurring pattern.

separation — the distance, expressed in angular measure, between two celestial bodies.

September equinox — that instant when the Sun, moving southerly, crosses the equatorial plane of the Earth; one of two points where the ecliptic and celestial equator meet, having coordinates RA = 12^h, Dec. = $0°$. NOTE this point is the "autumn" equinox only in the northern hemisphere of Earth.

setting — the disappearance of a celestial object below the horizon due to the rotation of a planet or satellite.

setting circles — movable scales on an equatorial mount which, when aligned, allow the observer to manually find celestial objects from their right ascensions and declinations.

shadow bands — faint ripples of light sometimes seen on flat, light-colored surfaces just before and just after totality.

short-period comet — any comet with an orbital period of less than 200 years.

sidereal clock — a clock which keeps sidereal time.

sidereal period — an amount of time measured with reference to the stars.

sidereal time — the local hour angle of the vernal equinox; also defined as the right ascension of a (real or hypothetical) star on the local meridian. Sometimes known as "star time."

silvering — an archaic term once used in reference to the depositing of a very thin layer of a silver compound onto the front surface of a telescope mirror.

Small Magellanic Cloud — an irregular galaxy and a satellite of the Milky Way; the second closest galaxy to the Milky Way, at a distance of about 240 000 light years away.

solar — of or pertaining to the Sun.

solar antapex — the point on the celestial sphere away from which the Sun and solar system are moving, relative to the stars in our vicinity. This point lies within the boundaries of the constellation of Columba, and has the approximate coordinates RA = 6^h, Dec. = $-30°$.

solar apex — the point on the celestial sphere towards which the Sun and solar system are moving, relative to the stars in our vicinity. This point lies within the boundaries of the constellation of Hercules, and has the approximate coordinates RA = 18^h, Dec. = $+30°$.

solar cycle — the variation in the activity of the Sun during an eleven-year period, most noticeable because of its effect on the number of sunspots visible on the photosphere.

solar eclipse — an eclipse at New Moon when the Moon covers the disk of the Sun, either totally or partially, for locations on Earth.

solar flare — a sudden burst of light particles (protons, electrons, etc.) and electromagnetic energy from the Sun's photosphere. In space solar flares contribute to the solar wind and on Earth they result in increased low-energy cosmic ray and aurora activity.

solar irradiance — the amount of solar energy falling upon a unit area of the surface of a planet or satellite; usually measured in watts per square meter.

solar system — everything which is dominated by the Sun's gravitational field. The solar system is made up of the Sun, the nine planets and their moons, and also minor bodies such as asteroids and comets. It was formed from the solar nebula 4.6 billion years ago.

solar telescope   a telescope whose design is dedicated to observing the Sun.

solar wind   energetic charged particles that flow radially outward from the solar corona, carrying mass and angular momentum away from the Sun.

solstice   either of two points on the ecliptic, lying at right ascension 6 hours (June solstice) and 18 hours (December solstice).

solstitial colure   the great circle passing through both celestial poles and intersecting the ecliptic at both solstices.

spectroscopy   the study of a celestial object's light, that light having been dispersed into its component wavelengths.

speculum   an archaic term once referring to a figured and highly-polished alloy of 80% copper and 20% tin from which primary mirrors for reflecting telescopes were made.

spherical aberration   the inability to focus axial and paraxial ray bundles that are parallel to the axis at a single point in the image plane, creating a fuzzy image.

spider   any support for the diagonal mirror of a Newtonian reflector.

spider diffraction   the cross-shaped pattern formed by a Newtonian reflector due to diffraction of light around the spider support for the secondary mirror.

spiral galaxy   a galaxy in which a central bulge of older stars is surrounded by a flattened galactic disk containing a spiral pattern of young, hot stars.

sporadic meteor   any meteor not belonging to a recognized meteor shower.

spring equinox   see March equinox.

star   a self-luminous sphere of gas that generates energy by means of nuclear fusion reactions in its core.

star atlas   a set of celestial maps showing stars and other celestial objects brighter than some designated magnitude, designed to aid the amateur astronomer in locating objects.

star catalog   a collection of data (no maps) about stars brighter than some designated magnitude.

star diagonal   see diagonal mirror.

star trails   a popular subject of amateur astrophotography where the camera is stationary with its shutter open for some period of time (usually minutes to hours) and which records curved images of stars as the Earth rotates.

stationary point   the position of a planet as it shifts from direct to retrograde motion, or from retrograde to direct, and thus appears motionless.

stellar parallax   the apparent angular displacement of a star or other celestial object that results from the revolution of the Earth about the Sun; numerically, this is the angle subtended by one astronomical unit at the distance of the particular object. This differs from the solar parallax, which is the apparent displacement of the Sun as seen from two (generally widely separated) places on Earth.

Strehl ratio   a measurement of the amount of light put into the peak of the image spot in an actual telescope, compared to that put in the spot of a perfect telescope; a Strehl ratio of 100% would constitute a perfect telescope.

summer solstice   see June solstice.

Sun   the star at the center of the solar system, around which the Earth and other planets orbit; classified as a G2V (dwarf), approximately 4.6 billion years old, containing 2×10^{30} kg of material.

sungrazer   a comet which passes very close to the Sun; it is often destroyed in the process.

sunrise	the moment when the upper limb of the Sun appears above the horizon as a result of the rotation of the Earth.
sunset	the moment when the upper limb of the Sun disappears below the horizon as a result of the rotation of the Earth.
sunspot	a temporarily cooler region on the photosphere of the Sun caused by magnetic field variations.
sunspot cycle	a period of approximately 11 years during which the number of sunspots observed is at a maximum.
supercluster	a cluster of clusters of galaxies which stretches for hundreds of millions of light years. So far about fifty have been observationally identified.
supergiant	a star with a higher luminosity and a larger radius than a giant of the same spectral classification. It will have typically one hundred times the luminosity of a giant. It will almost certainly become a supernova.
superior conjunction	the position of an inferior planet when it is in conjunction with the Sun and on the side of the Sun opposite the Earth.
superior planet	any planet whose orbit around the Sun is further out than that of the Earth: Mars, Jupiter, Saturn, Uranus, Neptune or Pluto.
supernova	a catastrophic explosion which causes a star to explode.
supernova remnant	an expanding cloud of gas created by a supernova.
surface	for a terrestrial planet or a satellite, the boundary between the planet itself and its atmosphere; for a Jovian planet, this may refer to the boundary between the atmosphere and the solid core deep within or it may refer to the "optical" surface, beginning at a depth where the atmosphere becomes opaque.
surface brightness	the luminosity of any celestial object (or of the sky itself) per unit area, often expressed as magnitude per square arcsecond.

symmetrical	an eyepiece exhibiting symmetry in its element configuration. The most common symmetricals are Plossl eyepieces, which comprise two identical achromats facing one another.
synchronous rotation	the condition arising when the rotational period of a satellite is equal to its orbital period.
synodic period	the average time between successive conjunctions of two planets as seen from the Earth.
syzygy	the lineup of the Sun, the Earth and either the Moon or a planet. For the Moon, this occurs at New Moon and Full Moon. For a planet, it occurs at conjunction and opposition.
tail	the most distinctive feature of a comet, which generally points away from the Sun; can appear in a variety of shapes and lengths.
telecompressor	an optical device which multiplies the focal length of a telescope by a factor which is less than one, thus reducing the focal length, usually to improve the photographic speed of a system.
tele-extender	another name for a Barlow lens.
telescope	a device used to collect, focus and magnify light.
terminator	the boundary between the sunlit and dark hemispheres of the Moon or other solar system object. At the terminator, either sunrise or sunset is occurring.
terrestrial	of or pertaining to the Earth.
terrestrial planets	the four small inner planets: Mercury, Venus, Earth, Mars; named due to their similarities in size and composition with Earth.
third contact	during a total eclipse, the moment that total obscuration of the Sun or Moon ends; the instant totality ends; during a transit, the moment the disk of the planet makes contact with the edge of the Sun, during egress.

tilt of axis — the angle between the axis of a planet or satellite and its orbital plane.

time zone — any of 24 areas into which the Earth's surface is divided so as to provide standard time.

total eclipse — an eclipse of the Sun or Moon in which (1) the Sun is totally covered by the Moon or (2) the Moon is totally immersed within the umbral shadow of the Earth.

total magnitude — *see* integrated magnitude.

transient lunar phenomena (TLP) — also known as lunar transient phenomena (LTP); temporary glows, flashes, color or brightness changes, etc., which occur on the surface of the Moon.

transit — (1) the passage of Mercury or Venus across the Sun's disk, as seen from Earth; (2) also known as upper culmination, this is the passage of a celestial body across an observer's meridian nearest to the zenith (to distinguish it from lower culmination).

truss-tube telescope — a telescope in which the tube is a light truss assembly, designed for quick and easy assembly.

twilight — the condition of solar illumination at a point on the Earth's surface before sunrise or after sunset when the Sun's zenith distance is 96° for civil twilight, 102° for nautical twilight, or 108° for astronomical twilight.

twinkling — *see* scintillation.

umbra — the dark inner region of a shadow cast by a solar system object illuminated by the Sun.

unit-magnification finder — a finder which operates without magnification.

Universal Time (UT) — also known as Greenwich Mean Time (GMT), time kept on the Greenwich meridian (longitude = 0°); in astronomical and navigational usage, UT often refers to a specific time called UT1, which is a measure of the rotation angle of the Earth as observed astronomically; in civil usage, UT refers to a time scale called Coordinated Universal Time (UTC), which is the basis for the worldwide system of civil time, determined using highly precise atomic clocks.

upper culmination — *see* culmination.

variable star — a star whose luminosity varies by any means.

vernal equinox — the instant when the Sun, moving northerly, crosses the equatorial plane of the Earth; one of two points where the ecliptic and the celestial equator meet, having celestial coordinates of RA = 0^h, Dec. = 0°. NOTE: this point is also known as the March equinox and is the spring equinox only in the northern hemisphere of Earth.

vertical circle — any of an infinite number of great circles that pass through the zenith and intersect the horizon at right angles.

vignetting — the darkening of the outer areas of an image relative to the center of the image.

visual binary — a double star which appears to be binary star system because of a chance alignment.

visual magnitude — *see* apparent magnitude.

waning — the time when the illuminated portion of the Moon or a planet is becoming smaller.

wavelength — the measurement of the distance from crest to crest in transverse waves such as electromagnetic radiation; the shorter the wavelength, the greater the energy carried by the wave.

waxing — the time when the illuminated portion of the Moon or a planet is growing larger.

white dwarf — the remains of a stellar core following the cessation of nuclear fusion, composed of electron degenerate matter.

Wilson Effect — the effect whereby the limbward side of the penumbra of a sunspot appears foreshortened just before the spot disappears over the limb, first described by the Scottish astronomer Alexander Wilson in 1769.

y,d,h,m,s short for years, days, hours, minutes, seconds; thus $2^y298^d16^h33^m20^s = 2$ years, 298 days, 16 hours, 33 minutes and 20 seconds.

year generally defined as the time interval for the Earth or any planet to make one complete revolution around the Sun. Many other types of years are defined.

zenith the point on the celestial sphere which lies 90° from all points on the horizon; the overhead point lying on the observer's meridian; the point opposite the nadir.

zenith distance the angular distance of a celestial object from the zenith; equal to 90 degrees minus the altitude of the object.

zenithal hourly rate (ZHR) theoretical number of meteors per hour which would be visible if the observed shower were to come from the direction of the zenith under optimum observing conditions (no Moon, no clouds) and with a limiting visual magnitude of 6.5.

zero-power finder a common misnomer for a unit-magnification finder (which has a magnification of unity, not zero).

zodiac a band around the celestial sphere 18° in width and centered on the ecliptic.

zodiacal band a glow (fainter than the zodiacal light and greater in extent) sometimes seen (from a dark site) extending along the ecliptic.

zodiacal dust dust found in the plane of the inner solar system, which is believed to have been created during asteroid collisions and from the breaking up of comets.

zodiacal light a glow, visible to the eye from a dark location, caused by dust particles spread along the ecliptic plane.

zone of avoidance a band around the sky, centered on the Milky Way, in which galaxies are obscured by the Milky Way's dust.

Index